Lecture Notes in Mathematics 1558

Thomas J. Bridges   Jacques E. Furter

Singularity Theory and Equivariant Symplectic Maps

Springer-Verlag
Berlin Heidelberg New York
London Paris Tokyo
Hong Kong Barcelona
Budapest

Authors

Thomas J. Bridges
Jacques E. Furter
Mathematics Institute
University of Warwick
Coventry CV4 7AL, Great Britain

Mathematics Subject Classification (1991): 58C27, 58F14, 58F05, 58F22, 58F36, 39A10, 70Hxx

ISBN 3-540-57296-1 Springer-Verlag Berlin Heidelberg New York
ISBN 0-387-57296-1 Springer-Verlag New York Berlin Heidelberg

Printed in Germany

2146/3140-543210 - Printed on acid-free paper

Table of Contents

1. Introduction

The study of orbits of symplectic maps under iteration has a rich history motivated by basic questions in symplectic geometry and their importance in applications – celestial mechanics, plasma physics, accelerator dynamics, condensed matter physics and fluid flow, for example. The present work is a research monograph that considers particular interesting questions about symplectic maps on $\mathbb{R}^{2n}$ with $n \geq 1$. The monograph consists of three parts: a general theory for bifurcating period-q points of equivariant symplectic maps, introduction of a singularity theory framework for equivariant gradient bifurcation problems which is used to classify singularities of bifurcating period-q points of symplectic maps and thirdly a compendium – much of which is contained in a sequence of appendices – of basic questions and results for symplectic maps on $\mathbb{R}^{2n}$ and their generating functions.

Symmetry arises in two ways in the analysis of period-q points. A period-q orbit of a symplectic map has a natural cyclic ($\mathbb{Z}_q$) symmetry; that is, an orbit can start at any of the q distinct points that make up the period-q orbit. Period-q points can therefore be characterized as fixed points of a $\mathbb{Z}_q$ equivariant map on $\mathbb{R}^{nq}$. When the symplectic map is itself equivariant we call this a spatial symmetry (that is, it is independent of discrete time). A general framework to account for the two types of symmetry is introduced in Chapters 2 and 8 respectively.

When the map depends on parameters the periodic points will undergo bifurcations as the parameters are varied. To classify the bifurcations we introduce a singularity theory framework in Chapter 3 with special cases treated in Chapters 4 and 6. These results are self-contained and have general application to the classification of bifurcations for equivariant gradient maps. In particular, our results provide a complete theory when the symmetry group is discrete.

Throughout the monograph various questions arise that are of a basic nature such as the signature and linear stability question for period-q points on configuration space, discrete-time conservation laws, linear normal forms, generating functions and reversibility, multiple resonant Floquet multipliers, dynamical equivalence and so forth. Each of these topics is treated independently in the appendices.

The classic theory on the bifurcation of period-q points of area-preserving maps depending on a single parameter – generic bifurcation of periodic points – is due to Meyer [1970,1971] with extension to the reversible area-preserving case by Rimmer [1978,1983]. We extend the basic theory in a number of significant ways. The framework of Meyer does not extend easily to symplectic maps on $\mathbb{R}^{2n}$ with $n \geq 2$. We introduce a framework for the bifurcation of period-q points based on "Lagrangian generating functions". Let $h(x, x')$ be a smooth function of $x, x' \in \mathbf{R}^n$ and suppose $\det h_{xx'}(x, x') \neq 0$ then the relations (the notation is described in Chapter 2)

$$\begin{aligned} y' &= h_{x'}(x, x') \\ y &= -h_x(x, x') \end{aligned} \tag{1.1}$$

generate, implicitly, a symplectic map $\mathbf{T} : (x, y) \mapsto (x', y')$.

The idea of Lagrangian generating functions for symplectic maps is relatively new, having been introduced in the late 1970's (Percival [1980], Aubry [1983]). However, because of their associated variational structure – an orbit of the symplectic map $\mathbf{T}$ generated by $h(x, x')$ corresponds to a critical point of the action $\mathrm{W}(x) = \sum_j h(x^j, x^{j+1})$ – Lagrangian generating functions have turned out to be extremely useful for proving existence of invariant circles, for converse KAM theory and for transport theories (reviewed in MacKay [1986] and Meiss [1991]). Most of the above results are for the area-preserving case. It is even more recent that Lagrangian generating functions have been extended to symplectic maps on $\mathbf{R}^{2n}$, $n \geq 2$ (Bernstein & Katok [1987], MacKay, Meiss & Stark [1989], Kook & Meiss [1989], Meiss [1991]). For period-q points of symplectic maps on $\mathbf{R}^{2n}$ we use Lagrangian generating functions to reduce the existence question to a problem of equivariant critical point theory. We concentrate in this work on $q \geq 3$, although in general q can be any natural number.

The idea is to reduce the question of existence of bifurcating solutions to a normal form on some low (lowest) dimensional space. For symplectic maps on $\mathbf{R}^{2n}$, period-q points correspond to critical points of a functional – the action functional – on an nq-dimensional space. We decompose the nq-dimensional space into an m-dimensional space (typically $m = 2$ or with symmetry $m = 2n$) on which the functional is "singular" and an $nq - m$ dimensional space on which the functional is nondegenerate. Because of the presence of *spatial* (Σ) and *temporal* ($\mathbb{Z}_q$) symmetries the action functional is $\Sigma \times \mathbb{Z}_q$-invariant. The Equivariant Splitting Lemma is used to decompose the action functional into a nondegenerate quadratic functional on $\mathbf{R}^{nq-m}$ and a reduced functional on $\mathbf{R}^m$ *that inherits the symmetry of the full problem*. This decomposition is basic to our analysis and with modest assumptions the idea extends to symplectic maps of arbitrary dimension, to symmetric maps with Σ a compact Lie group and to various degeneracies of interest such as the collision of multipliers.

In other words, the existence of bifurcating period-q points is in one-to-one correspondence with critical points of a $\Sigma \times \mathbb{Z}_q$-invariant functional on $\mathbf{R}^m$ and when $\Sigma = \mathrm{id}$ $m = 2$; in particular, the dimension of the reduced space, m, is independent of the configuration space dimension, n, and the sequence space dimension, q. In Chapter 2 a generalization of Meyer's theorem is proved on the generic bifurcation and stability of periodic points in one-parameter symplectomorphisms on $\mathbf{R}^{2n}$. In our framework the question reduces to the existence of bifurcating critical points of an $\mathbb{Z}_q$-equivariant gradient map on $\mathbf{R}^2$ dependent on a single parameter.

In subsequent chapters the theory for bifurcating period-q points is extended to include

(a) the case where the symplectomorphism depends on more than one parameter and we classify all possible singularities of bifurcating period-q points up to codimension 2 (with a precise definition of codimension given in Chapter 3).

(b) In 4D-symplectomorphisms, the effect of a collision of multipliers of opposite signature on bifurcating period-q points (Chapters 9-10).

(c) The effect of spatial symmetry on the bifurcation of period-q points (Chapter 8).

In all cases we "project" the action functional, via the Equivariant Splitting Lemma, to a reduced $\Sigma \times \mathbb{Z}_q$-equivariant gradient map on $\mathbb{R}^m$. In other words the problem can be reduced to a question in singularity theory.

Suppose we have the simplest case: $m = 2$, $\Sigma = \mathrm{id}$ and the map depends on a single parameter λ. From a singularity theory point of view, given a $\mathbb{Z}_q$-equivariant gradient map on $\mathbb{R}^2$, $\nabla_x f(x, \lambda)$, obtained from a reduction of the action functional for period-q points, satisfying say

$$\nabla_x f(0,0) = 0 \quad \text{and} \quad \det \mathrm{Hess}_x f(0,0) = 0\,,$$

what is the simplest "normal form" "equivalent" to $f(x, \lambda)$? What equivalence relation is most suitable to the origin of the problem, that is bifurcating period-q points ? Is the germ finitely determined ? What are all possible perturbations of f (universal unfolding) ?

There are various "off the shelf" singularity theory frameworks available. For example Mather's classic contact-equivalence, without any distinguished parameter and without a gradient structure, has been used successfully by Hummel [1979] to classify $\mathbb{Z}_q$-equivariant maps on $\mathbb{R}^2$ (with an eye towards classifying singularities of periodic points of (non-symplectic) diffeomorphisms). There is catastrophe theory – right equivalence – which is natural for maps with a gradient structure. And there is singularity theory for bifurcation problems – distinguished parameter singularity theory – of Golubitsky and Schaeffer. In a similar vein the recent developments of Mond & Montaldi [1991] put special emphasis on the role of parameters and paths in parameter space.

Periodic points occur generically in one-parameter families of symplectic maps in the neighborhood of an elliptic fixed point. Therefore it is natural to include a distinguished parameter in the singularity theory framework. (There are a number of other arguments in favor of this. For example if the one-parameter represents energy of a Hamiltonian system and a bifurcation is associated with increasing energy the equivalence relation should preserve this orientation.) It is also of interest to preserve the gradient structure. Therefore the singularity theory framework we use is equivariant contact-equivalence with a distinguished parameter that preserves the gradient structure.

Here, the difficulty is that the set of such contact-equivalences does not form a nice algebraic structure; in particular, its elements depend on which gradient they are applied to. But we can circumvent the difficulties and present a new theory, in Chapter 3, in complete generality for Γ-equivariant gradient maps on $\mathbf{R}^m$ with Γ a discrete group.

Then in Chapter 4 the singularity theory is applied to $\mathbf{Z}_q$-equivariant gradient maps on $\mathbf{R}^2$ with a distinguished parameter. This is the basic result necessary for the classification of the singularities of bifurcating period-q points.

In Chapters 5 and 7 the singularities of two-parameter area-preserving maps are considered; that is, when certain coefficients, in the normal form for generic (one-parameter) period-q points, go through zero the structure of the bifurcating period-q points changes. In particular there are folds, symmetry-breaking bifurcations and secondary bifurcations of periodic points.

It should be noted as well that rigorous results on (linear) stability are obtained in the case where the reduced functional is on $\mathbf{R}^2$ using a generalization of the MacKay-Meiss formula (see Appendix B) and projecting onto the normal form space. In this way we are able to track the effect of degenerate bifurcations, in 2-parameter maps, on the stability configurations.

In Chapter 8 the effect of spatial symmetries on bifurcating period-q points is considered. As far as we are aware this is the first work on equivariant symplectomorphisms with nontrivial spatial symmetries; in particular continuous symmetries. This is surprising as equivariant symplectic maps are quite important in applications. Simple examples giving rise to equivariant symplectic maps can be obtained by parametrically forcing some of the classical problems in mechanics such as the spherical pendulum and the rigid body. In Section 8.4 we show for example how the periodic points of an $\mathbf{O}(2)$-equivariant symplectic map are related to the dynamics of a parametrically forced spherical pendulum. A phase space of at least 4 dimensions (2D-configuration space) is necessary for non-trivial configuration space symmetries. An area-preserving map for example has only a one-dimensional configuration space: hence, the only nontrivial spatial symmetries are equivalent to $\mathbf{Z}_2$ in which case the bifurcations are not significantly different from those of a reversible area-preserving map.

Let $\mathbf{T} : (x, y) \mapsto (x', y')$ be a Σ-equivariant symplectic map with a Σ-invariant fixed point and suppose there is a bifurcation of period-q points from the fixed point. Then a basic question is: how much of $\Sigma \times \mathbf{Z}_q$ is inherited by the branches of bifurcating period-q points? This is a question of spontaneous symmetry breaking. A related but more abstract question is: given a spatial symmetry Σ, how many distinct families of period-q points can we expect to bifurcate? We will take two approaches to this problem in Chapter 8. First, using a minimal amount of information about the Σ-equivariant symplectomorphism (spatial-temporal symmetry and gradient structure) and suitable non-degeneracy hypotheses, we use topological results – results based on Γ-length of sets of Bartsch [1992b], Bartsch & Clapp [1990] (among others) or equivariant Lusternik-Schnirelman category – to obtain lower bounds on the number of geometrically distinct branches of bifurcating period-q points and their symmetry group. This idea is reminiscent of the classical "Weinstein-Moser" theory for periodic orbits of continuous-time Hamiltonian systems and generalizes to the subharmonic case a theorem of Montaldi, Roberts & Stewart [1988] on bifurcation of periodic solutions from symmetric equilibria of Hamiltonian vectorfields.

When the spatial symmetry Σ acts absolutely irreducibly on the configuration space of the symplectic map, the results on bifurcation of period-q points can be sharpened. In particular, for every isotropy subgroup $\Pi \subset \Sigma \times \mathbf{Z}_q$ with Weyl group $W\Pi \stackrel{\text{def}}{=} N\Pi/\Pi$ (where $N\Pi$ is the normalizer of Π in $\Sigma \times \mathbf{Z}_q$) there exist *at least* $\mathrm{cat}_{W\Pi}\mathrm{Fix}\,\Pi$ $W\Pi$-orbits of bifurcating period-q points whose symmetry group is at least Π ($\mathrm{cat}_{W\Pi}\mathrm{Fix}\,\Pi$ denotes the $W\Pi$-category of the unit sphere in $\mathrm{Fix}\,\Pi$).

This result is straightforward to apply. For example, it is used in Section 8.3 to classify bifurcating period-q points of $\mathbf{O}(2)$-equivariant symplectomorphisms; in particular, for each $q \geq 3$ there are 3 geometrically distinct conjugacy classes. In cases where $\dim(\mathrm{Fix}\,\Pi) = 2$ we can restrict to the fixed-point subspace and again apply the singularity theory of Chapter 4 to obtain more precise information on the branches of period-q points. This program is carried out for $\Sigma = \mathbf{O}(2)$ in Section 8.3.

When a symplectomorphism has a continuous symmetry then we expect an associated conserved quantity. For symplectic maps with a Lagrangian generating function it is particularly easy to prove – via a discrete version of Noether's Theorem – that every continuous spatial symmetry generates a conserved quantity. This result is proved in Appendix E. When applied to the $\mathbf{O}(2)$-equivariant map on $\mathbf{R}^4$ is shows that, generically, every orbit of the map lies on an invariant submanifold that is equivalent to the interior of a solid torus! This leads to interesting geometric structures in the phase space. In the case of the parametrically forced spherical pendulum the conserved quantity of the symplectic map is a discrete analog of the angular momentum of the pendulum. For the $\mathbf{O}(2)$-equivariant symplectic map we give results for the period-q points only, but they are quite suggestive about other interesting dynamics – invariant circles that ride on the group orbit, drift of structures transverse to the group orbit, etc. – of equivariant symplectic maps.

One of the original motivations for this study was to understand the collision of multipliers singularity in symplectic maps. The collision of multipliers of opposite signature is one of the three ways that a one-parameter family of periodic orbits in a Hamiltonian system can lose stability and it is the most difficult to analyze because it involves the bifurcation of invariant tori. Just before a collision, when the multipliers are distinct and sufficiently irrational, we expect – by KAM theory – a surrounding invariant 2-torus (in the map) but, in some sense, the 2-torus collapses to a 1-torus (an invariant circle), at the collision, that may or may not vanish after the collision. Our idea is to treat the two-parameter problem and give a complete treatment of the collision of multipliers at rational points. We will see that for $\theta = \frac{2p}{q}\pi$, and q sufficiently large we get a rough idea about the structure of the irrational collision.

The collision of multipliers in symplectomorphisms is the analogue of the collision of purely imaginary eigenvalues in the vectorfield case (Meyer & Schmidt [1971], Sokol'skij [1974], van der Meer [1985], Bridges [1990,1991]). In fact, consider the linear normal form for the collision of purely imaginary eigenvalues in a Hamiltonian vectorfield,

$$\frac{d}{dt}\begin{pmatrix} q \\ p \end{pmatrix} = \mathbf{J}\mathbf{A}_0 \begin{pmatrix} q \\ p \end{pmatrix} \quad \text{with} \quad \mathbf{A}_0 = \begin{pmatrix} \mathbf{0} & -\theta\mathbf{J} \\ \theta\mathbf{J} & \epsilon\mathbf{I} \end{pmatrix}$$

with $\theta \in \mathbb{R}$ and $\epsilon = \pm 1$. Exponentiation of $\mathbf{J}\mathbf{A}_0$ then yields the linear normal form for a collision of multipliers in symplectic maps on $\mathbb{R}^4$ (see Appendix D),

$$\begin{pmatrix} x_{n+1} \\ y_{n+1} \end{pmatrix} = \mathbf{M}_0 \begin{pmatrix} x_n \\ y_n \end{pmatrix} \quad \text{with} \quad \mathbf{M}_0 = \exp(\mathbf{J}\mathbf{A}_0) = \begin{pmatrix} \mathbf{R}_\theta & \epsilon\mathbf{R}_\theta \\ \mathbf{0} & \mathbf{R}_\theta \end{pmatrix}$$

where $\mathbf{R}_\theta$ is the rotation matrix. Note that the value of θ (the frequency) is not particularly important in the vectorfield case but has great importance in the map case (rotation number).

An unfolding of $\mathbf{A}_0$ is given by

$$\mathbf{A}(\lambda,\alpha) = \begin{pmatrix} \alpha\mathbf{I} & -(\theta+\lambda)\mathbf{J} \\ (\theta+\lambda)\mathbf{J} & \epsilon\mathbf{I} \end{pmatrix}.$$

However exponentiation of $\mathbf{A}(\lambda,\alpha)$ leads to $\widehat{\mathbf{M}}(\lambda,\alpha) = \exp\big(\mathbf{J}\mathbf{A}(\lambda,\alpha)\big)$ which depends nonlinearly on α. We prove (Proposition 9.1) that an equivalent unfolding of $\mathbf{M}_0$ in $\mathbf{Sp}(4,\mathbb{R})$ is

$$\mathbf{M}(\lambda,\alpha) = \begin{pmatrix} (1+\epsilon\,\alpha)\mathbf{R}_{\theta+\lambda} & \epsilon\mathbf{R}_{\theta+\lambda} \\ \alpha\mathbf{R}_{\theta+\lambda} & \mathbf{R}_{\theta+\lambda} \end{pmatrix}.$$

The 2-parameter matrix $\mathbf{M}(\lambda,\alpha)$ is in general position for a collision of multipliers at rational points (for a collision at irrational points λ can be taken to be zero. On the other hand a more complete analysis of the structure of the irrational collision is obtained with the two parameters.) and is the starting point for our theory. In particular, for (λ,α) sufficiently small we study bifurcating period-q points of the map $z_{n+1} = \mathbf{M}(\lambda,\alpha)\, z_n + h.o.t.$. Linear stability results for the bifurcating period-q points near a collision turn out to be difficult but we make some headway on this problem in Sections 9.5 and 9.6. A sufficient condition for linear instability is obtained in Section 9.5. In Section 9.6 a linear stability theory for bifurcating period-4 points – in the unfolding of the collision at $\pm i$ – is presented. Here the interesting result is that there is always a secondary irrational, small-angle collision along one of the globally connected branches.

Of crucial importance for the study of multiplier collision is the signature of multipliers and we obtain new results on signature in configuration space for Lagrangian generating functions and this material is recorded in Appendix B.

Surprisingly there has not been much work on the bifurcation of symplectomorphisms near a collision of multipliers. The linear theory (normal form and signature theory) has been well understood for some time but the nonlinear problem is difficult because it involves the bifurcation of invariant tori in the map. Recent results of Bridges & Cushman [1993] and Bridges, Cushman & MacKay [1993] use normal form theory and Poisson reduction to obtain a geometrical picture in the phase space for the irrational collision. There are interesting numerical calculations on the bifurcations near irrational and rational collisions by Pfenniger [1985,1987]. For reversible (but non-symplectic) diffeomorphism the collision of multipliers singularity also occurs. Recent results on the collision of multipliers in one-parameter reversible maps have been reported by Sevryuk & Lahiri [1991] and Bridges, Cushman & MacKay [1993, Section 4].

Finally in Chapter 10 we consider equivariant symplectic maps with a collision of multipliers. We are mainly interested here in admissible spatial symmetries that do not prevent the basic collision in 4D-maps. The simplest additional symmetry is reversibility. However reversibility has little effect on the collision: extra $\mathbf{Z}_2$ in the normal form and extra symmetry in the sequences of period-q points, but the normal form and local geometric structure in the neighborhood of the collision is similar to the non-reversible case. On the other hand non-reversible symplectomorphisms admit a larger group of spatial symmetries which in turn leads to interesting structures in the neighborhood of a collision. Admissible spatial symmetries, for symplectomorphisms on $\mathbf{R}^4$, include $\mathbf{SO}(2)$, $\mathbf{Z}_m$, $m \geq 2$, and $\mathbf{Z}_2 \times \mathbf{Z}_2$. The group $\mathbf{SO}(2)$ has the most dramatic effect on the collision (surprisingly, it does not eliminate the collision) because of the presence of a continuous symmetry and we analyze the $\mathbf{SO}(2)$-symmetric collision in some detail.

Throughout, unless otherwise mentioned (see remarks in Appendix I), the symplectic form is taken to be the standard one:

$$\omega = \sum_{i=1}^{n} dx_i \wedge dy_i \quad \text{with} \quad (x, y) \in \mathbf{R}^{2n}$$

with unit symplectic operator $\mathbf{J}_{2n} = \mathbf{J} \otimes \mathbf{I}_n$. When $n = 1$ we drop the subscript:

$$\mathbf{J} \stackrel{\text{def}}{=} \mathbf{J}_{2n}\Big|_{n=1} = \begin{pmatrix} 0 & 1 \\ -1 & 0 \end{pmatrix} \quad \text{and} \quad \mathbf{I} \stackrel{\text{def}}{=} \mathbf{I}_2 \,.$$

The term symplectomorphism is now becoming standard, especially in the Russian literature: the terms symplectomorphism and symplectic map are used interchangeably throughout the text. The symplectic maps and generating functions are assumed to be as smooth as necessary. Our analyses are local: we work in the neighborhood of fixed points throughout (this avoids questions of global existence of generating functions and makes it convenient to work with germs) and have found it convenient to express the phase space and configuration space variables in Euclidean coordinates – rather than action-angle variables.

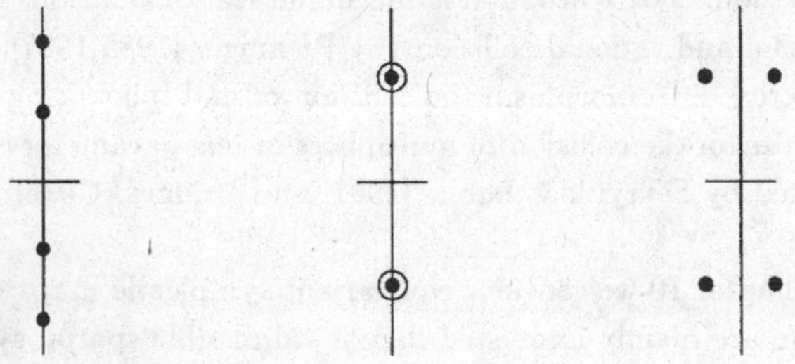

(a) Collision of purely imaginary eigenvalues in a Hamiltonian vectorfield

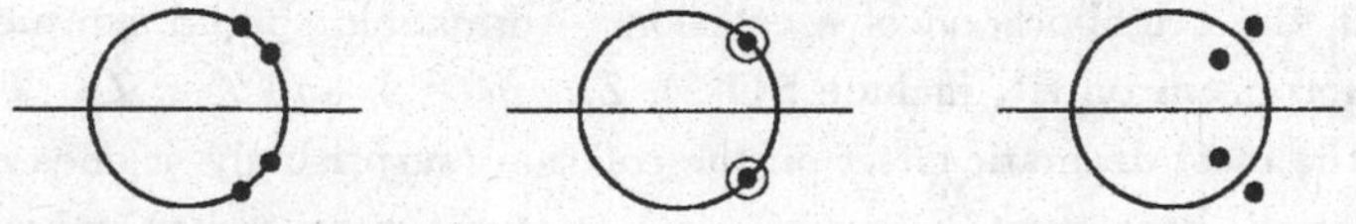

(b) Collision of multipliers of opposite signature in a symplectic map

Figure 1

2. Generic bifurcation of periodic points

Let $\mathbf{T} : (x, y, \Lambda) \mapsto (x', y')$ be a smooth family of symplectic maps on $\mathbf{R}^{2n}$ depending on a multidimensional parameter Λ and suppose that $\mathbf{T}$ has a fixed point (which without loss of generality can be taken to be the origin) when $\Lambda = 0$.

There are several ways for the parameters to affect $\mathbf{T}$ but, to fix the ideas, we have chosen the following "canonical" splitting of the parameter space:

$$\Lambda = (\hat{\lambda}, \beta)$$

where $\hat{\lambda}$ is a control (multi)parameter for the linear part in (x, y) of $\mathbf{T}$ at the origin, $\mathbf{DT}(\hat{\lambda}) \stackrel{\text{def}}{=} D_{(x,y)}\mathbf{T}(0, 0, \Lambda) \in \mathbf{Sp}(2n, \mathbf{R})$. The (multi)parameter β controls only the coefficients of the higher order terms of $\mathbf{T}$ and so is considered as a perturbation (*unfolding*) parameter.

Typically $\hat{\lambda}$ controls the eigenvalues of $\mathbf{DT}$ around the resonances provoking the bifurcations and so, generically, $\hat{\lambda}$ is unidimensional, denoted by λ. Otherwise $\hat{\lambda}$ is itself split into (λ, α) where λ is the *main* bifurcation parameter and α is another *perturbation* parameter (cf. Section 2.6 on the collision of multipliers). Although λ can conceivably be taken to be multidimensional, in this work we are going to consider only $\lambda \in \mathbf{R}$. Together, (λ, α) control the (multi)parameter unfolding of the linear part $\mathbf{DT}^o$ in the sense of Arnold [1988]. By perturbation/unfolding parameter we mean that we consider α, β as fixed with respect to the variation of λ.

Suppose that the origin is an elliptic fixed point of $\mathbf{T}$ when $\Lambda = 0$, say. In particular, for $\mathbf{DT}^o \stackrel{\text{def}}{=} \mathbf{DT}(0 \ldots 0)$ we assume that

$$\sigma(\mathbf{DT}^o) = \{e^{\pm i\theta_1}, \cdots, e^{\pm i\theta_n}\}, \quad \theta_j = 2\pi\rho_j \quad \text{and} \quad \rho_j \in (0, \tfrac{1}{2}),\ 1 \leq j \leq n. \qquad (2.1)$$

In that case, a simple application of the Implicit Function Theorem shows that locally (in a neighborhood of the origin) $\mathbf{T}$ has an unique family of fixed points, which we can consider to be parametrized by $\hat{\lambda}$. Generically for a one-parameter family each multiplier in $\sigma(\mathbf{DT}(\lambda))$ lies at an irrational point: $\rho_j \in (0, \frac{1}{2}) \setminus \mathbf{Q}$ for almost all values of λ. But, at particular values of λ, individual multipliers may lie at rational points.

Suppose that when $\lambda = 0$ one of the multipliers lies at a rational point and the remaining $n-1$ multipliers lie at irrational points. Without loss of generality, take

$$\rho_1 = \tfrac{p}{q} \in \mathbf{Q} \cap (0, \tfrac{1}{2}) \quad \text{and} \quad \rho_j \in (0, \tfrac{1}{2}) \setminus \mathbf{Q},\ 2 \leq j \leq n, \qquad \text{(H0)}$$

with $\frac{p}{q}$ in lowest terms and $q \geq 3$.

The assumption (2.1) is not a necessary condition for bifurcation of period-q points. The requirement that all multipliers lie on the unit circle is not essential but makes the discussion of stability worthwhile, as all periodic points in the neighborhood of a fixed point with a pair of multipliers outside the unit circle are, by continuity, unstable. In other words, the *existence* theory for period-q points is not affected by the presence of multipliers outside (or inside) the unit circle, although such periodic points will be unstable.

The hypothesis (H0) is introduced to simplify some technical details of the exposition. More general situations, such as additional multipliers at rational points or higher multiplicities, can be treated either with a "Weinstein-Moser"-type of theory (cf. Section 8.1) or adapted from the framework introduced in this chapter. For instance, if another multiplier is at a rational point, say $\frac{r}{s}$, our results hold true if $s > q$. On the other hand, there may be additional periodic points of period $s, \ldots, gcm(q, s)$. The problem is thus best treated in the space of period-$gcm(q, s)$ orbits (see Appendix J for mode interactions). Results on stability also hold true in this context if there are no multiplier collisions and no resonances.

In this chapter we introduce a framework for the bifurcation of periodic points and give a new proof of Meyer's theorem (Meyer [1970,1971]) on the bifurcation and stability of period-q points. In particular, in the one-parameter family of symplectic twist maps $\mathbf{T}(\lambda)$ with $\mathbf{DT}^o$ satisfying (2.1), with the hypothesis (H0), and assuming the multiplier passes through the rational point with nonzero speed, there is a bifurcation of period-q points associated with the multiplier with exponent $\theta_1 = \frac{2p}{q}\pi$.

2.1 Lagrangian variational formulation

To study the existence and stability of period-q points, the symplectic map will be identified with an underlying variational structure given by a generating function. In particular, what we will show (Proposition 2.1) is that at an elliptic point there always exists, at least locally, a Lagrangian generating function with an associated variational principle. That is, for symplectic maps in the neighborhood of an elliptic fixed point, there is a one-to-one correspondence between the orbits of the map and critical points of a Lagrangian generating function (see also remarks in Appendix G on twist maps) First we introduce some basic facts about Lagrangian generating functions for symplectic maps on $\mathbb{R}^{2n}$ following MacKay, Meiss & Stark [1989, Appendix A].

Let $\mathbf{T} : (x, y, \Lambda) \to (x', y')$ be a smooth Λ-parametrized family of symplectic maps on $\mathbb{R}^{2n}$; that is $\mathbf{DT}(x, y, \Lambda) \in \mathbf{Sp}(2n, \mathbb{R})$ and let $\pi_x, \pi_y : \mathbb{R}^{2n} \to \mathbb{R}^n$ be projection operators taking $\mathbb{R}^{2n}$ onto the first and second n components respectively: $\pi_x(x, y) = x$ and $\pi_y(x, y) = y$. Suppose

$$|\pi_x \cdot \mathbf{DT}(x, y, \Lambda) \cdot \pi_y| \neq 0, \tag{2.2}$$

where $|\mathbf{A}|$ denotes the determinant of the matrix $\mathbf{A}$, then there exists a Lagrangian variational formulation associated with $\mathbf{T}$. In particular, there exists a generating function $h(x, x', \Lambda)$ with

$$y = -\nabla_x h(x, x', \Lambda) \overset{\text{def}}{=} -h_1(x, x', \Lambda) \tag{2.3a}$$

$$y' = \nabla_{x'} h(x, x', \Lambda) \overset{\text{def}}{=} h_2(x, x', \Lambda) \tag{2.3b}$$

and in terms of h the condition (2.2) becomes

$$|h_{12}(x, x', \Lambda)| \neq 0 . \tag{2.4}$$

When h satisfies the condition (2.4), the relations (2.3) define implicitly a family of symplectic maps $(x, y, \Lambda) \mapsto (x', y')$ on $\mathbf{R}^{2n}$.

The generating function h can be used to define a gradient structure through the *action*:

$$\mathrm{W}_{(M,N)}(\mathbf{x}, \Lambda) = \sum_{i=M}^{N-1} h(x^i, x^{i+1}, \Lambda) \quad \text{where} \quad \mathbf{x} = \{x^i\}_{i=M}^{N}.$$

Note that, at fixed Λ, the orbit by $\mathbf{T}$ of the point $(x^M, y^M) \in \mathbf{R}^{2n}$ from time M to time N is the sequence $\{x^i, y^i\}$ defined by

$$(x^{i+1}, y^{i+1}) = \mathbf{T}(x^i, y^i, \Lambda) , \quad M \leq i \leq N-1 .$$

In terms of the action however the orbit corresponds to the sequence $\{x^i\}$ in *configuration space* for which the action is stationary for all finite segments (with respect to variations that fix x^M and x^N) (see also remarks in Appendix G on dynamical equivalence). Let $\{\xi^i\}$ be a variation with $\xi^M = \xi^N = 0$, then

$$\begin{aligned}
\left.\frac{d}{d\epsilon}\right|_{\epsilon=0} \mathrm{W}_{(M,N)}(\mathbf{x} + \epsilon\xi, \Lambda) &= \sum_{j=M}^{N-1} (h_1(x^i, x^{i+1}, \Lambda), \xi^i) + (h_2(x^i, x^{i+1}, \Lambda), \xi^{i+1}) \\
&= \sum_{j=M}^{N-1} (h_1(x^i, x^{i+1}, \Lambda) + h_2(x^{i-1}, x^i, \Lambda), \xi^i) \\
&= [\nabla_{\mathbf{x}} \mathrm{W}_{(M,N)}(\mathbf{x}, \Lambda), \xi] ,
\end{aligned}$$

where $(,)$, $[,]$ are the scalar products on $\mathbf{R}^n$ and the space of $(N-M-1)$-sequences, respectively. In particular, the action is stationary when $\nabla_{\mathbf{x}} \mathrm{W}_{(M,N)}(\mathbf{x}, \Lambda) = 0$ for all integers (M, N) with $M < N$ or when

$$h_1(x^i, x^{i+1}, \Lambda) + h_2(x^{i-1}, x^i, \Lambda) = 0 , \quad \forall i \in \mathbf{Z} . \tag{2.5}$$

However, from equation (2.3), $h_1(x^i, x^{i+1}, \Lambda) = -y^i$ and $h_2(x^{i-1}, x^i, \Lambda) = y^i$; that is, an orbit of the symplectic map generated by h corresponds to a path of stationary action. Conversely, given a function h satisfying the condition (2.4), the relations

(2.3) generate a symplectic map whose orbits correspond to sequences of stationary action. Proofs of the above statements can be found in MacKay, Meiss & Stark [1989, Appendix A].

If in fact there exists a generating function h (see Proposition 2.1), then in the neighborhood of an elliptic fixed point we can take, without loss of generality, $h^o = h_1^o = h_2^o = 0$. Then a general expression for the (smooth) Lagrangian generating function h is

$$\begin{aligned} h(x, x', \Lambda) &= \frac{1}{2}\begin{pmatrix} x \\ x' \end{pmatrix}\begin{bmatrix} \mathbf{A}(\hat{\lambda}) & -\mathbf{B}(\hat{\lambda}) \\ -\mathbf{B}(\hat{\lambda})^T & \mathbf{C}(\hat{\lambda}) \end{bmatrix}\begin{pmatrix} x \\ x' \end{pmatrix} + \hat{h}(x, x', \Lambda), \\ &= \tfrac{1}{2}(x, \mathbf{A}(\hat{\lambda})\, x) - (x, \mathbf{B}(\hat{\lambda})\, x') + \tfrac{1}{2}(x', \mathbf{C}(\hat{\lambda})\, x') + \hat{h}(x, x', \Lambda), \end{aligned} \tag{2.6}$$

where $\hat{h}(x, x', \Lambda)$ begins with terms of degree 3 in (x, x') and $\mathbf{A}$, $\mathbf{B}$, $\mathbf{C}$ are $n \times n$ matrices with $\mathbf{A}$, $\mathbf{C}$ symmetric and $\mathbf{B}$ satisfying

$$|\mathbf{B}| \neq 0\,. \tag{2.7}$$

With the hypothesis (2.7) the quadratic part of the generating function in (2.6) generates the linear symplectic map

$$\begin{pmatrix} x' \\ y' \end{pmatrix} = \begin{bmatrix} \mathbf{B}^{-1}\mathbf{A} & \mathbf{B}^{-1} \\ \mathbf{C}\mathbf{B}^{-1}\mathbf{A} - \mathbf{B}^T & \mathbf{C}\mathbf{B}^{-1} \end{bmatrix}\begin{pmatrix} x \\ y \end{pmatrix} = \mathbf{DT}(\hat{\lambda})\begin{pmatrix} x \\ y \end{pmatrix}. \tag{2.8}$$

Proposition 2.1. *Let* $\mathbf{T} : (x, y) \mapsto (x', y')$ *be a symplectic map on* $\mathbf{R}^{2n}$ *with an elliptic fixed point at the origin and suppose* $\sigma(\mathbf{DT}^o)$ *satisfies (2.1) with* distinct *multipliers. Then there exists, at least locally, a generating function* $h(x, x')$ *for* $\mathbf{T}$.

Proof. If the spectrum of $\mathbf{DT}^o$ has the form given in (2.1) then there exists a symplectic transformation $\widehat{\mathbf{S}} \in \mathbf{Sp}(2n, \mathbf{R})$ such that

$$\widehat{\mathbf{S}}^{-1}\mathbf{DT}^o\widehat{\mathbf{S}} = \widehat{\mathbf{M}}^o = \begin{bmatrix} \cos\theta_1 & & 0 & \epsilon_1 \sin\theta_1 & & 0 \\ & \ddots & & & \ddots & \\ 0 & & \cos\theta_n & 0 & & \epsilon_n \sin\theta_n \\ -\epsilon_1 \sin\theta_1 & & 0 & \cos\theta_1 & & 0 \\ & \ddots & & & \ddots & \\ 0 & & -\epsilon_n \sin\theta_n & 0 & & \cos\theta_n \end{bmatrix} \tag{2.9}$$

where ϵ_j is the signature of the multiplier $e^{i\theta_j}$ (see Appendices B and D). Without loss of generality we can suppose that $\mathbf{DT}^o$ is in normal form (2.9). Let h be of the form given in (2.6) with

$$\begin{aligned} \mathbf{A} = \mathbf{C} &= \operatorname{diag}(\epsilon_1 \cot\theta_1, \ldots, \epsilon_n \cot\theta_n), \\ \mathbf{B} &= \operatorname{diag}\left(\frac{\epsilon_1}{\sin\theta_1}, \ldots, \frac{\epsilon_n}{\sin\theta_n}\right). \end{aligned} \tag{2.10}$$

Then the quadratic part of h generates the linear map (2.9) and the condition (2.4) becomes

$$|h_{12}(x,x')| = |-\mathbf{B}+\hat{h}_{12}(x,x')| = |-\mathbf{B}|\cdot|\mathbf{I}_n-\mathbf{B}^{-1}\hat{h}_{12}(x,x')|,$$

but

$$|-\mathbf{B}| = (-1)^n\prod_{k=1}^{n}\frac{\epsilon_k}{\sin\theta_k}\neq 0,$$

and, for orbits $\{x^i\}$ in a sufficiently small neighborhood of the trivial fixed point,

$$|\mathbf{I}_n-\mathbf{B}^{-1}\hat{h}_{12}(x,x')|\neq 0,$$

completing the proof. ■

Therefore, for the bifurcation theory of period-q points, we can assume that $\mathbf{T}$ is generated by h.

Now let

$$\mathbf{X}_q^n = \{\{x^i\}\in(\mathbf{R}^n)^{\mathbf{Z}} \;:\; x^{i+q}=x^i,\quad \forall i\in\mathbf{Z}\}. \tag{2.11}$$

Then a period-q point of the symplectic map $\mathbf{T}$ is a sequence

$$(x^{i+1},y^{i+1}) = \mathbf{T}(x^i,y^i,\Lambda)\quad\text{with}\quad \{x^i,y^i\}\in\mathbf{X}_q^n\times\mathbf{X}_q^n.$$

The idea is to use the variational structure to obtain period-q points. In particular let

$$\mathrm{W}_q(\mathbf{x},\Lambda)\stackrel{\text{def}}{=}\mathrm{W}_{(M,N)}(\mathbf{x},\Lambda)\Big|_{\mathbf{X}_q^n} = \sum_{j=1}^{q}h(x^i,x^{i+1},\Lambda), \tag{2.12}$$

then period-q points of the symplectic map $\mathbf{T}$ generated by h are critical points of W_q in $\mathbf{X}_q^n$.

2.2 Linearization and unfolding

When the spectrum of $\mathbf{DT}^o$ is given by (2.1) and one of the multipliers lies at a rational point the matrix is not stable under perturbation in $\mathbf{Sp}(2n,\mathbf{R})$.

Proposition 2.2. *Suppose* $\mathbf{DT}^o\in\mathbf{Sp}(2n,\mathbf{R})$ *has the spectrum given by (2.1) and satisfies (H0). Then* $\mathbf{DT}^o$ *is of codimension 1 in* $\mathbf{Sp}(2n,\mathbf{R})$ *and a miniversal unfolding is given by* $\widehat{\mathbf{M}}^o$, *in (2.9), with* θ_1 *replaced by* $\theta_1-\lambda$, *which we denote by* $\widehat{\mathbf{M}}(\lambda)$.

Proof. The idea is to give a proof in the Lie algebra of $\mathbf{Sp}(2n,\mathbf{R})$. Note that

$$\widehat{\mathbf{A}}^o\stackrel{\text{def}}{=}\ln\widehat{\mathbf{M}}^o = \mathbf{J}\otimes\mathbf{D}^o\quad\text{with}\quad\mathbf{D}^o=\mathrm{diag}(\epsilon_1\theta_1,\ldots,\epsilon_n\theta_n).$$

When $\theta_1 = 2\pi\rho_1$ and ρ_1 lies at a rational point, any small perturbation of $\mathbf{D}^o$ will perturb the multiplier to an irrational point whereas a multiplier at an irrational point is stable under perturbation. Therefore an unfolding of $\mathbf{D}^o$ in $\mathbf{sp}(2n,\mathsf{R})$ is given by perturbing θ_1 to $\theta_1 - \lambda$. Then exponentiation of $\mathbf{J} \otimes \mathbf{D}(\lambda)$ results in $\widehat{\mathbf{M}}(\lambda)$. ■

To revise (2.10) to take account of the unfolding of the rational multiplier θ_1, take $\mathbf{A}(\lambda)$ and $\mathbf{B}(\lambda)$ to be $\mathbf{A}$ and $\mathbf{B}$ given in (2.10) with θ_1 replaced by $\theta_1 - \lambda$. This means that we have thus chosen the direction of variation of our main bifurcation parameter λ. In practical terms there is no problem in changing coordinates to adapt to this situation. Such a coordinate transformation will in general bring λ into the higher order terms which usually already depend on the perturbation parameter β. Then the period$-q$ points of the map $\mathbf{T}$ are critical points of

$$\begin{aligned} \mathrm{W}_q(\mathbf{x},\Lambda) &= \sum_{j=1}^{q} h(x^j, x^{j+1}, \Lambda) \\ &= \sum_{j=1}^{q} \tfrac{1}{2}\langle x^j, \mathbf{A}(\lambda)\, x^j\rangle - \langle x^j, \mathbf{B}(\lambda)\, x^{j+1}\rangle + \tfrac{1}{2}\langle x^{j+1}, \mathbf{A}(\lambda)\, x^{j+1}\rangle \\ &\qquad + \sum_{j=1}^{q} \hat{h}(x^j, x^{j+1}, \Lambda) \\ &= \tfrac{1}{2}\langle \mathbf{x}\,,\, \mathbf{L}(\lambda)\mathbf{x}\,\rangle + N(\mathbf{x},\Lambda) \end{aligned} \tag{2.13}$$

where $\langle\, ,\, \rangle$ is the scalar product on $\mathbf{X}_q^n$ and $N(\mathbf{x},\Lambda) = \sum_{j=1}^{q} \hat{h}(x^j, x^{j+1}, \Lambda)$ begins with terms of degree three. With the symbol $\otimes$ denoting the usual tensor – or Kronecker – product, the linear operator $\mathbf{L}(\lambda)$ takes the interesting form

$$\mathbf{L}(\lambda) = 2\,\mathbf{I}_q \otimes \mathbf{A}(\lambda) - (\Gamma_q + \Gamma_q^T) \otimes \mathbf{B}(\lambda) \tag{2.14}$$

where

$$\Gamma_q = \begin{pmatrix} 0 & 1 & 0 & \cdots & 0 \\ 0 & \ddots & \ddots & \ddots & \vdots \\ \vdots & \ddots & 0 & \ddots & 0 \\ 0 & & \ddots & \ddots & 1 \\ 1 & 0 & \cdots & 0 & 0 \end{pmatrix}. \tag{2.15}$$

The matrix Γ_q is the *fundamental circulant matrix* (Davis [1979]) on R^q. It is a $q \times q$ permutation matrix satisfying $\Gamma_q^q = \mathbf{I}_q$ ($\mathbf{I}_q$ is the identity on R^q); in particular, it is the generator for a representation of the abelian group $\mathbf{Z}_q$ on R^q which will be of fundamental importance in the sequel. We will come back to a discussion of symmetries shortly but first we consider the spectral analysis of the operator $\mathbf{L}^o$.

Proposition 2.3. *Let* $\mathbf{A}$ *and* $\mathbf{B}$ *be* $n \times n$ *matrices; let* Γ_q *be the fundamental circulant matrix on* $\mathbf{R}^q$ *and let*

$$\omega = \exp(2\pi i/q) \qquad q \geq 3\,.$$

Then

$$\sigma(2\mathbf{I}_q \otimes \mathbf{A} - \Gamma_q \otimes \mathbf{B} - \Gamma_q^T \otimes \mathbf{B}^T) = \bigcup_{j=1}^{q} \sigma(2\,\mathbf{A} - \omega^{j-1}\mathbf{B} - \overline{\omega}^{j-1}\mathbf{B}^T).$$

Moreover, if $\mathbf{A}$ *is symmetric the spectrum is real.*

Proof. Let

$$\widehat{\mathbf{L}} = 2\mathbf{I}_q \otimes \mathbf{A} - \Gamma_q \otimes \mathbf{B} - \Gamma_q^T \otimes \mathbf{B}^T$$

and consider the eigenvalue problem

$$\widehat{\mathbf{L}}\, z = \eta\, z\,.$$

The idea is to block diagonalize $\widehat{\mathbf{L}}$ using the Fourier matrix $\mathbf{F}$ (Davis [1979, p.32]) which is defined by

$$\overline{\mathbf{F}} = \frac{1}{\sqrt{q}} \begin{pmatrix} 1 & 1 & 1 & \cdots & 1 \\ 1 & \omega & \omega^2 & \cdots & \omega^{q-1} \\ 1 & \omega^2 & \ddots & & \vdots \\ \vdots & \vdots & & \ddots & \vdots \\ 1 & \omega^{q-1} & \cdots & \cdots & \omega^{(q-1)(q-1)} \end{pmatrix} \qquad \text{with } \omega = \exp(i\tfrac{2\pi}{q})\,. \tag{2.16}$$

The Fourier matrix diagonalizes Γ_q as follows

$$\mathbf{F}\Gamma_q\overline{\mathbf{F}} = \Omega = \operatorname{diag}(1, \omega, \omega^2, \ldots, \omega^{q-1})\,. \tag{2.17}$$

Now let $z = \overline{\mathbf{F}} \otimes \mathbf{I}_n \cdot \hat{z} \otimes w$. Then the eigenvalue problem $\widehat{\mathbf{L}}z = \eta z$ can be written

$$[\mathbf{F} \otimes \mathbf{I}_n \cdot \widehat{\mathbf{L}} \cdot \overline{\mathbf{F}} \otimes \mathbf{I}_n - \eta \mathbf{I}_q \otimes \mathbf{I}_n]\, \hat{z} \otimes w = 0\,.$$

Then using (2.17)

$$\begin{aligned} \mathbf{F} \otimes \mathbf{I}_n \cdot \widehat{\mathbf{L}} \cdot \overline{\mathbf{F}} \otimes \mathbf{I}_n &= 2\mathbf{I}_q \otimes \mathbf{A} - \Omega \otimes \mathbf{B} - \overline{\Omega} \otimes \mathbf{B}^T \\ &= \bigoplus_{i=1}^{q} (2\mathbf{A} - \omega^{j-1}\mathbf{B} - \overline{\omega}^{j-1}\mathbf{B}^T) \end{aligned}$$

which is block diagonal. In general the eigenvalues of each $n \times n$ block are complex but when $\mathbf{A}$ is symmetric each of the blocks has real spectrum since

$$[2\,\mathbf{A} - \omega^{j-1}\mathbf{B} - \overline{\omega}^{j-1}\mathbf{B}^T] \qquad j = 1, \ldots, q$$

are each Hermitian. ■

Corollary 2.4. *Let* $\mathbf{A}^o$ *and* $\mathbf{B}^o$ *be as given by (2.10). Then*

$$\mathbf{L}^o = 2\,\mathbf{I}_q \otimes \mathbf{A}^o - (\Gamma_q + \Gamma_q^T) \otimes \mathbf{B}^o$$

has two zero eigenvalues with $\operatorname{Ker} \mathbf{L}^o = \operatorname{span}\{\xi_1, \xi_2\}$ *and*

$$\xi_1 = \sqrt{\frac{2}{q}} \begin{pmatrix} 1 \\ \cos\theta \\ \cos 2\theta \\ \vdots \\ \cos(q-1)\theta \end{pmatrix} \otimes \begin{pmatrix} 1 \\ 0 \\ \vdots \\ 0 \end{pmatrix} \quad \text{and} \quad \xi_2 = \sqrt{\frac{2}{q}} \begin{pmatrix} 0 \\ \sin\theta \\ \sin 2\theta \\ \vdots \\ \sin(q-1)\theta \end{pmatrix} \otimes \begin{pmatrix} 1 \\ 0 \\ \vdots \\ 0 \end{pmatrix}. \tag{2.18}$$

The remaining $nq-2$ *eigenvalues of* $\mathbf{L}^o$ *are nonzero and if* P_σ *denotes the product of the non-zero eigenvalues of* $\mathbf{L}^o$, *then*

$$\operatorname{sign} \mathsf{P}_\sigma = (-1)^n \left(\prod_{k=1}^{n} \epsilon_k \right)^q \tag{2.19}$$

where $\{\epsilon_j\}_{j=1}^n$ *is the set of the signatures of the* n *multipliers of* $\mathbf{DT}^o$ *in (2.1).*

Proof. Let

$$\mathbf{M}_j = 2\,\mathbf{A}^o - 2(\cos \tfrac{2(j-1)}{q}\pi)\,\mathbf{B}^o\,, \quad 1 \le j \le q\,, \tag{2.20}$$

then by Proposition 2.3, $\sigma(\mathbf{L}^o) = \cup_{j=1}^q \sigma(\mathbf{M}_j)$. Using (2.10),

$$\mathbf{M}_j = \operatorname{diag}\left[\frac{2\epsilon_1}{\sin\theta_1}(\cos\theta_1 - (\cos \tfrac{2(j-1)}{q}\pi)), \ldots, \frac{2\epsilon_n}{\sin\theta_n}(\cos\theta_n - (\cos \tfrac{2(j-1)}{q}\pi)) \right]$$

for $1 \le j \le n$ and the characteristic equation for each $\mathbf{M}_j$ is

$$|\mathbf{M}_j - \eta_j \mathbf{I}_n| = \prod_{k=1}^{n} \left[\frac{2\epsilon_k}{\sin\theta_k}(\cos\theta_k - (\cos \tfrac{2(j-1)}{q}\pi)) - \eta_j \right], \quad 1 \le j \le n\,.$$

Therefore $\mathbf{M}_{p+1}$ and $\mathbf{M}_{q-p+1}$ each have a simple zero eigenvalue (corresponding to $k=1$ since $\theta_1 = \frac{2p}{q}\pi$) and the eigenvector in each case is $e_1 \stackrel{\text{def}}{=} (1, 0, \cdots, 0) \in \mathbf{R}^n$. In terms of the original coordinates the eigenvectors are $\overline{\mathbf{F}}|_{p+1} \otimes e_1$ and $\overline{\mathbf{F}}|_{q-p+1} \otimes e_1$ which are complex conjugates whose real and imaginary parts yield (2.18).

The product of the eigenvalues of $\mathbf{L}^o$ is given by $\prod_{j=1}^q |\mathbf{M}_j|$. Suppose that q is odd and let $q = 2r+1$, $r \ge 1$, and use the fact that $\mathbf{M}_{q-j+2} = \mathbf{M}_j$ when $j \ge 2$. Then,

$$\begin{aligned} \prod_{j=1}^{q} |\mathbf{M}_j| &= |\mathbf{M}_1| \prod_{j=2}^{r+1} |\mathbf{M}_j| \prod_{j=r+2}^{2r+1} |\mathbf{M}_j| \\ &= |\mathbf{M}_1| \prod_{j=2}^{r+1} |\mathbf{M}_j| \prod_{j=2}^{r+1} |\mathbf{M}_{2r-j+3}| = |\mathbf{M}_1| \left(\prod_{j=2}^{r+1} |\mathbf{M}_j| \right)^2. \end{aligned}$$

For $|\mathbf{M}_1|$, note that

$$|\mathbf{M}_1| = 2^n\,|\mathbf{A}^o - \mathbf{B}^o| = \prod_{k=1}^{n} \frac{2\epsilon_k}{\sin\theta_k}(\cos\theta_k - 1) = (-2)^n \prod_{k=1}^{n} \epsilon_k \tan(\tfrac{1}{2}\theta_k).$$

But, by hypothesis, $\theta_k \in (0,1)$ and so sign $|\mathbf{M}_1| = (-1)^n \prod_{j=1}^{n} \epsilon_k$ when q is odd, proving (2.19) in the case where q is odd (when the zero eigenvalues in the above product are divided out).

When q is even, let $q = 2r$, then

$$\begin{aligned}\prod_{j=1}^{q} |\mathbf{M}_j| &= |\mathbf{M}_1|\,|\mathbf{M}_{r+1}| \prod_{j=2}^{r} |\mathbf{M}_j| \prod_{j=r+2}^{2r} |\mathbf{M}_j| \\ &= |\mathbf{M}_1|\,|\mathbf{M}_{r+1}| \prod_{j=2}^{r} |\mathbf{M}_j| \prod_{j=2}^{r} |\mathbf{M}_{2r-j+2}| \\ &= |\mathbf{M}_1|\,|\mathbf{M}_{r+1}| \left(\prod_{j=2}^{r} |\mathbf{M}_j|\right)^2 .\end{aligned}$$

But

$$|\mathbf{M}_{r+1}| = 2^n |\mathbf{A}^o + \mathbf{B}^o| = \prod_{k=1}^{n} \frac{2\epsilon_k}{\sin\theta_k}(1 + \cos\theta_k).$$

Therefore upon dividing out the 2 zero eigenvalues,

$$\operatorname{sign} \mathbf{P}_\sigma = \operatorname{sign}\left(|\mathbf{M}_1|\,|\mathbf{M}_{r+1}|\right) = (-1)^n \quad (q \text{ even})$$

proving (2.19) for the case when q is even. ∎

The sign of the product of the non-zero eigenvalues of $\mathbf{L}^o$ is used in the proof of reduced stability of the bifurcating period-q points.

2.3 Symmetries

Crucial to the subsequent analysis is the role of temporal and spatial symmetries. The operator $\Gamma_q \otimes \mathbf{I}_n$ generates an action for $\mathbb{Z}_q$ on $\mathbf{X}_q^n$ and the functional $\mathbf{W}_q$ is $\mathbb{Z}_q$-invariant; that is,

$$\begin{aligned}\mathbf{W}_q(\Gamma_q \otimes \mathbf{I}_n \cdot \mathbf{x}, \Lambda) &= \tfrac{1}{2}\langle \Gamma_q \otimes \mathbf{I}_n \cdot \mathbf{x},\, \mathbf{L}(\lambda)\Gamma_q \otimes \mathbf{I}_n \cdot \mathbf{x}\rangle + N(\Gamma_q \otimes \mathbf{I}_n \cdot \mathbf{x}, \Lambda) \\ &= \tfrac{1}{2}\langle \mathbf{x},\, \mathbf{L}(\lambda)\mathbf{x}\rangle + N(\mathbf{x}, \Lambda) \\ &= \mathbf{W}_q(\mathbf{x}, \Lambda) .\end{aligned}$$

It follows that period-q points of the symplectic map $\mathbf{T}$ generated by h are critical points of a $\mathbb{Z}_q$-invariant functional.

For consideration of reversible-symplectic maps (see Appendix F), we introduce a *reversor* on $\mathbf{X}_q^n$, denoted by $\mathcal{K}_q \otimes \mathbf{I}_n$. The reversor generates an action for the group $\mathbf{Z}_2^\kappa$ on $\mathbf{X}_q^n$; explicitly,

$$\mathbf{Z}_2^\kappa = \langle \mathcal{K}_q \otimes \mathbf{I}_n \rangle \quad \text{with} \quad \mathcal{K}_q = \begin{pmatrix} 1 & 0 & \cdots & 0 & 0 \\ 0 & & & & 1 \\ \vdots & & 0 & & 0 \\ 0 & & & & \vdots \\ 0 & 1 & 0 & \cdots & 0 \end{pmatrix}. \tag{2.21}$$

Clearly $\mathcal{K}_q$ satisfies $\mathcal{K}_q^2 = \mathbf{I}_q$. Note also that $\mathcal{K}_q$ is symmetric and moreover satisfies the interesting relation $\mathcal{K}_q = \overline{\mathbf{F}}^2$ where $\mathbf{F}$ is the Fourier matrix. Together $\Gamma_q \otimes \mathbf{I}_n$ and $\mathcal{K}_q \otimes \mathbf{I}_n$ generate an action on $\mathbf{X}_q^n$ for the dihedral group of $2q$ elements,

$$\mathbf{D}_q = \langle\, \Gamma_q \otimes \mathbf{I}_n \,,\, \mathcal{K}_q \otimes \mathbf{I}_n \,\rangle. \tag{2.22}$$

This follows from the identity $\mathcal{K}_q \Gamma_q \mathcal{K}_q = \Gamma_q^T$. In particular, period-$q$ points of a reversible-symplectic map $\mathbf{T}$ generated by h are critical points of a $\mathbf{D}_q$-invariant functional. Note that a necessary and sufficient condition for the *linear* operator

$$\mathbf{L} = \mathbf{I}_q \otimes (\mathbf{A} + \mathbf{C}) - \Gamma_q \otimes \mathbf{B} - \Gamma_q^T \otimes \mathbf{B}^T$$

to commute with the action of $\mathbf{D}_q$ is that $\mathbf{B} = \mathbf{B}^T$. Similar restrictions are necessary for $\mathbf{D}_q$-invariance of the higher-order terms of h. Note that the normal form (2.9) is also reversible as $\mathbf{B}$ is diagonal, but might have been obtained through a *nonreversible* symplectic transformation $\widehat{\mathbf{S}}$. Hence, in that case, the additional information we have gained from reversibility loses any meaning for the original system. It is only when we start off with a *reversible* map $\mathbf{T}$ that the action of $\mathbf{D}_q$ has full power (cf. Theorem 2.9, Chapter 7 and Appendix F).

The tensor product decomposition of the group action of $\mathbf{Z}_q$ and $\mathbf{D}_q$ reflects the spatio-temporal symmetry structure. The $\mathbf{Z}_q$-symmetry for example is a *temporal symmetry*. The fact that the generating function h does not have any particular spatial symmetry is reflected by the identity acting on $\mathbf{R}^n$ in the representations $\Gamma_q \otimes \mathbf{I}_n$ and $\mathcal{K}_q \otimes \mathbf{I}_n$. Spatial symmetries are associated with a group action on $\mathbf{R}^n$ that is independent of discrete time.

We say that the symplectic map generated by h has *spatial symmetry* Σ if

$$h(\sigma x, \sigma x', \Lambda) = h(x, x', \Lambda), \quad \forall \sigma \in \Sigma.$$

We consider orthogonal actions of Σ. In consequence, note that, as Σ acts diagonally on $\mathbf{R}^{2n}$, it also acts symplectically with respect to $\mathbf{J}_{2n}$. We refer to Appendix I for a discussion of the relation between symmetry and the symplectic form.

When the generating function has spatial symmetry Σ, the functional $\mathbf{W}_q$ on $\mathbf{X}_q^n$ is then $\Sigma \times \mathbf{Z}_q$-invariant (or $\Sigma \times \mathbf{D}_q$ if it is also $\mathbf{Z}_2^\kappa$-invariant, cf. Appendix K) where

$$\Sigma \times \mathbf{Z}_q = \{ \langle \Gamma_q^j \otimes \sigma \rangle \mid 0 \leq j \leq q-1, \ \sigma \in \Sigma \}.$$

The role of spatial symmetries, in particular *continuous* spatial symmetries, in the bifurcation of period-q points, and the dynamics in general, in symplectic maps is an interesting subject. The bifurcation of periodic points in equivariant symplectic maps is considered in Chapter 8. In the remainder of this chapter however, we will suppose that the given symplectic map has no particular spatial symmetry, and will restrict attention to temporal symmetries.

Proposition 2.5. *The actions of $\mathbf{Z}_q = \langle \Gamma_q \otimes \mathbf{I}_n \rangle$ and $\mathbf{Z}_2^\kappa = \langle \mathcal{K}_q \otimes \mathbf{I}_n \rangle$ on* $\operatorname{Ker} \mathbf{L}^o =$ $\operatorname{span}\{\xi_1, \xi_2\}$ *with ξ_1 and ξ_2 given by Corollary 2.4 are the standard actions on* $\mathbf{R}^2$:

$$\mathbf{Z}_q \Big|_{\operatorname{Ker} \mathbf{L}^o} = \left\langle \begin{pmatrix} \cos\theta & \sin\theta \\ -\sin\theta & \cos\theta \end{pmatrix} \stackrel{\text{def}}{=} \mathbf{R}_\theta \right\rangle, \quad \mathbf{Z}_2^\kappa \Big|_{\operatorname{Ker} \mathbf{L}^o} = \left\langle \begin{pmatrix} 1 & 0 \\ 0 & -1 \end{pmatrix} \stackrel{\text{def}}{=} \kappa \right\rangle. \tag{2.23}$$

Proof. Follows from

$$[\xi_1\, \xi_2]^T\, \Gamma_q \otimes \mathbf{I}_n\, [\xi_1\, \xi_2] = \mathbf{R}_\theta \quad \text{and} \quad [\xi_1\, \xi_2]^T\, \mathcal{K}_q \otimes \mathbf{I}_n\, [\xi_1\, \xi_2] = \kappa\,.$$

■

2.4 Normal form for bifurcating period-q points

We now study critical points of the functional $\mathbf{W}_q$ given in (2.10). By Corollary 2.4 the Hessian of $\mathbf{W}_q$ evaluated at $\mathbf{x} = \lambda = 0$, $\mathbf{L}^o$, is singular with a 2-dimensional kernel. The idea is to use the Equivariant Splitting Lemma (see Appendix A) to split the functional $\mathbf{W}_q$ into a non-degenerate part to which Morse theory applies and a singular part, defined on the nullspace of $\mathbf{L}^o$, to which singularity theory will be applied. First we establish a preliminary result.

Proposition 2.6. *Let $\widehat{\mathbf{W}}_q$ be a smooth real valued potential defined on $\mathbf{R}^2 \times \mathbf{R}^k$ and suppose that $\widehat{\mathbf{W}}_q$ is $\mathbf{Z}_q$-invariant with respect to the standard action on $\mathbf{R}^2$. Then there exists smooth function germs F, G such that*

$$\widehat{\mathbf{W}}_q(\chi, \Lambda) = F(u, v, \Lambda) + w\, G(u, v, \Lambda) \tag{2.24}$$

with

$$u = \chi_1^2 + \chi_2^2 \quad \text{and} \quad v + iw = (\chi_1 + i\chi_2)^q. \tag{2.25}$$

In addition, if $\widehat{\mathbf{W}}_q$ is $\mathbf{Z}_2^\kappa$-invariant, with $\mathbf{Z}_2^\kappa = \langle \kappa \rangle$ (cf. (2.23)), then $G \equiv 0$; that is,

$$\widehat{\mathbf{W}}_q(\chi, \Lambda) = F(u, v, \Lambda). \tag{2.26}$$

Proof. The functions u, v and w form a Hilbert basis for the $\mathbb{Z}_q$-invariants on $\mathbb{R}^2$. The expression (2.24) then follows from Schwartz's Theorem (Golubitsky, Stewart & Schaeffer [1988, p.43]) and the fact that there is a relation: $v^2 + w^2 = u^q$. In particular every smooth $\mathbb{Z}_q$-invariant map on $\mathbb{R}^2$ can be expressed in the form (2.24). If $\widehat{\mathrm{W}}_q$ is also $\mathbb{Z}_2^\kappa$-invariant note that

$$\kappa \cdot w(\chi_1, \chi_2) = \mathrm{Im}\,(\chi_1 - i\chi_2)^q = -w(\chi_1, \chi_2).$$

Therefore w^2 (and not w) is an invariant function but, using the relation $w^2 = u^q - v^2$, the expression (2.26) follows. ∎

The following Theorem is a generalization of Meyer's Theorem on the generic bifurcation of period-q points.

Theorem 2.7. *Suppose* $\mathbf{T}$ *is a smooth* Λ*-parametrized family of symplectic maps on* $\mathbb{R}^{2n}$ *with its linear part depending on one parameter* λ: $\mathbf{DT}(\lambda)(0,0) = \widehat{\mathbf{M}}(\lambda)$ *with* $\widehat{\mathbf{M}}(\lambda)$ *as given in Proposition 2.2. Then, for* $(\|\chi\|, |\Lambda|)$ *sufficiently small, there exists an unique* $\mathbb{Z}_q$*-orbit of bifurcating period-q points which are in one-to-one correspondence with critical points of a* $\mathbb{Z}_q$*-invariant functional on* $\mathbb{R}^2$ *given by*

$$\widehat{\mathrm{W}}_q(\chi, \Lambda) = F(u, v, \Lambda) + w\, G(u, v, \Lambda) \tag{2.27}$$

and satisfying

$$F(0,0,\Lambda) = F_u^o = 0 \quad \text{and} \quad F_{u\lambda}^o = \epsilon_1\,, \tag{2.28}$$

where ϵ_1 *is the signature of the rational multiplier* $\exp i\theta_1$.

Moreover, generically, $(\nabla_\chi \widehat{\mathrm{W}}_q \sim g\ ;\ \mathcal{K}_\lambda^{\mathbb{Z}_q})$ *(that is,* $\nabla_\chi \widehat{\mathrm{W}}_q$ *is* $\mathcal{K}_\lambda^{\mathbb{Z}_q}$*-contact equivalent to g, cf. Chapters 3 and 4), where*

$$g(z,\lambda) = \begin{cases} \epsilon_1 \lambda z + \bar{z}^2, & q = 3\,, \\ (\epsilon_1\lambda + m|z|^2)\,z - \frac{1}{2}(z^2 + \bar{z}^2)\,\bar{z}\,, & q = 4\,, \\ (\epsilon_1\lambda + \delta_1|z|^2)\,z + \bar{z}^{q-1}, & q \geq 5\,, \end{cases} \tag{2.29}$$

and $z = \chi_1 + i\chi_2$, $\delta_1 = \pm 1$ *and* $m \in \mathbb{R}_+ \setminus \{0,1\}$.

Proof. By Proposition 2.1 $\mathbf{T}$ is generated by h which is also smooth. Decompose $\mathbf{X}_q^n = \mathrm{Ker}\,\mathbf{L}^o \oplus \mathbf{U}$. Then any $\mathbf{x} \in \mathbf{X}_q^n$ can be expressed as

$$\mathbf{x} = \chi_1\xi_1 + \chi_2\xi_2 + \Upsilon \quad \text{with} \quad \Upsilon \in \mathbf{U}.$$

All the hypotheses of the Equivariant Splitting Lemma (Appendix A) have been satisfied. Therefore there exists a $\mathbb{Z}_q$-equivariant smooth map ϕ and a $\mathbb{Z}_q$-invariant smooth map $\widehat{\mathrm{W}}_q$ such that

$$\mathrm{W}_q(\chi_1\xi_1 + \chi_2\xi_2 + \phi(\Upsilon, \chi, \Lambda), \Lambda) = \tfrac{1}{2}\langle\, \Upsilon\,, \mathbf{L}^o\Upsilon\,\rangle + \widehat{\mathrm{W}}_q(\chi, \Lambda)$$

and by Proposition 2.5 the action of $\mathbf{Z}_q$ on $\operatorname{Ker} \mathbf{L}^o$ is the standard action. The expression (2.27) then follows from Proposition 2.6.

The expressions in (2.28) are verified in the following way. They depend only on the quadratic part of the functional W_q and

$$\begin{aligned} \tfrac{1}{2}\langle \mathbf{x}, \mathbf{L}(\lambda)\mathbf{x}\rangle &= \tfrac{1}{2}\langle \chi_1\xi_1 + \chi_2\xi_2 , \mathbf{L}(\lambda)(\chi_1\xi_1 + \chi_2\xi_2)\rangle + \cdots \\ &= \tfrac{1}{2}\lambda\langle \chi_1\xi_1 + \chi_2\xi_2 , \mathbf{L}^o_\lambda(\chi_1\xi_1 + \chi_2\xi_2)\rangle + \cdots \qquad (2.30) \\ &= \epsilon_1\lambda(\chi_1^2 + \chi_2^2) + \cdots \\ &= \epsilon_1\lambda u + \cdots , \end{aligned}$$

using the fact that

$$\mathbf{L}_\lambda = 2\mathbf{I}_q \otimes \mathbf{A}_\lambda - (\Gamma_q + \Gamma_q^T) \otimes \mathbf{B}_\lambda$$

and

$$\begin{aligned} \mathbf{A}^o_\lambda &= -\frac{\epsilon_1}{\sin^2\theta_1} \operatorname{diag}(1, 0, \ldots, 0) \\ \mathbf{B}^o_\lambda &= -\frac{\epsilon_1}{\sin^2\theta_1} \operatorname{diag}(\cos\theta_1, 0, \ldots, 0) \quad \text{where } \theta_1 = \tfrac{2p}{q}\pi . \end{aligned}$$

With (2.28) the gradient of (2.27), with complex notation $z = \chi_1 + i\chi_2$, becomes

$$\nabla_z \widehat{\mathbf{W}}_q(z, \Lambda) = (\epsilon_1\lambda + F^o_{uu}|z|^2 + \cdots)\, z + (F^o_v + iG^o + \cdots)\, \bar{z}^{q-1}$$

and in the one bifurcation parameter family it can be assumed that $F^o_{uu}\cdot(F^{o2}_v + G^{o2}) \neq 0$. The remainder of the proof is an application of singularity theory to $\widehat{\mathbf{W}}_q$. In the expression given in (2.29) we anticipate the singularity theory to be introduced in Chapters 3 and 4; in particular, the normal forms (2.29) are a direct consequence of Corollary 4.3. ∎

Although the above proof is new, the normal forms for bifurcating period-q points given in (2.29) are well-known and have been derived using other methods in Meyer [1970] and Arnold et al. [1988, Section 7.4] and the bifurcation diagrams are familiar. An important "corollary" of our result however is that the framework introduced here admits generalization in a number of interesting directions.

Because of the quadratic term in the normal form, the bifurcating period-3 points are unstable (see Stability Lemma I) but very often the symplectic map has an involution (not to be confused with reversibility) which changes the normal form for generic bifurcation of period-3 points. This fact has often been observed in numerical calculations using an even generating function for an area-preserving map. The following result indicates how such an involution enters the normal form.

Corollary 2.8. *Let* $\mathbf{T} : (x, y, \Lambda) \mapsto (x', y')$ *be a* Λ*-parametrized family of symplectic maps on* $\mathbf{R}^{2n}$ *satisfying the hypotheses of Theorem 2.7. Suppose moreover that h has a* $\mathbf{Z}_2^\gamma$-spatial *symmetry; that is,*

$$h(\gamma x, \gamma x', \Lambda) = h(x, x', \Lambda) \quad \text{with} \quad \gamma^2 = \mathbf{I}_n$$

and that γ *acts nontrivially on* $e_1 \in \mathbf{R}^{2n}$, *that is* $\mathbf{Z}_2^\gamma\big|_{\text{Ker } \mathbf{L}^\rho} = \langle -\mathbf{I} \rangle$. *Then if* $q = 3$ *the reduced functional is* $\mathbf{Z}_6$*-invariant. In particular, generically,* $(\nabla_\chi \widehat{\mathbf{W}}_3 \sim g\,;\, \mathcal{K}_\lambda^{\mathbf{Z}_6})$ *with*

$$g(z, \lambda) = (\epsilon_1 \lambda + \epsilon_2 |z|^2)\, z + \bar{z}^5$$

where ϵ_1 *is the signature of the multiplier and* $\epsilon_2 = \pm 1$ *is determined by the nonlinear terms in the generating function.*

Proof. Since $\mathbf{Z}_2^\gamma$ *commutes* with $\mathbf{Z}_3$ the reduced functional is $\mathbf{Z}_2 \times \mathbf{Z}_3$-invariant; but $\mathbf{Z}_2 \times \mathbf{Z}_3$ is isomorphic to $\mathbf{Z}_6$. The normal form g for $\nabla_\chi \widehat{\mathbf{W}}_3$ then follows from application of the singularity theory for $\mathbf{Z}_6$-equivariant gradient maps. ■

Corollary 2.8 can be extended to arbitrary q but is of less interest when $q \geq 4$. In fact when q is even, $-\mathbf{I} \in \mathbf{Z}_q$, and so the effect of the involution is vacuous. When $q \geq 5$ is odd the normal form is modified as in Corollary 2.8 and the parity p in $\theta = \frac{2p}{q}\pi$ must be considered. But the effect is not as dramatic. When $q = 3$, we obtain not only new *stable* points but the overall aspect of the bifurcation is altered, whereas when $q \geq 5$ we simply double the number of points (cf. Chapter 4).

The $\mathbf{Z}_2^\gamma$-symmetry in Corollary 2.8 is a *spatial* symmetry. Another involutive symmetry of great interest is that of reversibility which is a *temporal* symmetry. As previously mentioned, the linear map $\widehat{\mathbf{M}}(\lambda)$ given in Proposition 2.2 is also reversible. However a *nonlinear map* that is reversible-symplectic with its linearization $\mathbf{DT}(\lambda)(0,0)$ equal to $\widehat{\mathbf{M}}(\lambda)$ will have slightly different normal forms for the bifurcating period-q points. The following theorem generalizes Rimmer's Theorem (Rimmer [1983]) on the generic bifurcation of period-q points in reversible-symplectic maps.

Theorem 2.9. *Suppose* $\mathbf{T}$ *is a reversible-symplectic* Λ*-parametrized family of maps on* $\mathbf{R}^{2n}$ *with* $\mathbf{DT}(\lambda)(0,0) = \widehat{\mathbf{M}}(\lambda)$ *with* $\widehat{\mathbf{M}}(\lambda)$ *given in Proposition 2.2. Then the bifurcating period-q points are in one-to-one correspondence with critical points of a* $\mathbf{D}_q$*-invariant functional on* $\mathbf{R}^2$ *given by*

$$\widehat{\mathbf{W}}_q(\chi, \Lambda) = F(u, v, \Lambda)$$

with F satisfying (2.28). Moreover, generically, $(\nabla_\chi \widehat{W}_q \sim g\,;\, \mathcal{K}_\lambda^{\mathbf{D}_q})$ *where*

$$g(z,\lambda) = \begin{cases} \epsilon_1 \lambda\, z + \delta_1 \bar{z}^2, & q = 3\,, \\ (\epsilon_1 \lambda + m|z|^2)\, z - \frac{1}{2}\delta_2 (z^2 + \bar{z}^2)\, \bar{z}, & q = 4\,, \\ (\epsilon_1 \lambda + \epsilon_2 |z|^2)\, z + \delta_1 \bar{z}^{q-1}\,, & q \geq 5 \end{cases} \tag{2.31}$$

and $z = \chi_1 + i\chi_2$, $\epsilon_1, \epsilon_2, \delta_1, \delta_2 = \pm 1$ *and* $m \in \mathbf{R} \setminus \{\pm 1\}$.

Proof. Same proof as Theorem 2.7, but with use of the singularity theory for $\mathbf{D}_q$-equivariant gradient maps given in Chapters 4 and 6. ■

The only difference between the zero codimension $\mathbf{D}_q$ and $\mathbf{Z}_q$-normal forms with one bifurcation parameter, λ, is that there are more equivalence classes in the $\mathbf{D}_q$-case but only trivially so, in the sense that there are additional signs in the normal forms but no essential difference in the bifurcation diagrams.

However, in the degenerate case (additional parameters) differences do show up in the bifurcation diagrams. Examples are given in Chapters 5 and 7.

Nevertheless, in all (also generic) cases, there is a nontrivial difference in the phase portraits for the reversible and nonreversible (symplectic) case. In particular the involution $\mathbf{Z}_2^\kappa$ and its conjugates result in symmetry planes in the phase space. In addition the sequences of period-q orbits have additional symmetry. This is discussed further in Chapter 7.

The fact that the coefficient $F^o_{u\lambda}$ in the reduced functional $\widehat{W}_q$ is the signature of the multiplier is extremely important. As we show in Section 2.6 (see also Appendix B) the signature goes to zero at an indefinite *collision* of multipliers. In particular the collision of multipliers at rational points is associated with the *collision singularity:*

$$F^o_{u\lambda} = 0\,.$$

2.5 Reduced stability of bifurcating periodic points

The idea of reduced stability is to obtain complete information on the linear stability of the bifurcating period-q points from the reduced functional $\widehat{W}_q$ on $\mathbf{R}^2$.

Lemma 2.10 (Stability Lemma I). *Let $\widehat{W}_q$ be the reduced functional governing the bifurcation of period-q points given by Theorem 2.7 or 2.9 and let* $\mathrm{Hess}_\chi \widehat{W}_q(\chi, \Lambda)$ *be the Hessian evaluated at an existing period-q point. Then for* $\|\chi\|$ *and* $|\Lambda|$ *sufficiently small the bifurcating period-q points are stable if* $|\mathrm{Hess}_\chi \widehat{W}_q(\chi, \Lambda)| > 0$ *and unstable if* $|\mathrm{Hess}_\chi \widehat{W}_q(\chi, \Lambda)| < 0$.

Remark. The result is actually more general than the situation presented by Theorems 2.7 and 2.9 in the sense that the Stability Lemma I is a reduced stability result for any non-degenerate bifurcating branch of period-q points provided we have only a pair of simple multipliers crossing a q-th root of unity (no resonances). In particular it is also applicable to the more complex bifurcations that take place in symplectic maps with the unfolding of degenerate nonlinearities (see Chapters 5 and 7).

The instability part of the result of Lemma 2.10 is also a simple consequence of the general Instability Lemma of Appendix M.

Proof. The idea of the proof is to use Floquet theory in configuration space (MacKay & Meiss [1983], Kook & Meiss [1989]) and then project onto $\operatorname{Ker}\mathbf{L}^o$.

Suppose $\{x^i\} \in (\mathsf{R}^n)^{\mathbf{Z}}$ is an orbit of the symplectic map generated by h. Then

$$h_1(x^i, x^{i+1}, \Lambda) + h_2(x^{i-1}, x^i, \Lambda) = 0\,, \quad \forall i \in \mathbf{Z}\,,$$

and the tangent orbit is determined from

$$\frac{d}{d\epsilon}\bigg|_{\epsilon=0} \left[h_1(x^i + \epsilon\xi^i, x^{i+1} + \epsilon\xi^{i+1}, \Lambda) + h_2(x^{i-1} + \epsilon\xi^{i-1}, x^i + \epsilon\xi^i, \Lambda) \right].$$

Therefore, with $\hat{\mathbf{A}}_i = \mathbf{A}_i + \mathbf{C}_i$,

$$\mathbf{A}_i = h_{11}(x^i, x^{i+1}, \Lambda)\,,\ \mathbf{C}_i = h_{22}(x^{i-1}, x^i, \Lambda)\,,\ \mathbf{B}_i = -h_{12}(x^i, x^{i+1}, \Lambda)\,, \tag{2.32}$$

a sequence $\{\xi^i\}$ in the tangent space to the orbit satisfies

$$-\mathbf{B}_{i-1}^T \xi^{i-1} + \hat{\mathbf{A}}_i \xi^i - \mathbf{B}_i \xi^{i+1} = 0\,, \quad \forall i \in \mathbf{Z}\,. \tag{2.33}$$

Now suppose the sequence $\mathbf{x} = \{x^i\} \in \mathbf{X}_q^n$. Then the governing equation for the sequence $\{\xi^i\}$ is a linear system with periodic coefficients and, by the discrete Floquet Theorem,

$$\xi^{i+q} = \mu\,\xi^i\,, \quad \forall i \in \mathbf{Z}\,, \tag{2.34}$$

where $\mu \in \mathbf{C}$ is the Floquet multiplier. Let $\xi = (\xi^1, \ldots, \xi^q)$, then combination of (2.33) and (2.34) results in the following eigenvalue problem for μ

$$\mathbf{M}(\mu) \cdot \xi = 0$$

where

$$\mathbf{M}(\mu) = \begin{bmatrix} \hat{\mathbf{A}}_1 & -\mathbf{B}_1 & \mathbf{0} & \cdots & \mathbf{0} & -\frac{1}{\mu}\mathbf{B}_q^T \\ -\mathbf{B}_1^T & \hat{\mathbf{A}}_2 & -\mathbf{B}_2 & \ddots & & \mathbf{0} \\ \mathbf{0} & -\mathbf{B}_2^T & \hat{\mathbf{A}}_3 & \ddots & \ddots & \vdots \\ \vdots & \ddots & \ddots & \ddots & \ddots & \mathbf{0} \\ \mathbf{0} & & \ddots & \ddots & \hat{\mathbf{A}}_{q-1} & -\mathbf{B}_{q-1} \\ -\mu\mathbf{B}_q & \mathbf{0} & \cdots & \mathbf{0} & -\mathbf{B}_{q-1}^T & \hat{\mathbf{A}}_q \end{bmatrix} \tag{2.35}$$

or, equivalently

$$\mathbf{M}(\mu) = \operatorname{Hess}_{\mathbf{x}} \mathbf{W}_q(\mathbf{x}, \Lambda) + (1-\mu)\,\mathbf{E}_{q1} \otimes \mathbf{B}_q + (1 - \tfrac{1}{\mu})\,\mathbf{E}_{q1}^T \otimes \mathbf{B}_q^T\,.$$

The matrix $\mathbf{E}_{q1}$ is a $q \times q$ matrix with 1 in the $(q,1)$-entry and zero otherwise. In particular $\mathbf{M}(1) = \mathrm{Hess}_{\mathbf{x}} \mathrm{W}_q(\mathbf{x}, \Lambda)$. Note however that the coefficients of $(1-\mu)$ and $(1-\frac{1}{\mu})$ are rank n matrices. Therefore $|\mathbf{M}(\mu)|$ will be a polynomial of at most degree n in μ and $\frac{1}{\mu}$ (see Appendix C for proof). In fact, since $|\mathbf{M}(\mu)| = |\mathbf{M}(\frac{1}{\mu})|$, the determinant will be a function of $\mu + \frac{1}{\mu}$.

For a period-q orbit there will be n pairs of multipliers associated with the symplectic map on $\mathbb{R}^{2n}$. However, for $\|\chi\|$ and $|\Lambda|$ sufficiently small, we can argue that the n multipliers will be perturbations of the case $(\chi, \Lambda) = (0,0)$, hence

$$\sigma(\mathbf{DT}^q) = \{e^{\pm iq\theta_1}, e^{\pm iq\theta_2}, \ldots, e^{\pm iq\theta_n}\}$$

when $\mathbf{DT}$ has the form given in (2.1). Since $\theta_2 \ldots \theta_n$ lie at irrational points, $\exp(iq\theta_j)$, $2 \le j \le n$, will each lie at some distinct point on the unit circle other than ± 1 and the multiplier $\exp(iq\theta_1)$ will lie at 1 (see Figure 2.1).

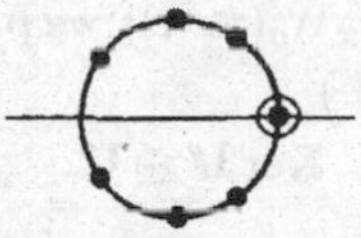

Figure 2.1: $\sigma(\mathbf{DT}^{oq})$

Therefore under sufficiently small perturbation the only multiplier that can become unstable is $\exp(iq\theta_1)$ as shown in Figure 2.2(a) and (b).

(a) (b)

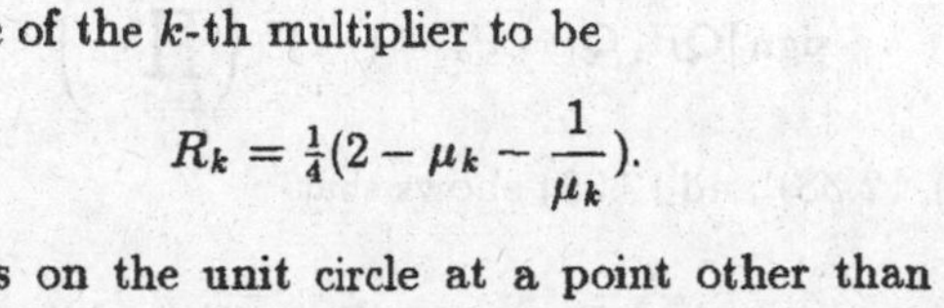

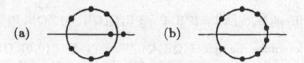

Figure 2.2: $\sigma(\mathbf{DT}(x,y)^q)$

Define the residue of the k-th multiplier to be

$$R_k = \tfrac{1}{4}(2 - \mu_k - \frac{1}{\mu_k}).$$

When a multiplier lies on the unit circle at a point other than ± 1, the residue satisfies $0 < R_k < 1$. For the configuration in Figure 2.1, $\{R_k\}_{k=2}^{n} \subset (0,1)$ and in Figures 2.1 and 2.2(a),(b), R_1 is respectively 0, negative and positive (and less than 1) respectively. Therefore when $\|\chi\|$ and $|\Lambda|$ are sufficiently small, the bifurcating period-q orbit is stable if $\prod_{k=1}^{q} R_k > 0$ and unstable if $\prod_{k=1}^{q} R_k < 0$.

Now Kook & Meiss [1989, p.77] have shown that for a symplectic map on $\mathbf{R}^{2n}$ generated by h the product of the residues has the following form:

$$\prod_{k=1}^{n} R_k = \left(-\frac{1}{4}\right)^n \frac{|\mathrm{Hess}_{\mathbf{x}} \mathbf{W}_q(\mathbf{x}, \Lambda)|}{\prod_{j=1}^{q} |\mathbf{B}_j|}. \tag{2.36}$$

However $\mathbf{B}_j = \mathbf{B} + (\mathbf{B}_j - \mathbf{B})$ and $\|\mathbf{B}_j - \mathbf{B}\| \to 0$ as $(\|\chi\|, |\Lambda|) \to 0$. Therefore

$$\operatorname{sign} \prod_{j=1}^{q} |\mathbf{B}_j| = \operatorname{sign} |\mathbf{B}|^q = \left(\prod_{k=1}^{n} \epsilon_k\right)^q \quad (\text{ for } (\|\chi\|, |\Lambda|) \text{ sufficiently small})$$

where ϵ_k is the signature of the k-th multiplier in the linear map $\mathbf{DT}$. Therefore (2.36) reduces to

$$\operatorname{sign}\left(\prod_{k=1}^{n} R_k\right) = (-1)^n \left(\prod_{k=1}^{q} \epsilon_k\right)^q \operatorname{sign} |\mathrm{Hess}_{\mathbf{x}} \mathbf{W}_q(\mathbf{x}, \Lambda)|. \tag{2.37}$$

To determine the sign of $|\mathrm{Hess}_{\mathbf{x}} \mathbf{W}_q(\mathbf{x}, \Lambda)|$, we partition $\mathrm{Hess}_{\mathbf{x}} \mathbf{W}_q(\mathbf{x}, \Lambda)$ according to the splitting

$$\operatorname{Ker} \mathbf{L}^o \oplus \mathbf{U}.$$

Let $\mathbf{P} = \mathbf{P}_1 + \mathbf{P}_2$ with $\mathbf{P}_j = \langle \xi_j, \cdot \rangle \, \xi_j$ $(j = 1, 2)$ and $\mathbf{Q} = \mathbf{I}_{nq} - \mathbf{P}$. Then

$$\begin{aligned} |\mathbf{H}_x| \stackrel{\text{def}}{=} |\mathrm{Hess}_{\mathbf{x}} \mathbf{W}_q(\mathbf{x}, \Lambda)| &= \begin{vmatrix} \mathbf{PH}_x\mathbf{P} & \mathbf{PH}_x\mathbf{Q} \\ \mathbf{QH}_x\mathbf{P} & \mathbf{QH}_x\mathbf{Q} \end{vmatrix} \\ &= |\mathbf{QH}_x\mathbf{Q}| \, |\mathbf{PH}_x\mathbf{P} - \mathbf{PH}_x\mathbf{Q}(\mathbf{QH}_x\mathbf{Q})^{-1}\mathbf{QH}_x\mathbf{P}|. \end{aligned} \tag{2.38}$$

But it follows from the definition of $\widehat{\mathbf{W}}_q$ that

$$|\mathrm{Hess}_{\chi} \widehat{\mathbf{W}}_q(\chi, \Lambda)| = |\mathbf{PH}_x\mathbf{P} - \mathbf{PH}_x\mathbf{Q}(\mathbf{QH}_x\mathbf{Q})^{-1}\mathbf{QH}_x\mathbf{P}|.$$

It remains to determine the sign of $|\mathbf{QH}_x\mathbf{Q}|$. When $(\chi, \Lambda) = (0, 0)$, $\mathbf{QH}_x\mathbf{Q} = \mathbf{QL}^o\mathbf{Q}$ which is $\mathbf{L}^o$ restricted to the complement of $\operatorname{Ker} \mathbf{L}^o$. Therefore, using Corollary 2.4, for (χ, Λ) sufficiently small,

$$\operatorname{sign} |\mathbf{QH}_x\mathbf{Q}| = \mathbf{P}_\sigma = (-1)^n \left(\prod_{k=1}^{n} \epsilon_k\right)^q. \tag{2.39}$$

Combining (2.37), (2.38) and (2.39) shows that

$$\operatorname{sign} \prod_{k=1}^{n} R_k = \operatorname{sign} |\mathrm{Hess}_{\chi} \widehat{\mathbf{W}}_q(\chi, \Lambda)|$$

for $\|\chi\|$ and $|\Lambda|$ sufficiently small. ■

Application of Stability Lemma I to the bifurcating period-3 points recovers the well-known fact that generic period-3 points are unstable. The stability properties for period-4 points are summarized in Table 2.1. For $q \geq 5$ there are two families of bifurcating period-q points one of which is unstable and the other stable. Further analysis of stability properties is given in Chapters 5 and 7 where there is more than one parameter resulting in more complex bifurcations and stability assignments.

Family	Branch	sign $\lvert \mathrm{Hess}_\chi \widehat{W}_q(\chi, \Lambda)\rvert$
I	$\epsilon_1\lambda + (m-1)\chi_1^2 = 0$ $\chi_2 = 0$	$\mathrm{sign}\,(m-1)$
II	$\epsilon_1\lambda + 2m\,\chi_1^2 = 0$ $\chi_1 = \pm\chi_2$	$-\mathrm{sign}\, m$

Table 2.1: Branches of period-4 points for $(\epsilon_1\lambda + m|z|^2)\,z - (\mathrm{Re}\, z^2)\overline{z}$

2.6 4D-symplectic maps and the collision singularity

In the area-preserving case, singularities in the bifurcation of period-q points are associated with certain coefficients in the nonlinear terms passing through zero. These singularities are considered in Chapters 5 and 7. With an increase in the configuration space to two-dimensions, corresponding to 4D-maps, a number of new and interesting questions arise in the bifurcation of period-q points:

(a) the collision of multipliers of opposite signature at rational points,

(b) the bifurcation of periodic points for equivariant symplectic maps; in particular, symplectic maps with continuous spatial symmetries.

Both of these subjects will be considered in the sequel. Here we introduce some preliminary facts about 4D-maps, in particular, the role of signature in the collision of multipliers that is central to the later analysis.

Let $n = 2$ and consider a smooth generating function at an elliptic point with $h^o = h_1^o = h_2^o = 0$,

$$h(x, x') = \tfrac{1}{2}(x, \mathbf{A}x) - (x, \mathbf{B}x') + \tfrac{1}{2}(x', \mathbf{C}x') + \hat{h}(x, x') \qquad (2.40)$$

where $\hat{h}(x, x')$ begins with terms of degree 3, $\mathbf{A}$ and $\mathbf{C}$ are symmetric 2×2 matrices and $\mathbf{B}$ is a general 2×2 matrix satisfying $|\mathbf{B}| \neq 0$.

The quadratic part of the generating function $h(x, x')$ in (2.40) corresponds to the map

$$\begin{pmatrix} x' \\ y' \end{pmatrix} \stackrel{\mathrm{def}}{=} \mathbf{DT}\begin{pmatrix} x \\ y \end{pmatrix} = \begin{bmatrix} \mathbf{B}^{-1}\mathbf{A} & \mathbf{B}^{-1} \\ \mathbf{C}\mathbf{B}^{-1}\mathbf{A} - \mathbf{B}^T & \mathbf{C}\mathbf{B}^{-1} \end{bmatrix}\begin{pmatrix} x \\ y \end{pmatrix}, \qquad (2.41)$$

from which it follows that

$$\begin{aligned} \operatorname{Tr} \mathbf{DT} &= \operatorname{Tr}\left(\mathbf{B}^{-1}(\mathbf{A}+\mathbf{C})\right) \\ \operatorname{Tr}\left(\mathbf{DT}\circ\mathbf{DT}\right) &= \operatorname{Tr}\left(\mathbf{B}^{-1}(\mathbf{A}+\mathbf{C})\right)^2 - 2\operatorname{Tr}\left(\mathbf{B}^{-1}\mathbf{B}^T\right). \end{aligned} \tag{2.42}$$

The reduced characteristic equation for $\mathbf{DT}$ formed from $|\mathbf{DT}-\mu\mathbf{I}_4| = 0$ is given by

$$\rho^2 - \operatorname{Tr}\left(\mathbf{B}^{-1}(\mathbf{A}+\mathbf{C})\right)\rho + |\mathbf{B}^{-1}(\mathbf{A}+\mathbf{C})| - \frac{|\mathbf{B}-\mathbf{B}^T|}{|\mathbf{B}|} = 0 \tag{2.43}$$

where $\rho = \mu + \mu^{-1}$. Let $a = \operatorname{Tr}\mathbf{DT}$ and $2b = (\operatorname{Tr}\mathbf{DT})^2 - \operatorname{Tr}(\mathbf{DT}\circ\mathbf{DT})$ then (2.43) becomes

$$\rho^2 - a\rho + b - 2 = 0$$

and the familiar stability diagram for 4D-symplectic maps is recovered and is shown in Figure 2.3 (Howard & MacKay [1987, p.1041]). The stable region is the closed region in the center where both (pairs) of multipliers lie on the unit circle. Passage through the lines $b+2 = 2a$ or $b+2 = -2a$ results in a multiplier passing through $+1$ or -1 (period doubling) respectively. Both of these bifurcations are familiar in area-preserving maps. Of great interest is the loss of stability associated with a collision of multipliers along the curve $b = 2 + \frac{1}{4}a^2$ ($a \in (-4,4)$). The collision of multipliers of opposite signature at rational points will be treated in Chapter 9.

The required results on signature in configuration space are given in Appendix B. In particular, it is demonstrated (Proposition B.3) that the signature of a simple multiplier on the unit circle: $e^{i\theta}$, $\theta \in (0,\pi)$, is given by

$$\sigma = -\operatorname{sign}\operatorname{Im}\left(e^{i\theta}(\overline{\zeta},\mathbf{B}\zeta)\right) \stackrel{\text{def}}{=} \operatorname{sign} Q \tag{2.44}$$

where $\mathbf{B} = h^o_{12}$ and ζ satisfies

$$(\mathbf{A}+\mathbf{C} - e^{i\theta}\mathbf{B} - e^{-i\theta}\mathbf{B}^T)\zeta = 0. \tag{2.45}$$

Now suppose that $\sigma(\mathbf{DT}^o) = \{e^{\pm i\theta}, e^{\pm i\phi}\}$ with $\theta,\phi \in (0,\pi)$ but ϕ at an irrational point, $\theta = \frac{2p}{q}\pi$ and that $\mathbf{T}$ generated by h. If the hypotheses of Theorem 2.7 are met, we know that generically there exists a bifurcating branch of period-q points. It is also clear that the signature is important, it appears as a critical coefficient in the reduced functional (see equation (2.28)). We give a direct proof of the role of the signature in order to show that the coefficient vanishes (signature function goes to zero) if a multiplier of opposite signature collides with the basic multiplier.

Proposition 2.11. *Let* $\mathbf{T} : (x, y, \Lambda) \mapsto (x', y')$ *be a smooth* Λ*-parametrized family of symplectic maps on* $\mathbf{R}^4$ *generated by* h*. Suppose that the spectrum of* $\mathbf{DT}^o$ *is* $\{e^{\pm i\theta}, e^{\pm i\phi}\}$ *with* $\theta = \frac{2p}{q}\pi$*,* ϕ *at an irrational point and* $\theta^o_\lambda \neq 0$*. Then*

$$\operatorname{sign}\left(\frac{\partial^2 \widehat{\mathrm{W}}_q}{\partial\lambda\partial u}\right) = \operatorname{sign}(\theta^o_\lambda Q) \tag{2.46}$$

where $u = \chi_1^2 + \chi_2^2$ *and* $\widehat{\mathrm{W}}_q$ *is the reduced* $\mathbb{Z}_q$*-invariant functional given in (2.27).*

Suppose $\mathbf{T}$ *depends also on a second parameter* α *and that* ϕ *is of opposite signature to* θ*. If* $\alpha \to 0$ *results in* $\phi \to \theta$ *then* $\widehat{\mathrm{W}}^o_{q,\lambda u} \to 0$*, the "rational collision singularity".*

Proof. Consider the quadratic part of the functional W_q,

$$\tfrac{1}{2}\langle \mathbf{x}, \mathbf{L}(\lambda)\mathbf{x}\rangle = \tfrac{1}{2}\langle \mathbf{x}, \mathbf{L}^o\mathbf{x}\rangle + \tfrac{1}{2}\lambda\langle \mathbf{x}, \mathbf{L}^o_\lambda\mathbf{x}\rangle + \cdots$$

or with $\mathbf{x} = \chi_1\xi_1 + \chi_2\xi_2 + \Upsilon$,

$$\tfrac{1}{2}\langle \mathbf{x}, \mathbf{L}(\lambda)\mathbf{x}\rangle = \tfrac{1}{2}\lambda\langle \chi_1\xi_1 + \chi_2\xi_2\,, \mathbf{L}^o_\lambda(\chi_1\xi_1 + \chi_2\xi_2)\rangle + \cdots. \tag{2.47}$$

However from the definition of $\mathbf{L}$

$$\mathbf{L}_\lambda = \mathbf{I}_q \otimes (\mathbf{A}_\lambda + \mathbf{C}_\lambda) - \Gamma_q \otimes \mathbf{B}_\lambda - \Gamma_q^T \otimes \mathbf{B}_\lambda^T.$$

The derivatives of the basic (**A**,**B**,**C**) matrices can be obtained from the eigenvalue problem (2.45),

$$(\mathbf{A}_\lambda + \mathbf{C}_\lambda - e^{i\theta}\mathbf{B}_\lambda - e^{-i\theta}\mathbf{B}_\lambda^T)\zeta - i\,\theta_\lambda(e^{i\theta}\mathbf{B} - e^{-i\theta}\mathbf{B}^T)\zeta$$
$$+(\mathbf{A} + \mathbf{C} - e^{i\theta}\mathbf{B} - e^{-i\theta}\mathbf{B}^T)\zeta_\lambda = 0.$$

The adjoint eigenvector of (2.45) is simply $\overline{\zeta}$ since the matrix is Hermitian, therefore

$$\begin{aligned}(\overline{\zeta}\,, (\mathbf{A}_\lambda + \mathbf{C}_\lambda - e^{i\theta}\mathbf{B}_\lambda - e^{-i\theta}\mathbf{B}_\lambda^T)\zeta) &= i\,\theta_\lambda\,(\overline{\zeta}\,, (e^{i\theta}\mathbf{B} - e^{-i\theta}\mathbf{B}^T)\zeta)\\ &= -2\,\theta_\lambda \operatorname{Im}(e^{i\theta}(\overline{\zeta}\,, \mathbf{B}\zeta))\\ &= 2\,\theta_\lambda\,Q\end{aligned} \tag{2.48}$$

using the definition of the signature function Q given in (2.44). Now using the fact that

$$\operatorname{Ker}\mathbf{L}^o = \{\xi_1, \xi_2\} \quad \text{where} \quad \xi_1 + i\xi_2 = \frac{1}{\sqrt{q}}\begin{pmatrix} 1 \\ e^{i\theta} \\ e^{i2\theta} \\ \vdots \\ e^{i(q-1)\theta}\end{pmatrix} \otimes \zeta\,,$$

it follows that

$$\mathbf{L}_\lambda(\xi_1 + i\xi_2) = \frac{1}{\sqrt{q}} \begin{pmatrix} 1 \\ e^{i\theta} \\ e^{i2\theta} \\ \vdots \\ e^{i(q-1)\theta} \end{pmatrix} \otimes \left(\mathbf{A}_\lambda + \mathbf{C}_\lambda - e^{i\theta}\mathbf{B}_\lambda - e^{-i\theta}\mathbf{B}_\lambda^T\right)\zeta\,,$$

and

$$\begin{aligned}\langle\, \xi_1 - i\xi_2\,,\, \mathbf{L}_\lambda(\xi_1 + i\xi_2)\,\rangle &= (\,\overline{\zeta}\,,\, (\mathbf{A}_\lambda + \mathbf{C}_\lambda - e^{i\theta}\mathbf{B}_\lambda - e^{-i\theta}\mathbf{B}_\lambda^T)\zeta\,) \\ &= 2\theta_\lambda\, Q\end{aligned}$$

using (2.48). Expansion of the real and imaginary parts results in

$$\langle\, \chi_1\xi_1 + \chi_2\xi_2\,,\, \mathbf{L}_\lambda(\chi_1\xi_1 + \chi_2\xi_2)\rangle = \tfrac{1}{2}(\chi_1^2 + \chi_2^2)\langle\, \xi_1 - i\xi_2\,,\, \mathbf{L}_\lambda(\xi_1 + i\xi_2)\,\rangle.$$

Therefore the functional in (2.47) reduces to

$$\tfrac{1}{2}\langle\, \mathbf{x}\,,\, \mathbf{L}(\lambda)\,\mathbf{x}\,\rangle = \tfrac{1}{2}\lambda\, \theta_\lambda^o\, Q\, (\chi_1^2 + \chi_2^2) + \cdots\,. \tag{2.49}$$

By hypothesis $\theta_\lambda^o \neq 0$. In other words the signature of the multiplier at the rational point (the sign of Q) determines the sign of the coefficient of λu in the reduced functional. It is shown in Appendix B that at a collision of multipliers of opposite signature $Q \to 0$. ■

The central observation of Proposition 2.11 is the role of the collision singularity. When two multipliers of opposite signature collide at a rational point, *the dimension of the nullspace of the linear operator* $\mathbf{L}^o$ *does not increase*; the effect of the collision is to cause a singularity in the parameter structure instead. Therefore the normal form for bifurcating period-q points at a rational collision is a $\mathbb{Z}_q$-equivariant gradient map on $\mathbb{R}^2$ but with the collision singularity. We will come back to this point in Chapter 9 after we have explored some other singularities and introduced the necessary singularity theory.

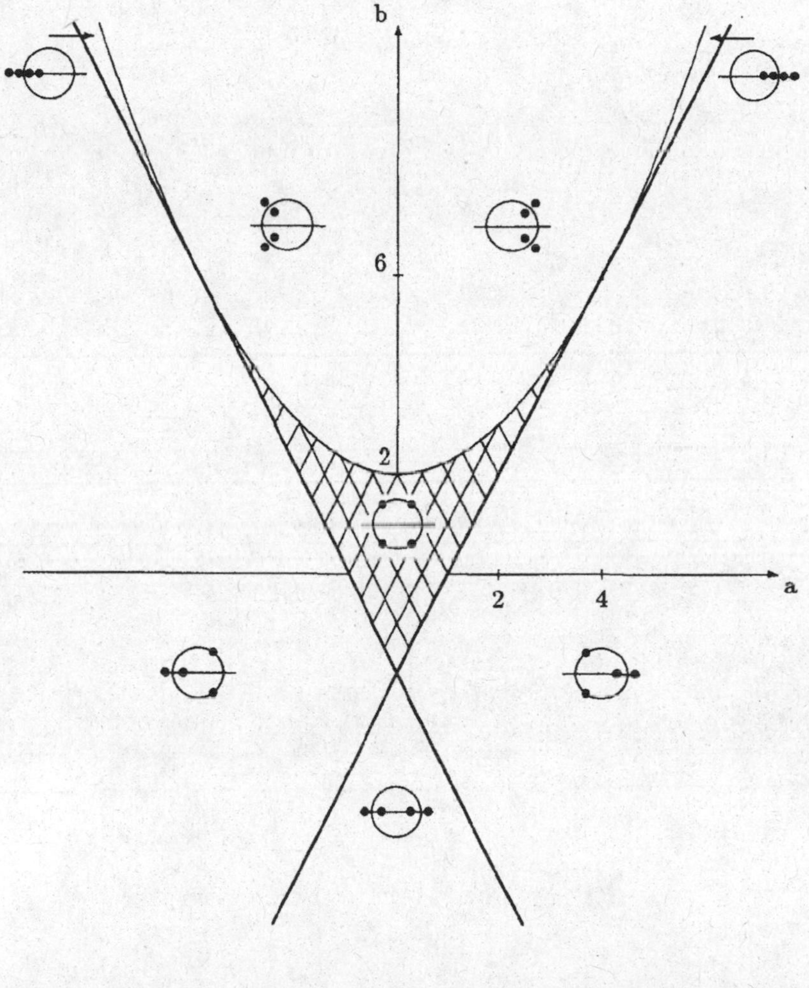

Figure 2.3: Zone of stability (instability) of the 4D-symplectic maps and the positions of the multipliers

3. Singularity Theory for Equivariant Gradient Bifurcation Problems

In this chapter we introduce a singularity theory framework for Γ-equivariant gradient bifurcation problems. Let $f : \mathbf{R}^n \times \mathbf{R} \to \mathbf{R}$ be a Γ-invariant germ with Γ acting orthogonally on $\mathbf{R}^n$ and trivially on parameters. Then $\nabla_x f$ is a Γ-equivariant map and we say that

$$\nabla_x f(x,\lambda) = 0 \qquad \text{with} \quad \nabla_x f(0,0) = 0 \quad \text{and} \quad |\nabla_x^2 f(0,0)| = 0 \tag{3.1}$$

is a Γ-equivariant gradient bifurcation problem. The function f is the potential of the bifurcation problem and $\lambda \in \mathbf{R}$ is the bifurcation parameter. In bifurcation problems the changes of the solution set of $\nabla_x f(x,\lambda) = 0$ as λ changes are of interest; that is, it is important to preserve the λ-slices of the zero-set. In other words, it is required to preserve both the gradient structure and certain directions in parameter space and this dual requirement leads to the greatest technicalities in the theory.

It would seem that a natural singularity theory framework for such germs is equivariant right equivalence for potentials with some special consideration for the parameter λ. To be precise define $\mathcal{E}^{\Gamma}_{o,\lambda}$ to be the set of λ-dependent Γ-invariant germs,

$$\mathcal{E}^{\Gamma}_{o,\lambda} = \{\, h(x,\lambda) \mid h : (\mathbf{R}^{n+1}, 0) \to \mathbf{R} \ \text{ and } \ h(\gamma x, \lambda) = h(x,\lambda) \ , \ \forall \gamma \in \Gamma \,\} .$$

Then a natural equivalence relation on $\mathcal{E}^{\Gamma}_{o,\lambda}$ is Γ-right equivalence preserving λ-slices

$$f(x,\lambda) = g\big(X(x,\lambda), \Lambda(\lambda)\big) \tag{3.2}$$

where $f, g \in \mathcal{E}^{\Gamma}_{o,\lambda}$ and $(I, X, \Lambda) \in \mathcal{K}^{\Gamma}_{\lambda}$. $\mathcal{K}^{\Gamma}_{\lambda}$ is a contact group to be defined shortly (see paragraph including equation (3.3)). We will refer to the equivalence in (3.2) as Γ-right equivalence for bifurcation problems and denote the group associated with it by $\mathcal{R}^{\Gamma}_{\lambda}$. Although the $\mathcal{R}^{\Gamma}_{\lambda}$-theory has an elegant simplicity, it turns out to inadequate.

It is readily demonstrated that if $f, g \in \mathcal{E}^{\Gamma}_{o,\lambda}$ and f is $\mathcal{R}^{\Gamma}_{\lambda}$-equivalent to g then $\nabla_x f$ is contact equivalent to $\nabla_x g$ (simply differentiate (3.2)). The problem is that the converse is not true in general. There are two distinct difficulties with the $\mathcal{R}^{\Gamma}_{\lambda}$-theory which involve different levels of complexity.

The first (simple) obstruction is associated with the difference between contact equivalence ($\mathcal{K}^{\Gamma}$) of gradients and right equivalence ($\mathcal{R}^{\Gamma}$) of potentials and does not involve the distinguished parameter. In particular $\mathcal{R}^{\Gamma}$-equivalence can introduce *moduli* (parameters that cannot be scaled away by a smooth change of coordinates) that are irrelevant in the context of $\mathcal{K}^\Gamma$-equivalence. This is easily seen in the following example of Hummel [1979, p.155].

Let $u = z\bar{z}$ and $v = \frac{1}{2}(z^3 + \bar{z}^3)$ be the $\mathbf{D}_3$-invariants on $\mathbf{R}^2$ (see Proposition 2.6 or Chapter 6) and consider the $\mathbf{D}_3$-invariant potential $\phi_m = u^2 + m v^2$. In the setting of $\mathbf{D}_3$-right equivalence ($\mathcal{R}^{\mathbf{D}_3}$), positive m are real moduli and so for positive $m \neq n$ ϕ_m is *not* $\mathcal{R}^{\mathbf{D}_3}$-equivalent to ϕ_n. However, introduction of simple scalings on the domain and range shows that, for all positive m, $\nabla_x \phi_m = 4u\, z + 6mv\, \bar{z}^2$ is $\mathbf{D}_3$-contact equivalent to $4u\, z + 6v\, \bar{z}^2$ (the gradient of the potential $\phi_1 = u^2 + v^2$).

The second, more fundamental, difficulty with $\mathcal{R}^\Gamma_\lambda$-equivalence is that the single change of coordinates (X, Λ) in (3.2) is too restrictive when the parameter λ is distinguished. In particular, singularities of infinite codimension arise immediately, even in the simplest context.

For example, suppose Γ is trivial and consider the potential $f_1(x, \lambda) = \frac{1}{4}x^4 + \frac{1}{2}\lambda x^2$ which corresponds to the pitchfork $\nabla_x f_1 = x^3 + \lambda x$. Using distinguished parameter contact equivalence ($\mathcal{K}_\lambda$) the pitchfork is codimension 2 but in the $\mathcal{R}_\lambda$-context f_1 is of *infinite* codimension! A similar, somewhat more immediate fact is that $x^4 + x^3 + \lambda x$ is $\mathcal{K}_\lambda$-contact equivalent to $x^3 + \lambda x$ but, at the potential level, $\frac{1}{5}x^5 + \frac{1}{4}x^4 + \frac{1}{2}\lambda x^2$ is *not* $\mathcal{R}_\lambda$-equivalent to $\frac{1}{4}x^4 + \frac{1}{2}\lambda x^2$.

The modal obstruction is more of a nuisance than a serious difficulty: it introduces unnecessary distinctions in the classification. The second difficulty is fundamental however as it precludes finite codimension. In the case of the pitchfork the difficulties can be circumvented by using left-right equivalence ($\mathcal{A}^\Gamma_\lambda$) with parameters. The classical theory of left-right equivalence ($\mathcal{A}$; independent change of coordinates at the source and image) also preserves the zero-set of the gradient (recall that we are only interested in controlling what happens to the zero-set of the gradient). The correct results for the pitchfork are recovered if we use $\mathcal{A}^\Gamma_\lambda$-equivalence (that is λ-parametrized left-right equivalence) for *potentials*. On the other hand, this is again not a satisfactory solution for a general approach. For instance the $\mathcal{A}^\Gamma_\lambda$-codimension tends to increase much more rapidly than the $\mathcal{K}^\Gamma_\lambda$-codimension, and actually, in most cases we still cannot get finite $\mathcal{A}^\Gamma_\lambda$-codimension. This can be easily checked on problems with two space variables without symmetry or on most problems with nontrivial symmetries.

Another approach is to use the concept of paths in the parameter space. The basic idea was suggested in Golubitsky & Schaeffer [1979] and has been extended and applied to some bifurcation problems in $\mathbf{R}^2$ (without symmetry) by Zuppa [1984]. In those two works the idea was to consider the $\mathcal{R}$-universal unfolding F of $f_o \stackrel{\text{def}}{=} f(x, 0)$ and $Z(F)$, the zero-set of its gradient with respect to the space-variable x. Then bifurcation problems are represented as lifts on $Z(F)$ of one dimensional paths through the parameter space of F. Thus the last step consists of defining an equivalence relation for paths that respects the contact equivalence of the gradient bifurcation problems. This approach has the great advantage of nicely organizing the results. There is a hierarchy of potentials – independent of the number of distinguished parameters – and for each potential, a hierarchy of paths. But this too has drawbacks: it still does not solve the problem of the appearance of unwanted moduli in the classification of the potentials (f_o) and of greater importance: the calculations associated with path equivalence rapidly become very complicated. Moreover, in the equivariant setting the difficulties appear at low codimension.

For these reasons we take a different (hybrid) approach. After reviewing (in Section 3.1) the classical methods for contact equivalence for bifurcation diagrams of GSS II [1988], we develop, in Section 3.2, a theory based on the "gradient" part of the tangent spaces used in the classical approach. Although there is not an underlying group of change of coordinates, the principal results go through.

We are able to define and compute the *(gradient)* $\mathcal{G}^{\Gamma}_{\lambda}$*-universal unfolding.* In some sense it corresponds to the "gradient" part of the $\mathcal{K}^{\Gamma}_{\lambda}$-universal unfolding of $\nabla_x f$ (Universality Theorem). In a second step we tackle some aspects of the recognition problem in particular when Γ is a finite group.

When Γ is finite, it is possible to redefine the path formulation in terms of the *(gradient)* $\mathcal{G}^{\Gamma}$*-universal unfolding* of the potential f_o. Our idea is to consider paths in that modified parameter space. The theory for such paths is developed in Section 3.4, where we use ideas of Mond & Montaldi [1991]. In Mond & Montaldi the algebraic base for a similar program to tackle multi-parameter ($\lambda \in \mathbf{R}^k$) $\mathcal{K}_\lambda$-contact equivalence without gradient structure or symmetry is laid down. A consequence of our adaptation of the path formulation is Theorem 3.3 which states that under mild assumptions finite *gradient* codimension is equivalent to finite $\mathcal{K}^{\Gamma}_{\lambda}$-codimension. Hence, in practical terms, we can use the $\mathcal{K}^{\Gamma}_{\lambda}$-recognition theory, simply imposing that the normal forms should be gradients. We shall also be using the path formulation for the nice organization it provides (cf. Chapter 4 on classification of $\mathbf{Z}_q$-equivariant gradient bifurcation problems). Most of the proofs are to be found in Section 3.5. Before proceeding to the general theory some notation and preliminary results are introduced.

3.1 Contact Equivalence and Gradient Maps

The classic equivalence relation for bifurcation problems in $\mathcal{E}^{\Gamma}_{\lambda}$ is equivariant contact equivalence where

$$\mathcal{E}^{\Gamma}_{\lambda} = \{\ h(x,\lambda) \mid h : (\mathbf{R}^{n+1},0) \to \mathbf{R}^n \ \text{ and } \ \gamma h(x,\lambda) = h(\gamma x,\lambda)\ ,\ \forall \gamma \in \Gamma\ \}.$$

Germs that are also gradients form a subset of $\mathcal{E}^{\Gamma}_{\lambda}$ which we define by

$$\mathcal{E}^{\Gamma}_{\nabla,\lambda} = \{\ \nabla_x f(x,\lambda) \mid f \in \mathcal{E}^{\Gamma}_{o,\lambda}\ \}.$$

Associated with the sets $\mathcal{E}^{\Gamma}_{o,\lambda}$, $\mathcal{E}^{\Gamma}_{\lambda}$ and $\mathcal{E}^{\Gamma}_{\nabla,\lambda}$ are the following basic facts.

(a) $\mathcal{E}^{\Gamma}_{o,\lambda}$ is a finitely generated ring with a unique maximal ideal $m^{\Gamma}_{x,\lambda}$. When clear from the context we shall denote the maximal ideal simply by m. The potentials we consider are elements of $m^{\Gamma}_{x,\lambda}$.

(b) $\mathcal{E}^{\Gamma}_{\lambda}$ is a finitely generated module over $\mathcal{E}^{\Gamma}_{o,\lambda}$ and a vector space over $\mathbf{R}$.

(c) $\mathcal{E}^{\Gamma}_{\nabla,\lambda} \subset \mathcal{E}^{\Gamma}_{\lambda}$ is a real vector subspace of $\mathcal{E}^{\Gamma}_{\lambda}$ and has a module structure only over the ring $\mathcal{O}_{\lambda}$ of functions in λ.

In general, the ring of functions over some space of parameters α will be denoted by $\mathcal{O}_{\alpha}$.

Consider the set of Γ-commuting matrices

$$\mathcal{M}^{\Gamma}_{\lambda} = \{\, T : (\mathbf{R}^{n+1}, 0) \to \mathbf{M}_n(\mathbf{R}) \mid T(\gamma x, \lambda)\gamma = \gamma T(x, \lambda)\ ,\ \forall \gamma \in \Gamma \,\}.$$

Note that $\mathcal{M}^{\Gamma}_{\lambda}$ is a finitely generated $\mathcal{E}^{\Gamma}_{o,\lambda}$-module. By definition, $\mathcal{M}^{\Gamma}_{\lambda,o}$ denotes the subset of $\mathcal{M}^{\Gamma}_{\lambda}$ formed by the elements T where T^o is in the connected component of the identity.

The *contact group* $\mathcal{K}^{\Gamma}_{\lambda}$ is by definition the set of triples (T, X, Λ) where $T \in \mathcal{M}^{\Gamma}_{\lambda,o}$, $X \in (m\mathcal{E}_{\lambda})^{\Gamma}$, $X^o_x \in \mathcal{M}^{\Gamma}_{\lambda,o}$ and $\Lambda : (\mathbf{R}, 0) \to (\mathbf{R}, 0)$ is required to satisfy $\Lambda^o_{\lambda} > 0$. The contact group $\mathcal{K}^{\Gamma}_{\lambda}$ acts in a natural way in $(m\mathcal{E}_{\lambda})^{\Gamma}$ by

$$(T, X, \Lambda)\cdot f = T(x, \lambda)\ f\big(X(x, \lambda), \Lambda(\lambda)\big). \tag{3.3}$$

By definition two elements $f, g \in (m\mathcal{E}_{\lambda})^{\Gamma}$ are said to be *contact equivalent* if they belong to the same orbit of the action (3.3). Contact equivalence is also an equivalence relation on $\mathcal{E}^{\Gamma}_{\nabla,\lambda}$ but for $f \in \mathcal{E}^{\Gamma}_{\nabla,\lambda}$, $(T, X, \Lambda)\cdot f$ is not necessarily in $\mathcal{E}^{\Gamma}_{\nabla,\lambda}$ for arbitrary $(T, X, \Lambda) \in \mathcal{K}^{\Gamma}_{\lambda}$. Therefore some modification of the usual techniques is necessary in order to describe the contact classes inside $\mathcal{E}^{\Gamma}_{\nabla,\lambda}$. Before proceeding to address this central question we record additional algebraic details necessary for the theory of contact equivalence.

Basic among the algebraic objects is the *tangent space* at $h \in (m\mathcal{E}_{\lambda})^{\Gamma}$ of its $\mathcal{K}^{\Gamma}_{\lambda}$-orbit defined by

$$\mathcal{T}(h, \mathcal{K}^{\Gamma}_{\lambda}) = \{\, Th + h_x X + h_{\lambda}\Lambda \mid T \in \mathcal{M}^{\Gamma}_{\lambda},\ X \in (m\mathcal{E}_{\lambda})^{\Gamma},\ \Lambda^o = 0 \,\}.$$

For computational convenience, a $\mathcal{E}^{\Gamma}_{o,\lambda}$-submodule $\mathcal{RT}(h)$ of $\mathcal{T}(h, \mathcal{K}^{\Gamma}_{\lambda})$ is often used instead (GSS I/II [1985/1988]). When $\mathcal{E}^{\Gamma}_{\lambda}/\mathcal{T}(h, \mathcal{K}^{\Gamma}_{\lambda})$ is of finite dimension as a real vector space we say that h is of finite $\mathcal{K}^{\Gamma}_{\lambda}$-codimension. When in fact $\mathcal{T}(h, \mathcal{K}^{\Gamma}_{\lambda})$ is of finite codimension then

(a) there exists a "simple" polynomial in the contact class; that is, *the normal form* of h,

(b) the elements of the class are defined by a finite set of equalities and inequalities involving only a fixed finite part of their Taylor series.

Perturbations of elements in $\mathcal{E}_\lambda^\Gamma$ are described using the concept of an unfolding. Given $h \in (m\mathcal{E}_\lambda)^\Gamma$, an *unfolding* H of h with k parameters is a germ of a Γ-equivariant map $(\mathbf{R}^{n+1+k}, 0) \to \mathbf{R}^n$ such that $H(x,\lambda,0) = h(x,\lambda)$. Let

$$\mathcal{E}_{\lambda,un}^\Gamma(k) = \{\, H : (\mathbf{R}^{n+1+k},0) \to \mathbf{R}^n \mid H \text{ is } \Gamma-\text{equivariant},\ H(\cdot,\cdot,0) \in \mathcal{E}_\lambda^\Gamma \,\}$$

and $\mathcal{E}_{\nabla,\lambda,un}^\Gamma(k)$ be the subset of $\mathcal{E}_{\lambda,un}^\Gamma(k)$ where H is a gradient map. More generally, the subscript ∇ indicates that we consider the elements of the space which are also gradients in x. The group $\mathcal{K}_\lambda^\Gamma$ extends naturally to $\mathcal{K}_{\lambda,un}^\Gamma(k)$ by considering the quadruple (T, X, Λ, Φ) acting on $\mathcal{E}_{\lambda,un}^\Gamma(k)$ where (T, X, Λ) is a k-parameter unfolding of an element in $\mathcal{K}_\lambda^\Gamma$ and $\Phi : (\mathbf{R}^k, 0) \to (\mathbf{R}^k, 0)$ is a diffeomorphism. The action of (T, X, Λ, Φ) on $\mathcal{E}_{\lambda,un}^\Gamma(k)$ is simply

$$(T, X, \Lambda, \Phi) \cdot H = T(x,\lambda,\alpha)\; H\big(X(x,\lambda,\alpha), \Lambda(\lambda,\alpha), \Phi(\alpha)\big). \tag{3.4}$$

Contact equivalence actually induces two relations on unfoldings. In addition to (3.4) we can also compare two unfolding with different numbers of parameters but the same $h \in \mathcal{E}_\lambda^\Gamma$. In particular, let $H \in \mathcal{E}_{\lambda,un}^\Gamma(k)$ and $G \in \mathcal{E}_{\lambda,un}^\Gamma(l)$ be two unfoldings of $h \in \mathcal{E}_\lambda^\Gamma$. We say that G *maps into* H if there is $(T, X, \Lambda, \Phi) \in \mathcal{K}_{\lambda,un}^\Gamma(l)$, unfolding of the identity and $\Psi : (\mathbf{R}^l, 0) \to (\mathbf{R}^k, 0)$ such that $G = (T, X, \Lambda, \Psi \circ \Phi) \cdot H$. The important property here is versality. H is said to be a *versal* unfolding of h if any other unfolding G of h maps into H. In particular the versal unfoldings with the minimal number of parameters, say r, are called *universal*. The universal unfoldings of h are all equivalent in the sense that they belong to the same $\mathcal{K}_{\lambda,un}^\Gamma(r)$-orbit.

The construction of an universal unfolding uses the (extended) tangent space corresponding to $\mathcal{K}_{\lambda,un}^\Gamma(1)$,

$$T_e(h, \mathcal{K}_\lambda^\Gamma) = \{\, Th + h_x X + h_\lambda \Lambda \mid T \in \mathcal{M}_\lambda^\Gamma,\ X \in \mathcal{E}_\lambda^\Gamma,\ \Lambda : (\mathbf{R},0) \to \mathbf{R} \,\}. \tag{3.5}$$

We use that space essentially to form the quotient $\mathcal{E}_\lambda^\Gamma / T_e(h, \mathcal{K}_\lambda^\Gamma) \stackrel{\text{def}}{=} \mathcal{N}_e(h, \mathcal{K}_\lambda^\Gamma)$; that is, the *extended normal space.* More generally, we will define normal spaces for all the tangent spaces we consider. If both the ambient space and the tangent space are clear from the context, we shall denote the associated normal space simply by exchanging T for $\mathcal{N}$.

Fundamental results on the normal space $\mathcal{N}_e(h, \mathcal{K}_\lambda^\Gamma)$ (GSS II [1988, p.212]) are:

(a) $\dim_{\mathbf{R}} \mathcal{N}_e(h, \mathcal{K}_\lambda^\Gamma) < \infty$ if and only if h is of finite codimension.

(b) If $\dim_{\mathbf{R}} \mathcal{N}_e(h, \mathcal{K}_\lambda^\Gamma) = r < \infty$, the universal unfoldings of h have r parameters and are equivalent to a polynomial. If $\{h_i\}_{i=1}^r \subset \mathcal{E}_\lambda^\Gamma$ is a base for $\mathcal{N}_e(h, \mathcal{K}_\lambda^\Gamma)$ then $h + \sum_{i=1}^r \alpha_i h_i$ is a universal unfolding of h.

The tangent spaces defined so far have only a structure of $\mathcal{O}_\lambda$-modules, being the sum of $\mathcal{E}_{o,\lambda}^\Gamma$-modules and $\mathcal{O}_\lambda$-modules. We denote by a hat superscript the $\mathcal{E}_{o,\lambda}^\Gamma$-modules obtained by forgetting about the Λ-part (*restricted* tangent spaces), for example

$$\widehat{\mathcal{T}}(h, \mathcal{K}_\lambda^\Gamma) = \{\, Th + h_x X \mid T \in \mathcal{M}_\lambda^\Gamma,\ X \in (m\mathcal{E}_\lambda)^\Gamma \,\}.$$

3.2 Fundamental Results

Let Γ be any compact Lie group we can suppose to act orthogonally on $\mathbf{R}^n$ and let $\nabla_x f$ be a gradient bifurcation problem in $\mathcal{E}_\lambda^\Gamma$. The construction of the *gradient extended tangent space* of $\nabla_x f$ is as follows. Define

$$\mathcal{T}_{\nabla,e}(\nabla_x f, \mathcal{K}_\lambda^\Gamma) = \mathcal{T}_e(\nabla_x f, \mathcal{K}_\lambda^\Gamma) \cap \mathcal{E}_{\nabla,\lambda}^\Gamma\,.$$

As we have previously mentioned, $\mathcal{T}_{\nabla,e}$ has only a structure of $\mathcal{O}_\lambda$-module but note that, unlike $\mathcal{T}_e$, it has no a-priori nice $\mathcal{E}_{o,\lambda}^\Gamma$-submodule like $\widehat{\mathcal{T}}_e$. In particular, we cannot find any useful finitely generated submodules. Its normal space in $\mathcal{E}_{\nabla,\lambda}^\Gamma$ is

$$\mathcal{N}_{\nabla,e}(\nabla_x f, \mathcal{K}_\lambda^\Gamma) \stackrel{\text{def}}{=} \mathcal{E}_{\nabla,\lambda}^\Gamma / \mathcal{T}_{\nabla,e}(\nabla_x f, \mathcal{K}_\lambda^\Gamma)\,.$$

Note that $\mathcal{N}_{\nabla,e}$ is a subspace of $\mathcal{N}_e$. The dimension of $\mathcal{N}_{\nabla,e}$ as a real vector space is the *gradient codimension* of $\nabla_x f$, denoted by $\mathrm{cod}_\nabla(\nabla_x f, \mathcal{K}_\lambda^\Gamma)$.

An alternative characterization of the above expressions is obtained by working in the space of potentials $\mathcal{E}_{o,\lambda}^\Gamma$. Define

$$\begin{aligned}
&\text{(a)} \quad \nabla : (m\mathcal{E}_{o,\lambda})^\Gamma \to \mathcal{E}_\lambda^\Gamma \text{ by } f \mapsto \nabla_x f,\ \nabla \text{ is an } \mathcal{O}_\lambda\text{-isomorphism onto } \mathcal{E}_{\nabla,\lambda}^\Gamma,\\
&\text{(b)} \quad a_e^\Gamma(f) \stackrel{\text{def}}{=} \{\, g \in \mathcal{E}_{o,\lambda}^\Gamma \mid \widehat{\mathcal{T}}_e(g, \mathcal{R}_\lambda) \subset \widehat{\mathcal{T}}_e(f, \mathcal{R}_\lambda) \,\},\\
&\text{(c)} \quad \mathcal{T}_e(f, \mathcal{G}_\lambda^\Gamma) \stackrel{\text{def}}{=} a_e^\Gamma(f) + \mathcal{T}_e(f, \mathcal{R}_\lambda^\Gamma), \qquad (3.6)\\
&\text{(c)} \quad \mathcal{N}_e(f, \mathcal{G}_\lambda^\Gamma) \stackrel{\text{def}}{=} (m\mathcal{E}_{o,\lambda})^\Gamma / \mathcal{T}_e(f, \mathcal{G}_\lambda^\Gamma),
\end{aligned}$$

where it is to be recalled that,

$$\widehat{\mathcal{T}}_e(f, \mathcal{R}_\lambda) = \{\, <\nabla_x f, X> \mid X \in \mathcal{E}_\lambda \,\} \quad \text{for} \quad f \in \mathcal{E}_{o,\lambda}^\Gamma\,,$$

where $<,>$ is the scalar product on $\mathbf{R}^n$.

Remark. Note that, for technical reasons, we deliberately took the *non*-equivariant formulation for (b). As can be seen from the proof of Theorem 3.1 (p.49), we want to prove that $g \in a_e^\Gamma(f)$ is equivalent to the existence of a matrix T (which can be made Γ-equivariant by averaging) such that $\nabla_x g = T\,\nabla_x f$. At the present time, it is not clear to us how to prove the result in all the cases if one takes the Γ-invariant formulation ($\mathcal{R}_\lambda^\Gamma$; it is even not clear whether it is actually always true).

Technically the result we want is the following. Consider $\{X_i\}_{i=1}^r$ to be a set of generators for $\mathcal{E}_\lambda^\Gamma$. Form the matrix X such that $X_{ij}(x,\lambda) = (X_i)_j(x,\lambda)$. Then the fact that $\widehat{\mathcal{T}}_e(g, \mathcal{R}_\lambda^\Gamma) \subset \widehat{\mathcal{T}}_e(f, \mathcal{R}_\lambda^\Gamma)$ means that there exist a Γ-equivariant matrix A (depending on (x, λ)) such that

$$X\, \nabla_x g = A\, X\, \nabla_x f\,.$$

To be able to get the conclusion we want that this last relation implies the existence of a matrix T such that $\nabla_x g = T\,\nabla_x f$. Note that if the Γ-action is trivial then $X = \mathbf{I}_n$ and the result is trivial.

Clearly $a_e^\Gamma(f)$, $\mathcal{T}_e(f, \mathcal{G}_\lambda^\Gamma)$ and $\mathcal{N}_e(f, \mathcal{G}_\lambda^\Gamma)$ are in general only $\mathcal{O}_\lambda$-modules, although $a_c^\Gamma(f)$ and $\mathcal{T}_e(f, \mathcal{G}_\lambda^\Gamma)$ are rings. The next proposition shows that the "gradient part" of $\mathcal{T}_e(\nabla_x f, \mathcal{K}_\lambda^\Gamma)$ and its codimension are invariant under $\mathcal{K}_\lambda^\Gamma$-equivalence, that is, well-defined in our context. The proof is in Section 3.5.

Theorem 3.1. (a) $\nabla\big(\mathcal{T}_e(f, \mathcal{G}_\lambda^\Gamma)\big) = \mathcal{T}_{\nabla,e}(\nabla_x f, \mathcal{K}_\lambda^\Gamma)$.
(b) *The gradient codimension is an invariant of the $\mathcal{K}_\lambda^\Gamma$-equivalence.*
(c) *As a corollary of (a), $\nabla\big(\mathcal{N}_e(f, \mathcal{G}_\lambda^\Gamma)\big) = \mathcal{N}_{\nabla,e}(\nabla_x f, \mathcal{K}_\lambda^\Gamma)$.*

When the gradient codimension of $\nabla_x f$ is finite, say m, let $\{\nabla_x h_i\}_{i=1}^m$ be a spanning basis for $\mathcal{N}_{\nabla,e}$. Then we define a $\mathcal{G}_\lambda^\Gamma$-unfolding of $\nabla_x f$ by

$$\nabla_x F(x, \lambda, \alpha) = \nabla_x f(x, \lambda) + \sum_{i=1}^{m} \alpha_i\, \nabla_x h_i(x, \lambda). \tag{3.7}$$

The next theorem explores the universal properties of $\nabla_x F$.

Theorem 3.2 (Universality Theorem). *Let $\nabla_x f$ be of finite gradient codimension and $\nabla_x F$ a $\mathcal{G}_\lambda^\Gamma$-universal unfolding of $\nabla_x f$ as defined in (3.7). Then*
(a) *(Gradient Versality) Let $\nabla_x G(x, \lambda, \delta)$ be an unfolding of $\nabla_x f$ with p-parameters, $\delta \in \mathbb{R}^p$, then $\nabla_x G$ $\mathcal{K}_\lambda^\Gamma$-maps into $\nabla_x F$.*
(b) $\mathrm{cod}_\nabla(\nabla_x f, \mathcal{K}_\lambda^\Gamma)$ *is the minimal number of parameters needed for an unfolding of $\nabla_x f$ to be gradient versal.*
(c) *Any other gradient versal unfolding of $\nabla_x f$ having $\mathrm{cod}_\nabla(\nabla_x f, \mathcal{K}_\lambda^\Gamma)$-parameters is $\mathcal{K}_\lambda^\Gamma$-isomorphic to $\nabla_x F$.*

The proof is to be found in Section 3.5. The theorem takes care of the unfolding theory. We can now turn our attention to the recognition problem. It is not immediately clear that we can also transpose the classical theory in the same generality that was obtained above for the unfolding theory. In the classical theory it is possible, using unipotent subgroups of $\mathcal{K}_\lambda^\Gamma$, to get – in a systematic way – powerful estimates of the set of higher order terms (terms which can be removed in the Taylor expansion of any element in a $\mathcal{K}_\lambda^\Gamma$-equivalence class). In the case that $\nabla_x f$ is of finite

$\mathcal{K}_\lambda^\Gamma$-codimension (non-gradient situation) we know that the following $\mathcal{O}_\lambda$-module of higher order terms is also of finite codimension (that is calculable)

$$\mathcal{P}(\nabla_x f, \mathcal{K}_\lambda^\Gamma) = \{\, h \in \mathcal{E}_\lambda^\Gamma \mid ((g+h) \sim \nabla_x f \,;\, \mathcal{K}_\lambda^\Gamma) \,,\; \forall g \text{ with } (g \sim \nabla_x f \,;\, \mathcal{K}_\lambda^\Gamma) \,\}\,.$$

(cf. Chapter XIV, GSS II [1988, p.204...] or this work p.47).

We can thus define the following $\mathcal{O}_\lambda$-module

$$\mathcal{P}_\nabla(\nabla_x f, \mathcal{K}_\lambda^\Gamma) = \mathcal{P}(\nabla_x f, \mathcal{K}_\lambda^\Gamma) \cap \mathcal{E}_{\nabla,\lambda}^\Gamma \,,$$

which we call the set of *gradient higher-order terms*. When Γ is not finite, the $\mathcal{K}_\lambda^\Gamma$-codimension of $\nabla_x f$ is not in general finite even if the gradient codimension is (cf. the examples in Montaldi, Roberts & Stewart [1991]). But when Γ is finite we can use the set of gradient higher-order terms provided the following assumption holds. We say that $f_o \in \mathcal{E}_o$ satisfies (H1) if

$$\mathrm{cod}_\nabla(\nabla_x f_o, \mathcal{K}) < \infty \,. \tag{H1}$$

As a consequence of the path formulation (cf. next section), we have

Theorem 3.3. *Suppose that Γ is finite and that f_o satisfies (H1), then*

$$\mathrm{cod}(\nabla_x f, \mathcal{K}_\lambda^\Gamma) < \infty \quad \textit{iff} \quad \mathrm{cod}_\nabla(\nabla_x f, \mathcal{K}_\lambda^\Gamma) < \infty \,.$$

Remark. Although (H1) is sufficient, we conjecture that it is not a necessary hypothesis for the conclusion of Theorem 3.3 to hold true (but still with a finite Γ). In Section 3.3 we shall discuss some alternative sufficient conditions.

When Γ is finite, Theorem 3.3 justifies the following methodology for the classification of problems with $f_o \in \mathcal{E}_o$ satisfying (H1). We can work in the context of $\mathcal{K}_\lambda^\Gamma$-equivalence and, for the *recognition problem*, we simply need to insure that the normal forms are gradients. The question of higher-order terms is treated in the classical way using the set of gradient higher-order terms. And then, for the *universal unfolding problem*, we compute $\mathcal{N}_{\nabla,e}$, which can be done by adapting the classical computations of $\mathcal{N}_e$.

3.3 Potentials and Paths

Suppose that $\mathrm{cod}_\nabla(\nabla_x f, \mathcal{K}_\lambda^\Gamma) < \infty$, it is then easy to see that $\mathrm{cod}_\nabla(\nabla_x f_o, \mathcal{K}^\Gamma)$ is also finite. There is a result analogous to the Universality Theorem for $\nabla_x f_o$.

Proposition 3.5. *Let* $\{\nabla_x h_i\}_{i=1}^l$ *be a base for* $\mathcal{N}_{\nabla,e}(\nabla_x f_o, \mathcal{K}^\Gamma)$*, then*

$$\nabla_x F(x,\alpha) = \nabla_x f_o(x) + \sum_{i=1}^{l} \alpha_i \nabla_x h_i(x)$$

is a gradient universal unfolding of $\nabla_x f_o$*.*

Proof. Follows the same line as the proof of Theorem 3.2. ■

More generally, the theorems and definitions of Section 3.2 transfer to this situation (without distinguished parameter) with the necessary adjustments.

The *path formulation* takes the following form. We can consider $\nabla_x f$ as a one-parameter unfolding of $\nabla_x f_o$. From Proposition 3.5, it maps into $\nabla_x F$, that is

$$\nabla_x f(x,\lambda) = T(x,\lambda)\, \nabla_x F\big(X(x,\lambda), \phi(\lambda)\big)$$

where $(T, X, I, \phi) \in \mathcal{K}^\Gamma_{un}(1)$ identified with $\mathcal{K}^\Gamma_\lambda$. This shows that $\nabla_x f$ is $\mathcal{K}^\Gamma_\lambda$-equivalent to $\nabla_x F(\cdot, \phi)$. Similarly any gradient unfolding of $\nabla_x f$, say $\nabla_x G$, is equivalent as an unfolding to $\nabla_x F(\cdot, \Phi)$, where Φ is an unfolding of ϕ. This is because $\nabla_x G$ is not only an unfolding of $\nabla_x f_o$ but also of $\nabla_x f$. Hence when $\nabla_x G$ maps into $\nabla_x F$ it can be made to respect $\nabla_x f$.

The problem is now to study the possible potentials and their universal gradient unfoldings and to define an equivalence relation on the space of paths and the unfoldings that respect the $\mathcal{K}^\Gamma_\lambda$-equivalence of the gradients. This last analysis is carried out in Section 3.4 for finite groups. In the proof of the results concerning the path formulation we shall need the hypothesis (H1). For the remainder of this section we are going to concentrate on the potentials f_o.

The a priori classification of the potentials f_o and their $\mathcal{G}^\Gamma$-universal unfoldings requires $\mathcal{K}^\Gamma$-theory for the gradients. However, it is useful to first consider directly $\mathcal{A}^\Gamma$-equivalence for the potentials. This has several advantages. First, at "low" codimension, it results in the same classification as $\mathcal{G}^\Gamma$ (in the equivariant setting we know that $\mathcal{R}^\Gamma$-equivalence results in too fine a classification). A second advantage is that for the common classes of the $\mathcal{A}^\Gamma$ and $\mathcal{G}^\Gamma$-classifications, we can directly compare any germs and unfoldings of those classes by changes of coordinates in the potential setting and so reduce calculation.

The action of the group $\mathcal{A}^\Gamma$ on $\mathcal{E}^\Gamma_o$ is given by

$$g_o = H \circ f_o \circ X\ ,$$

where $X \in \mathcal{R}^\Gamma$ and $H : (\mathbf{R}, 0) \to (\mathbf{R}, 0)$ is a diffeomorphism. The associated tangent spaces necessary for the classification are:

$$\begin{aligned} \mathcal{T}_e(f_o, \mathcal{A}^\Gamma) &= \mathcal{T}_e(f_o, \mathcal{R}^\Gamma) + \{\, a(f_o) \mid a : (\mathbf{R},0) \to \mathbf{R} \,\}\,, \\ \mathcal{T}(f_o, \mathcal{U}(\mathcal{A}^\Gamma)) &= \mathcal{T}(f_o, \mathcal{U}(\mathcal{R}^\Gamma)) + \{\, a(f_o) \mid a : (\mathbf{R},0) \to (\mathbf{R},0) \,\}\,. \end{aligned}$$

Note that the above tangent spaces are not $\mathcal{E}_o^\Gamma$-modules and so are not convenient to work with. In practical terms, we quotient first by the $\mathcal{R}^\Gamma$-part of the tangent space, and then use the second part to quotient again the result.

The next lemma indicates some alternative conditions to (H1).

Lemma 3.6. *Suppose* Γ *is a finite group and let* $h : \mathbf{R}^n \to \mathbf{R}$ *be a* Γ*-invariant potential. The following are equivalent and imply (H1)*

$$\begin{array}{lllll}(a) & \operatorname{cod}(h,\mathcal{R}^\Gamma)<\infty & \Leftrightarrow & \operatorname{cod}(h,\mathcal{A}^\Gamma)<\infty & \Leftrightarrow & \operatorname{cod}(h,\mathcal{K}^\Gamma)<\infty.\\ (b) & \operatorname{cod}(h,\mathcal{R})<\infty & \Leftrightarrow & \operatorname{cod}(h,\mathcal{A})<\infty & \Leftrightarrow & \operatorname{cod}(h,\mathcal{K})<\infty.\\ (c) & \operatorname{cod}(\nabla_x h,\mathcal{K})<\infty. & & & & \end{array}$$

Proof. It follows from Roberts [1986] that (a) is true and $(a) \Leftrightarrow (b)$. It is clear that $(b) \Rightarrow (c)$. For the reverse implication, passing to holomorphic germs, from Proposition 2.4(ii) of Wall [1981], $\operatorname{cod}(\nabla_x h, \mathcal{K}) < \infty$ if and only if $\operatorname{cod}(\nabla_x h, \mathcal{C}) < \infty$, where $\mathcal{C}$ denotes the action on the left by invertible matrices depending on the source x. Now denote by $\mathcal{J}(h)$ the jacobian ideal of h. Thus

$$\begin{aligned}\operatorname{cod}(\nabla_x h,\mathcal{C})<\infty &\Rightarrow \operatorname{cod}_{\mathbf{R}}(\ \{\ T\nabla_x h \mid T\in\mathcal{M}\ \}\ )<\infty \text{ in } \mathcal{E}\\ &\Rightarrow \operatorname{cod}_{\mathbf{R}}(\mathcal{J}(h),\ldots,\mathcal{J}(h))<\infty \text{ in } \mathcal{E}\\ &\Rightarrow \operatorname{cod}_{\mathbf{R}}(\mathcal{J}(h))<\infty \text{ in } \mathcal{E}_o\\ &\Rightarrow \operatorname{cod}(h,\mathcal{R})<\infty.\end{aligned}$$

■

The next lemma makes explicit a relation between $\mathcal{G}^\Gamma$ and $\mathcal{A}^\Gamma$-equivalence.

Lemma 3.7. $\operatorname{cod}(f_o,\mathcal{G}^\Gamma)\overset{\text{def}}{=}\operatorname{cod}_\nabla(\nabla_x f_o,\mathcal{K}^\Gamma)\leq \operatorname{cod}(f_o,\mathcal{A}^\Gamma)$.

(b) *If* f_o *is weighted homogeneous then*

$$\mathcal{T}_{\nabla,e}(\nabla_x f_o,\mathcal{K}^\Gamma)=\mathcal{T}_e(f_o,\mathcal{A}^\Gamma)=\mathcal{T}_e(f_o,\mathcal{R}^\Gamma).$$

Moreover, if f_o *is of (any) finite codimension (H1) is satisfied.*

Proof. (a) We assume that $\operatorname{cod}_\nabla(f_o,\mathcal{A}^\Gamma) < \infty$ (otherwise the result is obvious). Let $\{\nabla_x h_i\}_{i=1}^l$ be a base of $\mathcal{N}_e(f_o,\mathcal{A}^\Gamma)$. Via ∇, $\{\nabla_x h_i\}_{i=1}^l$ generates $\mathcal{N}_{\nabla,e}(\nabla_x f_o,\mathcal{K}^\Gamma)$ but it might not be independent. Hence the conclusion.

(b) Via the proof of Proposition 1.3 of Roberts [1986], it is enough to prove the result in the complexified situation. From Proposition 6 of Yau [1983], if $f_{o,\mathbf{C}}$ is weighted homogeneous $f_{o,\mathbf{C}} \in m\,\mathcal{J}(f_{o,\mathbf{C}})$ and $a(f_{o,\mathbf{C}}) \subset m\,\mathcal{J}(f_{o,\mathbf{C}})$, where $\mathcal{J}(f_{o,\mathbf{C}})$ denotes the Jacobian ideal of $f_{o,\mathbf{C}}$. Hence, by averaging over Γ, we get the equality of tangent spaces we are aiming for.

When the codimension of any of those tangent spaces is finite, so is $\operatorname{cod}(f_{o,\mathbf{C}})$, that is $\operatorname{cod}(\nabla_x f_{o,\mathbf{C}},\mathcal{K})$, hence $\operatorname{cod}_\nabla(\nabla_x f_o,\mathcal{K}) < \infty$, that is (H1). ■

We finish this section with a result that is needed for the discussion of the higher order terms for paths. We can choose a base $\{\nabla_x h_i\}_{i=1}^l$ of $\mathcal{N}_{\nabla,e}(\nabla_x f_o, \mathcal{K}^\Gamma)$ formed of homogeneous polynomials, which we order by increasing powers of x. For that base, consider the following $\mathcal{E}_o^\Gamma$-submodule of $\mathcal{M}^\Gamma \times \mathcal{E}^\Gamma$:

$$\begin{aligned}\mathcal{B} = \{\, (T, X) \mid T \in (m\mathcal{M})^\Gamma\,,\ \ & X \in (m^2\mathcal{M})^\Gamma \ \text{ such that} \\ & T\,\nabla_x f_o - (\nabla_x f_o)_x\, X \equiv 0 \ \text{ and} \\ & T\,\nabla_x h_k - (\nabla_x h_k)_x\, X \in \mathcal{E}_\nabla^\Gamma\,,\ \forall\, 1 \le k \le l \,\}.\end{aligned}$$

For any given $(T_o, X_o) \in \mathcal{B}$, we can form an $l\times l$-matrix $M_{(T_o,X_o)}$ that is constructed as follows. For $1 \le k \le l$, decompose $T_o\,\nabla_x h_k - (\nabla_x h_k)_x\, X_o$ modulo $T_{\nabla,e}(\nabla_x f_o, \mathcal{K}^\Gamma)$ in the base $\{\nabla_x h_i\}_{i=1}^l$ to get the vectors $\{\nu_k\}_{k=1}^l$. Then form the matrix $M_{(T_o,X_o)}$ whose k-th column is ν_k. By construction those matrices are upper triangular (with zero diagonal), hence nilpotent. This fact will be used in the proof of Lemma 3.16.

Remark (Linearized Stability). It is well-known that the (linearized) stability of a gradient bifurcation equation changes iff there is a secondary bifurcation (0 eigenvalue) because the Jacobian of the equation is given by the Hessian of the potential (symmetric matrix). This means that contact equivalence is perfectly adapted to deal with that situation as it preserves the sign of the determinant of the Hessian and can therefore keep track of zero eigenvalues (in the general non-gradient situation the problem is more difficult because Hopf bifurcation points are not in general invariant under coordinate changes).

3.4 Equivalence for paths

Assume Γ is a finite group. Consider $f_o \in \mathcal{E}_o^\Gamma$ satisfying (H1) and $F(x,\alpha)$ a given $\mathcal{G}^\Gamma$-universal unfolding of f_o – without loss of generality we can take f_o and F to be polynomials. And so, for a base $\{\nabla_x h_i\}_{i=1}^l$ of $\mathcal{N}_{\nabla,e}(\nabla_x f_o, \mathcal{K}^\Gamma)$ we may assume that

$$F(x,\alpha) = f_o(x) + \sum_{i=1}^{l} \alpha_i\, h_i(x). \tag{3.8}$$

Let $\mathcal{P}_l = \{\, \phi : (\mathbf{R}, 0) \to (\mathbf{R}^l, 0)\,\}$ be the space of one-parameter paths through the l-dimensional unfolding space. We consider the bifurcation problem given by $\nabla_x F(x, \phi(\lambda))$ with $\phi \in \mathcal{P}_l$. We can also define unfoldings; in particular, let

$$\mathcal{P}_{l,un}(q) = \{\, \Phi : \mathbf{R}^{1+q} \to \mathbf{R}^l \mid \Phi(\cdot, 0) \in \mathcal{P}_l \,\}$$

be the space of q-dimensional unfoldings. The objective here is to define coordinate changes on $\mathcal{P}_l$ and $\mathcal{P}_{l,un}(q)$ and furthermore to relate the coordinate changes on $\mathcal{P}_l$ to the coordinate changes for $\mathcal{K}_\lambda^\Gamma$-equivalence of the associated gradients.

A bifurcation diagram represented by a path ϕ is the slice of

$$Z(F) = \{ (x, \alpha) \mid \nabla_x F(x, \alpha) = 0 \}$$

above $\phi(\lambda)$; that is, those (x, α) where $\alpha = \phi(\lambda)$ for some λ. It is important to monitor the position of $\phi \in \mathbb{R}^l$ where bifurcation occurs in $Z(F)$. The natural way to proceed is to compare the position of the image of ϕ with the local bifurcation set

$$B = \{ \alpha \mid \exists x : \nabla_x F(x, \alpha) = 0 \ , \ |\mathrm{Hess}_x F| = 0 \ \}.$$

For technical reasons we are going to work on spaces of analytic functions (real or complex). The reason for this will become clear as we proceed. As we are interested in finitely determined problems the analyticity is not a restriction, since the determinacy and unfolding results then transfer from the analytic to the smooth case. We denote either $\mathbb{R}$ or $\mathbb{C}$ by $\mathbb{K}$. For some results (like some geometric criteria) we will need the complexification of some real analytic map f, and this will be denoted by $f_{\mathbb{C}}$. We retain the same notation for the function spaces and groups of diffeomorphisms defined heretofore. Therefore let F be defined on $\mathbb{K}^{n+l}$. F is also Γ-invariant with respect to the trivially extended action of Γ on $\mathbb{K}^{n+l}$. In general $Z(F)$ will not be a manifold, and so we introduce $\widetilde{F}$ which is the $\mathcal{G}$-universal unfolding (without symmetry) of f_o. For $\{\nabla_x h_{l+i}\}_{i=1}^{k}$ forming a base for the non-invariant part of the complement of $T_{\nabla,e}(\nabla_x f_o, \mathcal{K})$, it is easy to see that $\widetilde{F}$ can be chosen as

$$\widetilde{F}(x, \alpha, \beta) = F(x, \alpha) + \sum_{i=1}^{k} \beta_i \ h_{l+i}(x).$$

In a compact form, we denote (α, β) by $u \in \mathbb{K}^L$, $L = l + k$. As f_o is assumed to be in m^3 (cf. (3.1)), it is immediate that $Z(\widetilde{F})$ is an analytic manifold of dimension L.

Lemma 3.8. *There exists a Γ-action Γ_L on $\mathbb{K}^L$ such that, combined with the Γ-action Γ_n we already have on $\mathbb{K}^n$, it defines a Γ-diagonal action Γ_{n+L} on $\mathbb{K}^{n+L}$ by $\gamma_{n+L}(x, u) = (\gamma_n x, \gamma_L u)$ making $\nabla_x \widetilde{F}$ (Γ_n, Γ_{n+L})-equivariant, that is*

$$\nabla_x \widetilde{F}(\gamma_{n+L}(x, u)) = \gamma_n \ \nabla_x \widetilde{F}(x, u) \ , \ \forall \gamma \in \Gamma \ .$$

Proof. The Γ-action on $\mathbb{K}^n$ induces the following Γ-action Γ_* on $\mathcal{E}$:

$$(\gamma_* h)(x) = \gamma_n^{-1} h(\gamma_n x) \ .$$

Next, in the spirit of Slodowy [1978], we are going to introduce a Γ-action Γ_L on $\mathbb{K}^L$ which, combined with that action Γ_* on $\mathcal{E}$, will make $\nabla_x \widetilde{F}$ Γ-invariant.

Because $\nabla_x f_o$ is Γ_n-equivariant, $\mathcal{T}_e(\nabla_x f_o, \mathcal{K})$ is globally Γ_*-invariant. Moreover, $\mathcal{E}_\nabla$ is also globally left invariant by Γ_*. Hence Γ_* induces a well-defined action Γ_L

on the finite dimensional subspace $\mathcal{N}_e(\nabla_x f_o, \mathcal{K})$ (identified with $\mathbf{K}^L$). It is well-known (Vanderbauwhede [1982]) that we can choose the identification as to make Γ_L orthogonal. And so

$$\nabla_x \widetilde{F}(x,u) = \gamma_n^{-1} \nabla_x \widetilde{F}(\gamma_n x, \gamma_L u)\,, \quad \forall \gamma \in \Gamma.$$

This defines an orthogonal Γ-action on $\mathbf{K}^{n+L}$ by considering

$$\gamma_{n+L}(x,u) = (\gamma_n x, \gamma_L u)\,, \quad \forall \gamma \in \Gamma.$$

Clearly $Z(\widetilde{F})$ is globally invariant under Γ_{n+L}. It follows from the proofs of Proposition 2.4 and Corollary 4.5 in Slodowy [1978] that $\widetilde{F}|_{\mathbf{K}^n \times \mathrm{Fix}\,\Gamma_L} = F$. ■

Now consider the "catastrophe map" $\tilde{\pi} : Z(\widetilde{F}) \to \mathbf{K}^L$ induced from the natural projection $\pi_L : \mathbf{K}^{n+L} \to \mathbf{K}^L$. In the complex case, the theory of good representatives (Looijenga [1984, p.25,...]) can be used to obtain the following results:

(a) $(\pi_L, \tilde{\pi}_{\mathbf{C}})$ satisfy the properties of (2.7) (Looijenga, p.25), that is $\tilde{\pi}_{\mathbf{C}}$ defines an isolated singularity at 0. $\tilde{\pi}_{\mathbf{C}}$ is a proper, finite analytic map.

(b) Let $C_{\tilde{\pi}}$ be the critical set of $\tilde{\pi}_{\mathbf{C}}$ (that is the set of points where $\tilde{\pi}_{\mathbf{C}}$ is not a submersion), then $C_{\tilde{\pi}}$ is an analytic subset of $Z(\widetilde{F}_{\mathbf{C}})$ of dimension $L-1$.

(c) The discriminant $\widetilde{\Delta}_{\mathbf{C}} = \tilde{\pi}_{\mathbf{C}}(C_{\tilde{\pi}})$ is an hypersurface of $\mathbf{C}^L$.

(d) $\tilde{\pi}_{\mathbf{C}}$ defines an isolated complete intersection of dimension 0 at every point of $Z(\widetilde{F}_{\mathbf{C}})$.

Moreover it is easy to check that $\tilde{\pi}_{\mathbf{C}}$ is (Γ_{n+L}, Γ_L)-equivariant and so that $C_{\tilde{\pi}}$ and $\widetilde{\Delta}_{\mathbf{C}}$ are respectively Γ_{n+L}, Γ_L-invariant. We define $\Delta_{\mathbf{C}} = \widetilde{\Delta}_{\mathbf{C}} \cap \mathrm{Fix}\,\Gamma_L$. In the real case we define $C_{\tilde{\pi}}$, $\widetilde{\Delta}$, Δ as the real parts of their complex counterparts, in which case we expect $B \subset \Delta$ strictly, but the analytic objects we are going to define are the same in both (real and complex) cases.

It is now clear that the problem is to monitor the respective positions of Δ and the image of ϕ. For this, consider the following group $\mathcal{K}_\Delta$ of changes of coordinates on $\mathcal{P}_l$:

$$\phi(\lambda) = H\big(\lambda, \psi(\Lambda(\lambda))\big)\,, \quad \phi, \psi \in \mathcal{P}_l\,, \text{ where}$$

(a) $\Lambda : (\mathbf{K},0) \to (\mathbf{K},0)\,,\ \Lambda^o_\lambda > 0\,,$

(b) $H : (\mathbf{K}^{1+l},0) \to \mathbf{K}^l$ is the restriction on Fix Γ_L of $\widetilde{H} : (\mathbf{K}^{1+L},0) \to \mathbf{K}^L$ such that $(Id, \widetilde{H})$ is a local diffeomorphism at $(\mathbf{K}^{1+L},0)$ and $\widetilde{H}_u(0,0)$ is in the connected component of the identity.

(c) Moreover, we ask that $\widetilde{H}$ is $\widetilde{\Delta}$ preserving, that is

$$\widetilde{H}(\lambda,u) \in \widetilde{\Delta}\,, \ \forall \lambda\,, \ \forall u \in \widetilde{\Delta}\,.$$

For any analytic variety $V \subset \mathbf{K}^l$, a group – denoted $\mathcal{K}_V$, of changes of coordinates on $\mathcal{P}_l$ – has been introduced in Damon [1987]. In principle $\mathcal{K}_V$ is not required to preserve the origin. Nevertheless, $\mathcal{K}_V$ can be regarded as a subgroup of a contact group (Damon [1989]) having the additional property of preserving V in the image. For $\mathcal{K}$ it is well known that, without loss of generality, H can be taken as a matrix parametrized by the source. This is not possible for $\mathcal{K}_V$ however; in particular, it is essential to work with the full parametrized nonlinear changes of coordinates. On the other hand, Damon [1987] has shown that $\mathcal{K}_V$ is a geometric subgroup of $\mathcal{K}$ and that the singularity theory for $\mathcal{K}_V$ acting on $\mathcal{P}_l$ follows the usual pattern.

In our situation Δ is not, in principle, a reduced variety. For convenience we use the group $\mathcal{K}_\Delta$, defined in the previous paragraph, which is the restriction on Fix Γ_L of $\mathcal{K}_{\tilde{\Delta}}^{\Gamma_L}$, the group containing all $(H,\Lambda) \in \mathcal{K}_{\tilde{\Delta}}$ such that $H(\cdot,\lambda)$ is Γ_L-equivariant. We develop the explicit results we need in Section 3.5 where the proofs are given. In particular, our choice of Δ means that $\mathcal{K}_\Delta \subset \mathcal{K}$.

We are now in a position to state the fundamental result which gives conditions under which the local singularity theories for ϕ, using $\mathcal{K}_\Delta$-equivalence, and $\nabla_x F(\cdot,\phi)$ in $\mathcal{E}^\Gamma_{\nabla,\lambda}$, using $\mathcal{K}^\Gamma_\lambda$-equivalence, are similar (in ways that will be made precise).

Let us introduce the following linear map: let $\phi \in \mathcal{P}_l$, then $\omega_\phi : \mathcal{P}_l \to \mathcal{E}^\Gamma_{\nabla,\lambda}$ is defined by

$$\omega_\phi(\xi) = \frac{\partial}{\partial\tau}\bigg|_{\tau=0} \nabla_x F(x, \phi(\lambda) + \tau\xi(\lambda)) \,.$$

In other words, using the definition of F in (3.8), $\omega_\phi(\xi) = \sum_{i=1}^{l} \xi_i(\lambda)\ \nabla_x h_i(x)$. We denote by $\operatorname{Im}\omega_\phi$ the image of ω_ϕ. We can now state

Lemma 3.9 (Fundamental Lemma). *If* $\operatorname{cod}_\nabla\left(\nabla_x F(\cdot,\phi), \mathcal{K}^\Gamma_\lambda\right) < \infty$, *then*

(a) ω_ϕ *induces the following isomorphism of* $\mathcal{O}_\lambda$*-modules:*

$$\Omega_{\phi,e} \ :\ \mathcal{N}_e(\phi,\mathcal{K}_\Delta) \to \mathcal{N}_{\nabla,e}\left(\nabla_x F(\cdot,\phi), \mathcal{K}^\Gamma_\lambda\right) .$$

(b) ω_ϕ *is an isomorphism between* $\mathcal{T}\left(\nabla_x F(\cdot,\phi), \mathcal{U}(\mathcal{K}^\Gamma_\lambda)\right) \cap \operatorname{Im}\omega_\phi$ *and* $\mathcal{T}\left(\phi, \mathcal{U}(\mathcal{K}_\Delta)\right)$.

(c) ω_ϕ *is an isomorphism between* $\mathcal{T}\left(\nabla_x F(\cdot,\phi), \mathcal{K}^\Gamma_\lambda\right) \cap \operatorname{Im}\omega_\phi$ *and* $\mathcal{T}(\phi,\mathcal{K}_\Delta)$. *It also induces the following injective homomorphism of* $\mathcal{O}_\lambda$*-modules:*

$$\Omega_\phi \ :\ \mathcal{N}(\phi,\mathcal{K}_\Delta) \to \mathcal{N}_\nabla\left(\nabla_x F(\cdot,\phi), \mathcal{K}^\Gamma_\lambda\right) .$$

Moreover, $\dim \mathcal{N}(\phi,\mathcal{K}_\Delta) = \dim \mathcal{N}_e(\phi,\mathcal{K}_\Delta) + 1$.

The proof is given in Section 3.5. Note that we have to choose the following unipotent subgroups of $\mathcal{K}^\Gamma_\lambda$, (respectively $\mathcal{K}_\Delta$), for part (b) to hold true. Classically $\mathcal{U}(\mathcal{K}^\Gamma_\lambda)$ consists of changes of coordinates whose linear parts are the identity. To get ω_ϕ to be an isomorphism, we need to choose for $\mathcal{U}(\mathcal{K}_\Delta)$ the subgroup of $\mathcal{K}_\Delta$ whose linear parts are unipotent diffeomorphisms represented by upper triangular matrices. We can now state the basic result of this section.

Theorem 3.10 (Fundamental Theorem). *Under hypothesis (H1) and supposing that* $\operatorname{cod}(\phi, \mathcal{K}_\Delta) < \infty$,

(a) $\operatorname{cod}(\phi, \mathcal{K}_\Delta) < \infty$ *iff* $\operatorname{cod}(\nabla_x F(\cdot, \phi), \mathcal{K}^\Gamma_\lambda) < \infty$.

(b) $(\phi \sim \phi_1\,;\, \mathcal{K}_\Delta)$ *iff* $\nabla_x F(\cdot, \phi)$ *and* $\nabla_x F(\cdot, \phi_1)$ *belong to the same connected component of the intersection of a* $\mathcal{K}^\Gamma_\lambda$*-orbit with* $\mathcal{E}^\Gamma_{\nabla,\lambda}$.

(c) $\Phi \in \mathcal{P}_{l,un}(p)$ *is a* $\mathcal{K}_\Delta$*-versal unfolding of* ϕ *if and only if* $\nabla_x F(\cdot, \Psi)$ $\mathcal{K}^\Gamma_\lambda$*-maps into* $\nabla_x F(\cdot, \Phi)$ *for all unfoldings* Ψ *of* ϕ.

Let us now describe how these results are to be used. Different scenarios are possible. In some cases – for example, the simple singularities (Arnold [1972]) – it is possible to work directly in the path setting. However, it is necessary to compute the generators for the tangent spaces associated with $\mathcal{K}_\Delta$. In general, this is a lengthy computation, although feasible (in particular with the help of computer algebra). Alternatively the computations can be carried out in the gradient (that is $\mathcal{K}^\Gamma_\lambda$) setting where they are easier (at least in the first approximation). This seems true for our situation where the discriminants are complicated high-order polynomials. This is the approach we use to classify the $\mathbf{Z}_q$-equivariant gradient bifurcation problems in Chapter 4. We now describe in more detail the strategy we use (as a prototype for other classifications).

Step 1. Discussion of the potentials in $\mathcal{E}^\Gamma_o$.

(a) Classify under $\mathcal{A}^\Gamma$-equivalence the potentials up to some codimension k.

k can be determined by the codimensions of the problems of interest. For paths $\phi \in m^p_\lambda \mathcal{P}_l$ associated with a potential F with l parameters, note that $\operatorname{cod}(\phi, \mathcal{K}_\Delta) \geq \operatorname{cod}(\phi, \mathcal{K}) \geq pl - 1$. The classification gives us the normal forms F of the $\mathcal{A}^\Gamma$-universal unfoldings we are interested in.

(b) If necessary, adjust the classification to the $\mathcal{G}^\Gamma$-equivalence.

Step 2. Classification of paths.

(a) Low order terms.

First we use a low order $\mathcal{K}^\Gamma_\lambda$-change of coordinates to bring $\nabla_x f(x, \lambda)$ into $\nabla_x F(\cdot, \phi)$ plus higher order terms in (x, λ). In the first approximation it is not important whether the higher order terms are gradient or not. In particular this avoids the problem of having several connected components in the intersection of $\mathcal{E}^\Gamma_{\nabla,\lambda}$ and the $\mathcal{K}^\Gamma_\lambda$-orbit of $\nabla_x F(\cdot, \phi)$. Such a $\mathcal{K}^\Gamma_\lambda$-change of coordinates brings all the connected components into the one containing $\nabla_x F(\cdot, \phi)$.

(b) Higher order terms.

There are two kinds of higher order terms: those corresponding to the gradient theory, that is $(\phi + \text{h.o.t} \sim \phi\,;\, \mathcal{K}_\Delta)$ in $\mathcal{P}_l$, and those coming from the $\mathcal{K}^\Gamma_\lambda$-change of coordinates. Actually, as we use the $\mathcal{K}^\Gamma_\lambda$-calculations to estimate the higher order terms for the paths, both computations are done at the same time.

Let us describe the theory in the path setting. It follows the same line for $\mathcal{K}^{\Gamma}_{\lambda}$ and it can be found in GSS II [1988, Chapter XIV]. The set of higher order terms for ϕ is defined as:

$$\mathcal{P}(\phi,\mathcal{K}_{\Delta}) = \{\, \psi \in \mathcal{P}_l \mid \big((\phi_2+\psi)\sim\phi\,;\mathcal{K}_{\Delta}\big),\ \forall\phi_2 \text{ with } (\phi_2\sim\phi\,;\mathcal{K}_{\Delta})\,\}.$$

We refer to GSS II [1988, p.204] for more discussions on $\mathcal{P}(\nabla_x f,\mathcal{K}^{\Gamma}_{\lambda})$. Roughly speaking, it represents the terms that can be cut-off for any member of the equivalence class of ϕ. Another basic idea is the concept of an intrinsic part of a subspace of $\mathcal{P}_l$. Suppose $\mathcal{Q}\subset\mathcal{P}_l$ is a subspace, then $\operatorname{Intr}\{\mathcal{Q}\}\subset\mathcal{Q}$ is the largest subspace of $\mathcal{Q}$ invariant with respect to $\mathcal{K}_{\Delta}$-equivalence. The basic theorem along these lines is the following. Let us note that the same result holds for $\mathcal{K}^{\Gamma}_{\lambda}$.

Theorem 3.11 (following Bruce, du Plessis & Wall [1987]). *Let ϕ be of finite $\mathcal{K}_{\Delta}$-codimension, then* $\mathcal{P}(\phi,\mathcal{K}_{\Delta})\supset\operatorname{Intr}\{\mathcal{T}(\phi,\mathcal{U}(\mathcal{K}_{\Delta}))\}$.

From this result we can use ω_{ϕ} and the second part of the Fundamental Lemma to get estimates on $\mathcal{T}(\phi,\mathcal{U}(\mathcal{K}_{\Delta}))$ using the classical computations of $\mathcal{T}(\nabla_x F(\cdot,\phi),\mathcal{U}(\mathcal{K}^{\Gamma}_{\lambda}))$.

At the same time, we shall also get estimates to remove the nongradient higher order terms remaining after part (a). As usual, there is an interaction between the two procedures. We have to keep computing explicit changes of coordinates until we can apply the theory of part (b) to discard the higher order terms.

Step 3. Universal unfolding.

The last step is to obtain an unfolding Φ of ϕ such that $\nabla_x F(\cdot,\Phi)$ is a versal unfolding of $\nabla_x F(\cdot,\phi)$ in the sense of part (c) of the Fundamental Theorem. A base $\{\phi_i\}_{i=1}^{k}$ of $\mathcal{N}_e(\phi,\mathcal{K}_{\Delta})$ is computed using the isomorphism Ω_{ϕ}, $k=\operatorname{cod}(\phi,\mathcal{K}_{\Delta})$. Then

$$\Phi = \phi + \sum_{i=1}^{k} \alpha_i\,\phi_i$$

is the universal unfolding required.

3.5 Proofs

Proof of Theorem 3.1. (a) We have already shown that ∇ is $\mathcal{O}_{\lambda}$-linear and injective. What remains to be verified is that

$$\nabla(a^{\Gamma}_e(f))\subset\mathcal{T}_{\nabla,e}(\nabla_x f,\mathcal{K}^{\Gamma}_{\lambda}) \qquad\text{and}\qquad \nabla^{-1}(\mathcal{T}_{\nabla,e}(\nabla_x f,\mathcal{K}^{\Gamma}_{\lambda}))\subset\mathcal{T}_e(f,\mathcal{G}^{\Gamma}_{\lambda}).$$

It follows from (3.5), that the extended tangent space for $\nabla_x f$ is

$$\mathcal{T}_e(\nabla_x f,\mathcal{K}^{\Gamma}_{\lambda}) = \{\, T\,\nabla_x f + (\nabla_x f)_x\,X + (\nabla_x f)_{\lambda}\,\Lambda \mid (T,X,\Lambda)\in\mathcal{D}\,\}$$

where

$$\mathcal{D} = \{ (T, X, \Lambda) \mid T \in \mathcal{M}_\lambda^\Gamma ,\ X \in \mathcal{E}_\lambda^\Gamma ,\ \Lambda : (\mathbf{R},0) \to \mathbf{R} \}.$$

Moreover, simple manipulations show that

$$\begin{aligned} \mathcal{T}_e(\nabla_x f, \mathcal{K}_\lambda^\Gamma) &= \{ T\, \nabla_x f + \nabla(<\nabla_x f, X> + \Lambda f_\lambda) \mid (T,X,\Lambda) \in \mathcal{D} \} \\ &= \{ T\, \nabla_x f + \nabla(\mathcal{T}_e(f, \mathcal{R}_\lambda^\Gamma)) \mid (T,X,\Lambda) \in \mathcal{D} \}. \end{aligned}$$

Thus the proof of (a) will be complete if we can verify that

$$g \in a_e^\Gamma(f) \text{ iff } \exists\, T \in \mathcal{M}_\lambda^\Gamma \text{ such that } \nabla_x g = T\, \nabla_x f .$$

It is clear that $\nabla_x g = T\, \nabla_x f$ implies that $g \in a_e^\Gamma(f)$. Now let $g \in a_e^\Gamma(f)$. As the derivatives of g are in $\widehat{\mathcal{T}}_e(f, \mathcal{R}_\lambda)$, there exists a matrix $\widetilde{T} \in \mathcal{M}_\lambda$ such that $\nabla_x g = \widetilde{T}\, \nabla_x f$. Averaging $\widetilde{T}$ over Γ, we find that $\nabla_x g = T\, \nabla_x f$ for $T \in \mathcal{M}_\lambda^\Gamma$.

(b) The λ-part of the tangent spaces causes some technical difficulties. We are going to break the problem into two parts.

For the first part we need to show that if

$$\nabla_x g = (I, X_1, \Lambda_1) \cdot \nabla_x f , \quad \text{with} \quad (I, X_1, \Lambda_1) \in \mathcal{K}_\lambda^\Gamma$$

then

$$\mathcal{T}_{\nabla,e}(\nabla_x g, \mathcal{K}_\lambda^\Gamma) = (X_1, \Lambda_1)^* \mathcal{T}_{\nabla,e}(\nabla_x f, \mathcal{K}_\lambda^\Gamma),$$

that is, the gradient codimensions of $\nabla_x g$ and $\nabla_x f$ coincide. We denote by * the pull-back operation. For the second part, we show that the same result is true if $\nabla_x g = T_1\, \nabla_x f$ with $(T_1, I, I) \in \mathcal{K}_\lambda^\Gamma$.

For the first part, a straightforward computation shows that

$$\widehat{\mathcal{T}}_e(g, \mathcal{R}_\lambda) = (X_1, \Lambda_1)^* \widehat{\mathcal{T}}_e(f, \mathcal{R}_\lambda) \text{ and that } a_e^\Gamma(g) = (X_1, \Lambda_1)^* a_e^\Gamma(f).$$

Moreover, another calculation shows that

$$g_\lambda = (X_1, \Lambda_1)^* ((\Lambda_1)_\lambda f_\lambda \bmod \mathcal{T}_e(f, \mathcal{G}_\lambda^\Gamma)),$$

hence $\mathcal{T}_e(g, \mathcal{G}_\lambda^\Gamma) = (X_1, \Lambda_1)^* \mathcal{T}_e(f, \mathcal{G}_\lambda^\Gamma)$.

Similarly when $\nabla_x g = T_1\, \nabla_x f$, $\widehat{\mathcal{T}}_e(g, \mathcal{R}_\lambda) = \widehat{\mathcal{T}}_e(f, \mathcal{R}_\lambda)$ and $a_e^\Gamma(g) = a_e^\Gamma(f)$. Via ∇, it implies that $\widehat{\mathcal{T}}_{\nabla,e}(\nabla_x f, \mathcal{K}_\lambda^\Gamma) = \widehat{\mathcal{T}}_{\nabla,e}(\nabla_x f, \mathcal{K}_\lambda^\Gamma)$. Moreover, a verification shows that $\widehat{\mathcal{T}}_e(\nabla_x g, \mathcal{K}_\lambda^\Gamma) = T_1 \cdot \widehat{\mathcal{T}}_e(\nabla_x f, \mathcal{K}_\lambda^\Gamma)$ and, similarly, that $\mathcal{T}_e(\nabla_x g, \mathcal{K}_\lambda^\Gamma) = T_1 \cdot \mathcal{T}_e(\nabla_x f, \mathcal{K}_\lambda^\Gamma)$.

Using the operator *curl*, one can decompose $\mathcal{E}_\lambda^\Gamma$ into the direct sum: $\mathcal{E}_{\nabla,\lambda}^\Gamma \oplus \mathcal{E}_{c,\lambda}^\Gamma$ where $\mathcal{E}_{c,\lambda}^\Gamma$ corresponds to the space of nonzero curl functions (We work in any neighborhood of the origin).

One can similarly split the tangent spaces along the lines of the decomposition of $\mathcal{E}^{\Gamma}_{\lambda}$. First, as

$$\widehat{\mathcal{T}}_e(\nabla_x g, \mathcal{K}^{\Gamma}_{\lambda}) = \widehat{\mathcal{T}}_{\nabla,e}(\nabla_x g, \mathcal{K}^{\Gamma}_{\lambda}) \oplus \widehat{\mathcal{T}}_{c,e}(\nabla_x g, \mathcal{K}^{\Gamma}_{\lambda}) = \widehat{\mathcal{T}}_{\nabla,e}(\nabla_x f, \mathcal{K}^{\Gamma}_{\lambda}) \oplus \widehat{\mathcal{T}}_{c,e}(\nabla_x g, \mathcal{K}^{\Gamma}_{\lambda})$$

is isomorphic to

$$\widehat{\mathcal{T}}_e(\nabla_x f, \mathcal{K}^{\Gamma}_{\lambda}) = \widehat{\mathcal{T}}_{\nabla,e}(\nabla_x f, \mathcal{K}^{\Gamma}_{\lambda}) \oplus \widehat{\mathcal{T}}_{c,e}(\nabla_x f, \mathcal{K}^{\Gamma}_{\lambda}),$$

$\widehat{\mathcal{T}}_{c,e}(\nabla_x g, \mathcal{K}^{\Gamma}_{\lambda})$ is isomorphic to $\widehat{\mathcal{T}}_{c,e}(\nabla_x f, \mathcal{K}^{\Gamma}_{\lambda})$. Secondly, as the difference between the restricted and extented tangent spaces belong to $\mathcal{E}^{\Gamma}_{\nabla,\lambda}$,

$$\begin{aligned}\mathcal{T}_e(\nabla_x g, \mathcal{K}^{\Gamma}_{\lambda}) &= \mathcal{T}_{\nabla,e}(\nabla_x g, \mathcal{K}^{\Gamma}_{\lambda}) \oplus \widehat{\mathcal{T}}_{c,e}(\nabla_x g, \mathcal{K}^{\Gamma}_{\lambda}) \\ &= \left(\widehat{\mathcal{T}}_{\nabla,e}(\nabla_x f, \mathcal{K}^{\Gamma}_{\lambda}) + T_1 \cdot \mathcal{O}_{\lambda}(\nabla_x f)_{\lambda}\right) \oplus \widehat{\mathcal{T}}_{c,e}(\nabla_x g, \mathcal{K}^{\Gamma}_{\lambda}).\end{aligned}$$

As T_1 is invertible, $\mathcal{T}_e(\nabla_x g, \mathcal{K}^{\Gamma}_{\lambda})$ and $\mathcal{T}_e(\nabla_x f, \mathcal{K}^{\Gamma}_{\lambda})$ are isomorphic, hence

$$\mathcal{N}_e(\nabla_x g, \mathcal{K}^{\Gamma}_{\lambda}) = \left(\widehat{\mathcal{N}}_{\nabla,e}(\nabla_x f, \mathcal{K}^{\Gamma}_{\lambda}) / T_1 \cdot \mathcal{O}_{\lambda}(\nabla_x f)_{\lambda}\right) \oplus \widehat{\mathcal{N}}_{c,e}(\nabla_x g, \mathcal{K}^{\Gamma}_{\lambda})$$

is isomorphic to

$$\mathcal{N}_e(\nabla_x f, \mathcal{K}^{\Gamma}_{\lambda}) = \left(\widehat{\mathcal{N}}_{\nabla,e}(\nabla_x f, \mathcal{K}^{\Gamma}_{\lambda}) / \mathcal{O}_{\lambda}(\nabla_x f)_{\lambda}\right) \oplus \widehat{\mathcal{N}}_{c,e}(\nabla_x f, \mathcal{K}^{\Gamma}_{\lambda}).$$

As the two latter spaces ($\widehat{\mathcal{N}}_{c,e}$'s) are isomorphic,

$$\mathcal{N}_{\nabla,e}(\nabla_x g, \mathcal{K}^{\Gamma}_{\lambda}) = \widehat{\mathcal{N}}_{\nabla,e}(\nabla_x f, \mathcal{K}^{\Gamma}_{\lambda}) / T_1 \cdot \mathcal{O}_{\lambda}(\nabla_x f)_{\lambda}$$

is isomorphic to

$$\mathcal{N}_{\nabla,e}(\nabla_x f, \mathcal{K}^{\Gamma}_{\lambda}) = \widehat{\mathcal{N}}_{\nabla,e}(\nabla_x f, \mathcal{K}^{\Gamma}_{\lambda}) / \mathcal{O}_{\lambda}(\nabla_x f)_{\lambda}.$$

Therefore, the gradient codimension of $\nabla_x g$ and $\nabla_x f$ coincide. ∎

A useful ingredient for the proof of the Universality Theorem is the following lemma.

Lemma 3.12. *Under the assumptions of the Universality Theorem, consider any $\mathcal{G}^{\Gamma}_{\lambda}$-unfolding $\nabla_x H(x,\lambda,\delta)$ of $\nabla_x f$ with parameter space $\delta \in \mathbb{R}^p$. Then the normal space $\mathcal{N}_{\nabla,e,un}(\nabla_x H, \mathcal{K}^{\Gamma}_{\lambda,un}(p))$ is a finitely generated $\mathcal{O}_{\lambda,\delta}$-module.*

Proof. Via the isomorphism

$$\nabla : (m\mathcal{E}_o)^{\Gamma}_{\lambda,un}(p) \to \mathcal{E}^{\Gamma}_{\nabla,\lambda,un}(p),$$

$\mathcal{T}_{\nabla,e,un}(\nabla_x H, \mathcal{K}^{\Gamma}_{\lambda,un}(p))$ is the image of

$$\mathcal{T}_e(H, \mathcal{G}^{\Gamma}_{\lambda,un}(p)) = a^{\Gamma}_{e,un}(H) + \mathcal{T}_e(H, \mathcal{R}^{\Gamma}_{\lambda,un}(p)).$$

Therefore, the statement is verified if the following $\mathcal{O}_{\lambda,\delta}$-module,

$$\mathcal{W} = (m\mathcal{E}_o)^\Gamma_{\lambda,un}(p)/\mathcal{T}_{e,un}(H, \mathcal{G}^\Gamma_{\lambda,un}(p))$$

is finitely generated. However, $\mathcal{W}$ is trivially a finite $\mathcal{E}^\Gamma_{o,\lambda,un}(p)$-module and

$$\mathcal{W}_o = \mathcal{W}/m_{\lambda,\delta}\cdot\mathcal{W} = (m\mathcal{E}_o)^\Gamma/\mathcal{T}_e(f_o, \mathcal{G}^\Gamma)$$

is of finite dimension because f_o is of finite gradient codimension (without parameters) if f is (with parameters). Therefore, using Lemma 7.1 of GSS II [1988, p.234] it follows that $\mathcal{W}$ is a finitely generated $\mathcal{O}_{\lambda,\delta}$-module. ∎

Proof of the Universality Theorem. (a) This is the important part of the proposition. We use the classical proof that infinitesimal stability implies stability. As it is classical and long, it is not necessary to reproduce it here. A complete exposition can be found in GSS II [1988, p.238-241]. We can check that the same proof goes through if we can show the relation (7.13) in GSS II [1988, p.239]. Explicitly, let $\nabla_x H(x,\lambda,\delta)$ be any $\mathcal{G}^\Gamma_\lambda$-versal unfolding of $\nabla_x f$ with p parameters, $\delta \in \mathbf{R}^p$. The relation (7.13) in GSS II takes the form

$$\mathcal{E}^\Gamma_{\nabla,\lambda,un}(p) = \mathcal{T}_{\nabla,e,un}(\nabla_x H, \mathcal{K}^\Gamma_{\lambda,un}(p)) + \mathcal{O}_\delta < \frac{\partial H}{\partial \delta_1} \dots \frac{\partial H}{\partial \delta_p} > . \tag{3.9}$$

Define

$$\mathcal{W} = \mathcal{E}^\Gamma_{\nabla,\lambda,un}(p)/\mathcal{T}_{\nabla,e,un}(\nabla_x H, \mathcal{K}^\Gamma_{\lambda,un}(p)) \text{ and } \mathcal{W}_o = \mathcal{W}/m_\delta\cdot\mathcal{W}\,.$$

Obviously (3.9) follows if $\mathcal{W}$ is generated by $\{\frac{\partial H}{\partial \delta_i}\}_{i=1}^p$ as an $\mathcal{O}_\delta$-module. From Lemma 3.12, $\mathcal{W}$ is a finitely generated $\mathcal{O}_{\lambda,\delta}$-module. The proof now follows from the Equivariant Preparation Theorem as in Lemma 7.1 and Corollary 7.2 of GSS II [1988, p.234-36]. Because $\mathcal{W}_o = \mathcal{N}_{\nabla,e}(\nabla_x f, \mathcal{K}^\Gamma_\lambda)$ is a finite dimensional vector space generated by $\{\frac{\partial H}{\partial \delta_i}(x,\lambda,0)\}_{i=1}^p$, and so $\mathcal{W}$ is finitely generated by $\{\frac{\partial H}{\partial \delta_i}\}_{i=1}^p$ as an $\mathcal{O}_\delta$-module

(b,c) The proofs of these two remaining parts of the Universality Theorem follow exactly the classical pattern. We refer to Theorem 7.4 in GSS II [1988, p.241] and its proof. ∎

Proof of Theorem 3.3. If $\text{cod}(\nabla_x f, \mathcal{K}^\Gamma_\lambda) < \infty$, there exists an integer k such that $m^k \subset \mathcal{T}_e(\nabla_x f, \mathcal{K}^\Gamma_\lambda)$. And so $m^k_\nabla \overset{\text{def}}{=} m^k \cap \mathcal{E}^\Gamma_{\nabla,\lambda} \subset \mathcal{T}_{\nabla,e}(\nabla_x f, \mathcal{K}^\Gamma_\lambda)$. Hence

$$\text{cod}_\nabla(\nabla_x f, \mathcal{K}^\Gamma_\lambda) = \dim_{\mathbf{R}} \mathcal{N}_\nabla(\nabla_x f, \mathcal{K}^\Gamma_\lambda) \le \dim_{\mathbf{R}} \mathcal{E}^\Gamma_{\nabla,\lambda}/m^k_\nabla < \infty\,.$$

If $\text{cod}_\nabla(\nabla_x f, \mathcal{K}^\Gamma_\lambda) < \infty$ and f_o satisfy (H1), $\nabla_x f$ is represented by a potential F and a path ϕ such that $\text{cod}_\nabla(\nabla_x F(\cdot,\phi), \mathcal{K}^\Gamma_\lambda) < \infty$. From the Fundamental Lemma, part (a), it means that $\mathrm{cod}(\phi, \mathcal{K}_\Delta) < \infty$. And so, from part (a) of the Fundamental Theorem $\mathrm{cod}\big(\nabla_x F(\cdot,\phi), \mathcal{K}^\Gamma_\lambda\big) < \infty$ which implies that $\mathrm{cod}(\nabla_x f, \mathcal{K}^\Gamma_\lambda) < \infty$. ■

Before embarking on the remaining proofs, we turn now our attention to the $\mathcal{K}_\Delta$-theory. The group $\mathcal{K}_\Delta$ has been defined in the last section (see Section 3.4) and its action on $\mathcal{P}_l$ is

$$(H, \Lambda)\cdot\phi = H(\phi\circ\Lambda,\ \cdot\ )\,.$$

Associated with the space $\mathcal{P}_l$ and the group $\mathcal{K}_\Delta$ one has $\mathcal{P}_{l,un}(p)$, the space of p-parameter unfoldings, and $\mathcal{K}_{\Delta,un}(p)$, the group of isomorphisms of p-parameter unfoldings acting on $\mathcal{P}_l$. In addition, the notion of mapping between unfoldings with different numbers of parameters, hence of (uni)versality, also extends in the classical way.

Let us define the following :

$\mathcal{E}_{\alpha,\alpha} = \{\ \xi : (\mathbf{K}^l,0) \to \mathbf{K}^l\ \}$ and $\mathcal{E}_{u,u} = \{\ \xi : (\mathbf{K}^L,0) \to \mathbf{K}^L\ \}$,

$\mathcal{I}(\widetilde{\Delta}) = \{\ g \in \mathcal{O}_u \mid g|_{\widetilde{\Delta}} \equiv 0\ \}$, $\mathcal{I}^{\Gamma_L}(\widetilde{\Delta}) = \{\ g \in \mathcal{I}(\widetilde{\Delta}) \mid g$ is Γ_L-invariant$\}$,

$\widetilde{\mathcal{I}}(\Delta) = \{\ g \in \mathcal{O}_\alpha \mid \exists\ \tilde{g} \in \mathcal{I}^{\Gamma_L}(\widetilde{\Delta})$ such that $g = \tilde{g}|_{\mathrm{Fix}\,\Gamma_L}\ \}$,

$\Theta_{\widetilde{\Delta}} = \{\ \xi \in \mathcal{E}_{u,u} \mid \langle \xi, \nabla_u g\rangle \in \mathcal{I}(\widetilde{\Delta})\ ,\ \forall g \in \mathcal{I}(\widetilde{\Delta})\ \}$,

$\Theta^{\Gamma_L}_{\widetilde{\Delta}} = \{\ \xi \in \Theta_{\widetilde{\Delta}} \mid \xi$ is Γ_L-equivariant $\}$,

$\widetilde{\Theta}_\Delta = \{\ \xi \in \mathcal{E}_{\alpha,\alpha} \mid \exists\ \tilde{\xi} \in \Theta^{\Gamma_L}_{\widetilde{\Delta}}$ such that $\xi = \tilde{\xi}|_{\mathrm{Fix}\,\Gamma_L}\ \}$

$\mathrm{N}(\widetilde{\Theta}_\Delta) = \{\ \xi \in \widetilde{\Theta}_\Delta \mid \xi^o_\alpha$ is upper triangular $\}$.

Unlike the smooth real case, $\widetilde{\Theta}_\Delta$ and $\Theta^{\Gamma_L}_{\widetilde{\Delta}}$ are finitely generated $\mathcal{O}_\alpha$ (resp. $\mathcal{O}^{\Gamma_L}_u$)-modules of (equivariant) vector fields tangent to Δ (resp. $\widetilde{\Delta}$). The set of generators of $\widetilde{\Theta}_\Delta$ is denoted by $\{\mu_j\}_{j=1}^M$. For $\phi \in \mathcal{P}_l$, the $\mathcal{O}_\lambda$-module generated by $\{\mu_j \circ \phi\}_{j=1}^M$ is denoted by $\phi^*\widetilde{\Theta}_\Delta$.

The tangent spaces needed are the following $\mathcal{O}_\lambda$-submodules of $\mathcal{P}_l$:

$$\mathcal{T}(\phi, \mathcal{K}_\Delta) = m_\lambda < \phi_\lambda > +\ \phi^*(\widetilde{\Theta}_\Delta)$$
$$\mathcal{T}\big(\phi, \mathcal{U}(\mathcal{K}_\Delta)\big) = m_\lambda^2 < \phi_\lambda > +\ \phi^*(\mathrm{N}(\widetilde{\Theta}_\Delta))$$
$$\mathcal{T}_e(\phi, \mathcal{K}_\Delta) = < \phi_\lambda > +\ \phi^*(\widetilde{\Theta}_\Delta).$$

The 4 conditions for $\mathcal{K}_\Delta$ to be a geometric subgroup of $\mathcal{K}$ (in the sense of Damon [1984]) are satisfied. The main result for $\mathcal{K}_\Delta$-theory is then

Lemma 3.13. *Consider $\phi \in \mathcal{P}_l$ and $\Phi \in \mathcal{P}_{l,un}(p)$, a p-parameter unfolding of ϕ.*

(a) *(Theorem 9.3, Damon [1984, p.44]) The following are equivalent*

(i) Φ is a $\mathcal{K}_\Delta$-versal

(ii) $\mathcal{T}_e(\phi, \mathcal{K}_\Delta) + < \frac{\partial\Phi}{\partial\beta_1} \ldots \frac{\partial\Phi}{\partial\beta_p} >_{\mathbf{K}} = \mathcal{P}_l$.

(b) *(Theorem 10.2, Damon [1984, p.49])* $\mathrm{cod}(\phi, \mathcal{K}_\Delta) < \infty$ *iff ϕ has a versal unfolding iff ϕ is finitely determined.*

In the holomorphic case there is an alternative description of finite codimension in "geometrical" terms. We follow the line of Damon & Mond [1990]. Let us first define algebraic transversality. We say that $\phi_{\mathbf{C}} \in \mathcal{P}_l$ is algebraic-transverse at λ_0 to $\Delta_{\mathbf{C}}$ if $\phi_{\mathbf{C}}$ is transverse at $\phi_{\mathbf{C}}(\lambda_0)$ to the vector subspace of $\mathbf{C}^l$ generated by $\{\mu_j \circ \phi_{\mathbf{C}}\ (\lambda_0)\}_{j=1}^{M}$. We say that $\phi_{\mathbf{C}}$ is transverse to $\Delta_{\mathbf{C}}$ if it is transverse at every point of a punctured neighborhood of the origin. Remark that if $\phi_{\mathbf{C}}(\lambda_0) \notin \Delta_{\mathbf{C}}$, transversality is always satisfied. And so we get

Proposition 3.14. $\mathrm{cod}(\phi_{\mathbf{C}}, \mathcal{K}_\Delta) < \infty$ *iff* $\phi_{\mathbf{C}}$ *is transverse to* $\Delta_{\mathbf{C}}$.

Proof. Let $\widehat{M}(\lambda)$ be the matrix formed by the vectors $(\phi_{\mathbf{C}})_\lambda$, $\mu_j(\phi_{\mathbf{C}})$, $1 \le j \le M$. We claim that $\mathrm{cod}(\phi_{\mathbf{C}}, \mathcal{K}_\Delta) < \infty$ iff $\widehat{M}$ has rank l in a punctured neighborhood of the origin, that is iff $\phi_{\mathbf{C}}$ is transverse to $\Delta_{\mathbf{C}}$.

Denote by $\{e_i\}_{i=1}^{l}$ the standard base of $\mathbf{C}^l$. Then, $\mathrm{cod}(\phi_{\mathbf{C}}, \mathcal{K}_\Delta) < \infty$ iff there is an integer k such that $\lambda^k e_j \subset T(\phi_{\mathbf{C}}, \mathcal{K}_\Delta)$ for $1 \le j \le l$, iff there are $a_j \in (\mathcal{O}_\lambda)^{1+M}$ such that $\widehat{M} a_j(\lambda) = \lambda^k e_j$ for $1 \le j \le l$, iff $\widehat{M}$ has rank l for $\lambda \neq 0$. ■

We can now turn our attention to the proof of the Fundamental Lemma. First we need two preliminary results.

Lemma 3.15 (Liftable Vector Fields).

(a) $\tilde{\xi} \in \Theta_{\tilde{\Delta}}^{\Gamma_L}$ *iff there is a* Γ_{n+L}*-equivariant vector field* $\tilde{\mu}$ *on* $Z(\widetilde{F})$ *such that*

$$d\tilde{\pi}(\tilde{\mu}(x,u)) = \tilde{\xi}(\tilde{\pi}(x,u)) \ , \ \forall (x,u) \in Z(\widetilde{F}).$$

(b) *Moreover, in the context of part (a)* $\tilde{\xi}^o = 0$. *By local integration this means that* $\mathcal{K}_\Delta \subset \mathcal{K}$.

Proof. (a) As $\tilde{\pi}$ defines – at the origin – an isolated complete intersection singularity of dimension 0, from Theorem A in Terao [1983], we know that, without symmetry, $\tilde{\xi} \in \Theta_{\tilde{\Delta}}$ iff there is a vector field $\bar{\mu}$ on $Z(\widetilde{F})$ such that for all $(x,u) \in Z(\widetilde{F})$

$$d\tilde{\pi}(\bar{\mu}(x,u)) = \tilde{\xi}(\tilde{\pi}(x,u)).$$

We can now obtain the result through averaging. As Γ is a finite group, the formula

$$\tilde{\mu} = \frac{1}{|\Gamma|} \sum_{\gamma_i \in \Gamma} (\gamma_i)_{n+L}^{-1}\ \bar{\mu}((\gamma_i)_{n+L}(x,u))$$

gives the required Γ_{n+L}-equivariant vector field tangent to $Z(\widetilde{F})$.

(b) By writing $d\tilde{\pi}(\tilde{\mu}(x,u)) = \tilde{\xi}(\tilde{\pi}(x,u))$ in coordinates and putting $u = 0$, we find that

$$\sum_{i=1}^{L} \tilde{\xi}_i^o\ \nabla_x h_i \in T_{\nabla,e}(\nabla_x f_o, \mathcal{K}^\Gamma).$$

By the independence of the set $\{\nabla_x h_i\}_{i=1}^L$, it follows that $\tilde{\xi}^o = 0$. ∎

A consequence of Lemma 3.15 is:

Lemma 3.16. (a) $\omega_\phi{}^{-1}\Big(\widehat{T}_{\nabla,e}\big(\nabla_x F(\cdot,\phi),\mathcal{K}_\lambda^\Gamma\big)\Big) = \phi^*\widetilde{\Theta}_\Delta$.
(b) $\omega_\phi{}^{-1}\Big(\widehat{T}_\nabla\big(\nabla_x F(\cdot,\phi),\mathcal{U}(\mathcal{K}_\lambda^\Gamma)\big)\Big) = \phi^*(\mathrm{N}(\widetilde{\Theta}_\Delta))$.

Proof. (a) We shall proceed in three steps.

(i) Denote by ω_u the natural extension of ω_ϕ on $\mathcal{P}_L$, given by

$$\omega_u(\tilde{\xi}) = \frac{\partial}{\partial\tau}\Big|_{\tau=0} \nabla_x\widetilde{F}(\cdot,\tau\tilde{\xi}) = \sum_{i=1}^{L} \tilde{\xi}_i(u)\,\nabla_x h_i(x).$$

Denote by $\mathcal{K}_u^{\Gamma_{n+L}}$ the group of Γ_{n+L}-equivariant contact equivalences with u as parameters. First, let us prove the following, which is the equivalent result with u as parameters,

$$\omega_u^{-1}\Big(\widehat{T}_{\nabla,e}\big(\nabla_x\widetilde{F}(\cdot,u),\mathcal{K}_u^{\Gamma_{n+L}}\big)\Big) = \Theta_{\Delta}^{\Gamma_L}. \tag{3.10}$$

Suppose that for $\tilde{\xi}\in\Theta_{\Delta}^{\Gamma_L}$

$$\omega_u(\tilde{\xi}) = \tilde{\zeta}\in\widehat{T}_{\nabla,e}\big(\nabla_x\widetilde{F}(\cdot,u),\mathcal{K}_u^{\Gamma_{n+L}}\big). \tag{3.11}$$

This is equivalent to the existence of $T\in\mathcal{M}_u^{\Gamma_{n+L}}$ and $X\in\mathcal{E}_u^{\Gamma_{n+L}}$ such that

$$\tilde{\zeta} = T\,\nabla_x\widetilde{F} - (\nabla_x\widetilde{F})_x\,X.$$

Now, because $\tilde{\zeta} = (\nabla_x\widetilde{F})_u\,\tilde{\xi}$, denoting by D the total derivative, this is equivalent to $D(\nabla_x\widetilde{F})\cdot(X,\tilde{\xi}) = T\,\nabla_x\widetilde{F}$ iff $(X,\tilde{\xi})$ is a vector field tangent to $Z(\widetilde{F})$. Through averaging, $\tilde{\xi}$ can be made Γ_L-equivariant, that is, (3.11) is equivalent to $\tilde{\xi}\in\Theta_{\Delta}^{\Gamma_L}$.

(ii) The result is true if ϕ is an immersion. This is verified as follows. Denote by $\tilde{\phi}$ the trivial extension $(\phi,0)$ of ϕ onto $\mathbf{K}^L$. When ϕ is an immersion, there exists a diffeomorphism Φ on $\mathbf{K}^L$ such that $(\Phi\circ\tilde{\phi})(\lambda) = \big((\lambda,0),0\big)$. Via Φ, $\mathcal{M}_\lambda^\Gamma = \tilde{\phi}^*\mathcal{M}_u^{\Gamma_{n+L}}$ and $\mathcal{E}_\lambda^\Gamma = \tilde{\phi}^*\mathcal{E}_u^{\Gamma_{n+L}}$. Hence we would be finished if the following were true:

$$\omega_\phi{}^{-1}\Big(\widehat{T}_{\nabla,e}\big(\nabla_x F(\cdot,\phi),\mathcal{K}_\lambda^\Gamma\big)\Big) = \omega_\phi^{-1}\Big(\tilde{\phi}^*\widehat{T}_{\nabla,e}\big(\nabla_x\widetilde{F}(\cdot,u),\mathcal{K}_u^{\Gamma_{n+L}}\big)\Big) = \tilde{\phi}^*\omega_u^{-1}\Big(\widehat{T}_{\nabla,e}\big(\nabla_x\widetilde{F}(\cdot,u),\mathcal{K}_u^{\Gamma_{n+L}}\big)\Big).$$

However, $\tilde{\phi}^*$ and the ω's cannot be exchanged! Instead, consider $\xi\in\mathcal{E}_{\alpha,\alpha}$ and $\tilde{\xi}$, an extension of ξ to $\mathcal{E}_{u,u}$ (via Φ), such that

$$\omega_u(\tilde{\xi})\circ\tilde{\phi} = \omega_\phi(\xi)\in\widehat{T}_{\nabla,e}\big(\nabla_x F(\cdot,\phi),\mathcal{K}_\lambda^\Gamma\big) = \tilde{\phi}^*\widehat{T}_{\nabla,e}\big(\nabla_x\widetilde{F}(\cdot,u),\mathcal{K}_u^{\Gamma_{n+L}}\big).$$

Modulo $\tilde{\phi}^*$, $\omega_u(\tilde{\xi})$ is in the $\mathcal{K}_u^{\Gamma_n+L}$-tangent space, that is

$$\omega_u(\tilde{\xi}) \in \widetilde{\mathcal{T}}_{\nabla,e}\big(\nabla_x\widetilde{F}(\cdot,u),\mathcal{K}_u^{\Gamma_n+L}\big) + \mathcal{I}_{\tilde{\phi}}\cdot\mathcal{E}_{\nabla,u}^{\Gamma_n+L}\,,$$

where $\mathcal{I}_{\tilde{\phi}}$ is the ideal of functions vanishing on $\tilde{\phi}$. As in Lemma 3.12, as a consequence of the Equivariant Preparation Theorem,

$$\mathcal{E}_{\nabla,u}^{\hat{\Gamma}} = \operatorname{Im}\omega_u + \widehat{\mathcal{T}}_{\nabla,e}\big(\nabla_x\widetilde{F}(\cdot,u),\mathcal{K}_u^{\Gamma_n+L}\big)\,.$$

Hence, there exists $\nu \in \mathcal{E}_{u,u}$, with $\omega_u(\nu) \in \mathcal{I}_{\tilde{\phi}}\cdot\operatorname{Im}\omega_u$, such that

$$\omega_u(\tilde{\xi}) = \omega_u(\nu) \mod \widehat{\mathcal{T}}_{\nabla,e}\big(\nabla_x\widetilde{F}(\cdot,u),\mathcal{K}_u^{\Gamma_n+L}\big)\,.$$

Consider $\eta = \tilde{\xi} - \nu$, then

$$\omega_u(\eta) = \omega_u(\tilde{\xi}) - \omega_u(\nu) \in \widehat{\mathcal{T}}_{\nabla,e}\big(\nabla_x\widetilde{F}(\cdot,u),\mathcal{K}_u^{\Gamma_n+L}\big)\,.$$

From part (i), this implies that $\eta \in \Theta_{\tilde{\Delta}}^{\Gamma_L}$. Restricting to Fix Γ_L and composing with $\tilde{\phi}$ we find that $\eta\circ\tilde{\phi} = \xi$, hence

$$\xi \in \tilde{\phi}^*\Theta_{\tilde{\Delta}}^{\Gamma_L} = \phi^*\widetilde{\Theta}_\Delta\,.$$

(iii) The case when ϕ is not an immersion.
Define $\hat{\phi}:\mathbf{K}\to\mathbf{K}^{L+1}$ by $\hat{\phi}(\lambda) = (\tilde{\phi}(\lambda),\lambda)$ and let π_L be the natural projection $\mathbf{K}^{L+1}\to\mathbf{K}^L$. For $\hat{u}\in\mathbf{K}^{L+1}$, define the potential $\widehat{F}(x,\hat{u}) = \widetilde{F}(x,\pi_L\hat{u})$. We can follow the same theory as in (ii) above, the discriminant is now $\widehat{\Delta} = \widetilde{\Delta}\times\mathbf{K}$, with Γ_L acting trivially on $\mathbf{K}$. As $\hat{\phi}$ is an immersion

$$\omega_{\hat{\phi}}^{-1}\Big(\widehat{\mathcal{T}}_{\nabla,e}\big(\nabla_x\widehat{F}(\cdot,\hat{\phi}),\mathcal{K}_\lambda^\Gamma\big)\Big) = \hat{\phi}^*\widetilde{\Theta}_{\hat{\Delta}}^{\Gamma_L}.$$

Suppose $\xi\in\mathcal{P}_l$ is such that $\omega_u(\tilde{\xi}) \in \widehat{\mathcal{T}}_e\big(\nabla_xF(\cdot,\tilde{\phi}),\mathcal{K}_\lambda^\Gamma\big)$. From the definition of $\widehat{F}$ $\widehat{\mathcal{T}}_e\big(\nabla_xF(\cdot,\tilde{\phi}),\mathcal{K}_\lambda^\Gamma\big) = \widehat{\mathcal{T}}_e\big(\nabla_x\widehat{F}(\cdot,\hat{\phi}),\mathcal{K}_\lambda^\Gamma\big)$ and so there exists $\hat{\xi}\circ\hat{\phi}\in\hat{\phi}^*\widetilde{\Theta}_{\hat{\Delta}}^{\Gamma_L}$ such that $\omega_{\hat{\phi}}(\hat{\xi}\circ\hat{\phi}) = \omega_u(\tilde{\xi})$. Define $\eta(u,\lambda) = (\pi_L\hat{\xi})(u,\lambda,\lambda)$. Then

$$\tilde{\xi}(\lambda) = (\pi_L\hat{\xi})\circ\hat{\phi}\,(\lambda) = (\pi_L\hat{\xi})\big(\tilde{\phi}(\lambda),\lambda,\lambda\big) = \eta\circ\tilde{\phi}\,(\lambda).$$

As $\mathcal{I}(\widehat{\Delta}) = \mathcal{E}_u\cdot\mathcal{I}(\widetilde{\Delta})$, η is in $\Theta_{\tilde{\Delta}}^{\Gamma_L}$, restricting to Fix Γ_L implies that $\xi\in\tilde{\phi}^*\widetilde{\Theta}_\Delta$.

(b) We can use the line of proof in part (a) and work in the (x,u)-setting. As $\widehat{\mathcal{T}}_\nabla\big(\nabla_x\widetilde{F}(\cdot,u),\mathcal{U}(\mathcal{K}_u^{\Gamma_n+L})\big) \subset \widehat{\mathcal{T}}_{\nabla,e}\big(\nabla_x\widetilde{F}(\cdot,u),\mathcal{K}_u^{\Gamma_n+L}\big)$, from (3.11), we get that $\tilde{\xi}\in\Theta_{\tilde{\Delta}}^{\Gamma_L}$. Restricting to the fixed-point subspace of Γ_L, it remains to verify that $\xi^o = 0$ and that $\xi_\alpha^o = 0$. But, from Lemma 3.15(b), we already know that $\xi^o = 0$.

For the derivative ξ_α, consider the identity satisfied by ξ :

$$\sum_{i=1}^{l}\xi_i(\alpha)\,\nabla_xh_i = T\,\nabla_xF - (\nabla_xF)_x\,X\,. \tag{3.12}$$

From $\xi^o = 0$,

$$T(x,0)\,\nabla_x f_o(x) - (\nabla_x f_o)_x(x)\,X(x,0) \equiv 0\,.$$

Now differentiating (3.12) with respect to α and taking $\alpha = 0$, we get the system $(1 \le k \le l)$:

$$\begin{aligned}\sum_{i=1}^{l} \xi^o_{i\alpha_k}\, \nabla_x h_i &= T_{\alpha_k} \nabla_x f_o - (\nabla_x f_o)_x\, X_{\alpha_k} + T\, \nabla_x h_k - (\nabla_x h_k)_x\, X \\ &= T\, \nabla_x h_k - (\nabla_x h_k)_x\, X + \widehat{\mathcal{T}}_\nabla(\nabla_x f_o, \mathcal{K}^\Gamma)\,.\end{aligned}$$

From the discussion at the end of Section 3.3, ξ^o_α is an upper triangular matrix, hence $\xi \in \tilde{\phi}^*\mathrm{N}(\widetilde{\Theta}_\Delta)$. ∎

Proof of the Fundamental Lemma. The proof of part (a) consists of three steps.

(i) $\omega_\phi(\mathcal{T}_e(\phi, \mathcal{K}_\Delta)) \subset \mathcal{T}_{\nabla,e}(\nabla_x F(\cdot,\phi), \mathcal{K}^\Gamma_\lambda)$ and so $\Omega_{\phi,e}$ is a well-defined map.

(ii) $\Omega_{\phi,e}$ is surjective.

(iii) ${\omega_\phi}^{-1}\Big(\mathcal{T}_{\nabla,e}(\nabla_x F(\cdot,\phi), \mathcal{K}^\Gamma_\lambda)\Big) \subset \mathcal{T}_e(\phi, \mathcal{K}_\Delta)$ and so $\Omega_{\phi,e}$ is injective.

(i) From Lemma 3.16 we know that $\omega_\phi(\phi^*\widetilde{\Theta}_\Delta) \subset \mathcal{T}_{\nabla,e}(\nabla_x F(\cdot,\phi), \mathcal{K}^\Gamma_\lambda)$. Moreover it is clear that

$$\omega_\phi(\phi_\lambda) = \sum_{i=1}^{l} (\phi_\lambda)_i\, \nabla_x h_i = \frac{\partial}{\partial\lambda}\,(\nabla_x F(\cdot,\phi)) \in \mathcal{T}_{\nabla,e}(\nabla_x F(\cdot,\phi), \mathcal{K}^\Gamma_\lambda).$$

(ii) By definition $\Omega_{\phi,e}$ is surjective if

$$\mathcal{T}_{\nabla,e}(\nabla_x F(\cdot,\phi), \mathcal{K}^\Gamma_\lambda) + \omega_\phi(\mathcal{P}_l) = \mathcal{E}^\Gamma_{\nabla,\lambda}.$$

As $\mathrm{cod}_\nabla(\nabla_x F(\cdot,\phi), \mathcal{K}^\Gamma_\lambda))$ is finite, $\mathcal{W} \stackrel{\text{def}}{=} \mathcal{N}_{\nabla,e}(\nabla_x F(\cdot,\phi), \mathcal{K}^\Gamma_\lambda)$ is a finitely generated vector space, hence a finitely generated $\mathcal{O}_\lambda$-module. Moreover

$$\mathcal{W}/m_\lambda \mathcal{W} = \mathcal{N}_{\nabla,e}(\nabla_x f_o, \mathcal{K}^\Gamma) \tag{3.13}$$

and so is generated by $\{\nabla_x h_i\}^l_{i=1}$. We conclude that $\mathcal{N}_{\nabla,e}(\nabla_x F(\cdot,\phi), \mathcal{K}^\Gamma_\lambda)$ is spanned by $\omega_\phi(\mathcal{P}_l)$ as an $\mathcal{O}_\lambda$-module from (3.13) and the Equivariant Preparation Theorem (Theorem XIV.8.1, GSS II [1988, p.244])

(iii) Suppose

$$\omega_\phi(\xi) = \mu \in \mathcal{T}_e(\nabla_x F(\cdot,\phi), \mathcal{K}^\Gamma_\lambda).$$

We can split

$$\mathcal{T}_e(\nabla_x F(\cdot,\phi), \mathcal{K}^\Gamma_\lambda) = \widehat{\mathcal{T}}_e(\nabla_x F(\cdot,\phi), \mathcal{K}^\Gamma_\lambda) + \{\ \Lambda(\lambda)\ \frac{\partial}{\partial\lambda}(\nabla_x F(\cdot,\phi))\ |\ \Lambda : (\mathbf{K},0) \to \mathbf{K}\ \}\,,$$

and so we can accordingly split $\mu = \mu_1 + \mu_2$. It is then clear that there exists

$$\xi_2 \in \langle\phi_\lambda\rangle_{\mathbf{R}} \subset \mathcal{T}_e(\phi, \mathcal{K}_\Delta)$$

such that $\omega_\phi(\xi_2) = \mu_2$. From the linearity of ω_ϕ, $\mu_1 = \omega_\phi(\xi - \xi_2)$ and so from Lemma 3.16 (a) $\xi - \xi_2 \in \phi^*\widetilde{\Theta}_\Delta$, hence $\xi \in \mathcal{T}_e(\phi, \mathcal{K}_\Delta)$.

Part (b) is a simple consequence of Lemma 3.16(b). Again we can split

$$\mathcal{T}_\nabla(\nabla_x F(\cdot,\phi), \mathcal{U}(\mathcal{K}^\Gamma_\lambda)) = \widehat{\mathcal{T}}_\nabla(\nabla_x F(\cdot,\phi), \mathcal{U}(\mathcal{K}^\Gamma_\lambda)) + \\ \{ \Lambda(\lambda) \frac{\partial}{\partial\lambda}(\nabla_x F(\cdot,\phi)) \mid \Lambda : (\mathbf{K},0) \to (\mathbf{K},0),\ \Lambda^\circ_\lambda = 0 \}.$$

From Lemma 3.16(b), the pre-image of the first subspace lies in $\mathcal{T}(\phi, \mathcal{U}(\mathcal{K}_\Delta))$ and a computation shows that it is also true for the second subspace.

The proof of part (c) follows the same argument. From Lemma 3.16(a)

$$\omega_\phi{}^{-1}\Big(\widehat{\mathcal{T}}_\nabla(\nabla_x F(\cdot,\phi), \mathcal{K}^\Gamma_\lambda)\Big) = \phi^*\widetilde{\Theta}_\Delta.$$

As previously, the result is clear from the λ-part of the tangent space

$$\{ \Lambda(\lambda) \frac{\partial}{\partial\lambda}(\nabla_x F(\cdot,\phi)) \mid \Lambda : (\mathbf{K},0) \to (\mathbf{K},0) \}.$$

And so $\mathcal{T}(\phi, \mathcal{K}_\Delta) = \omega_\phi{}^{-1}\Big(\mathcal{T}_\nabla(\nabla_x F(\cdot,\phi), \mathcal{K}^\Gamma_\lambda)\Big)$. Moreover this means that Ω_ϕ is a well-defined injective homomorphism.

For the last statement of the lemma, it is clear that if $\phi_\lambda \notin \mathcal{T}(\phi, \mathcal{K}_\Delta)$ then dim $\mathcal{N}(\phi, \mathcal{K}_\Delta) = \dim \mathcal{N}_e(\phi, \mathcal{K}_\Delta) + 1$. If ϕ is an immersion, $\phi^\circ_\lambda \neq 0$ and so $\phi_\lambda \notin \mathcal{T}(\phi, \mathcal{K}) \supset \mathcal{T}(\phi, \mathcal{K}_\Delta)$. If ϕ is not an immersion, consider $\hat{\phi} = (\phi, \lambda)$ and $\hat{\Delta} = \Delta \times \mathbf{K}$. A simple computation of the tangent spaces associated with $\hat{\phi}$ shows that the dimensions of the normal spaces are respectively the same for ϕ and $\hat{\phi}$. Now $\hat{\phi}$ is an immersion and so

$$\dim \mathcal{N}(\phi, \mathcal{K}_\Delta) = \dim \mathcal{N}(\hat{\phi}, \mathcal{K}_{\hat{\Delta}}) = \dim \mathcal{N}_e(\hat{\phi}, \mathcal{K}_{\hat{\Delta}}) + 1 = \dim \mathcal{N}_e(\phi, \mathcal{K}_\Delta) + 1\,.$$

■

Lemma 3.17. *In the present framework, let $\Phi(\cdot,\delta) \in \mathcal{P}_{l,un}(p)$ be an unfolding of ϕ with $\delta \in \mathbf{R}^p$. Then $\mathcal{N}_{\nabla,e,un}(\nabla_x F(\cdot,\Phi), \mathcal{K}^\Gamma_{\lambda,un}(p))$ is a finitely generated $\mathcal{O}_{\lambda,\delta}$-module.*

Proof. Define

$$\mathcal{W} = \mathcal{E}^\Gamma_{o,\lambda,\delta} / \mathcal{T}_{e,un}(F(\cdot,\Phi), \mathcal{R}^\Gamma(p+1)).$$

$\mathcal{W}$ is trivially a finite $\mathcal{E}^\Gamma_{o,\lambda,\delta}$-module and

$$\mathcal{W}_o = \mathcal{W}/m_{\lambda,\delta}\cdot\mathcal{W} = \mathcal{E}^\Gamma_o / \mathcal{T}_e(f_o, \mathcal{R}^\Gamma)$$

is of finite dimension. From Lemma 7.1 of GSS II [1988, p.234], $\mathcal{W}$ is a finitely generated $\mathcal{O}_{\lambda,\delta}$-module, say generated by $\{h_i\}_{i=1}^r$.

We conclude as in the proof of Lemma 3.12. Recall that $l = \mathrm{cod}(f_o, \mathcal{G}^\Gamma) \leq r$, but, nevertheless consider $\nabla_x P(\cdot, \lambda, \delta) \in \mathcal{E}^\Gamma_{\nabla,\lambda,\delta}$. Because $\mathcal{W}$ is generated by $\{h_i\}_{i=1}^r$, we can write that

$$P(\cdot, \lambda, \delta) = \nabla_x F(\cdot, \Phi(\lambda, \delta))\, X(\cdot, \lambda, \delta) + \sum_{i=1}^{r} a_i(\lambda, \delta)\, h_i(\cdot).$$

Taking the gradient, we see that

$$\nabla_x P(\cdot, \lambda, \delta) - \sum_{i=1}^{r} a_i(\lambda, \delta)\, \nabla_x h_i \in \mathcal{T}_{\nabla,e,un}(\nabla_x F(\cdot, \Phi), \mathcal{K}^\Gamma_{\lambda,un}(q)).$$

And so the $\{\nabla_x h_i\}_{i=1}^r$ generate $\mathcal{N}_{\nabla,e,un}(\nabla_x F(\cdot, \Phi), \mathcal{K}^\Gamma_{\lambda,un}(q))$ as an $\mathcal{O}_{\lambda,\delta}$-module. ■

Proof of the Fundamental Theorem.

(a) Assume $\mathrm{cod}(\phi, \mathcal{K}_\Delta) < \infty$ and so $\mathrm{cod}(\phi_{\mathbf{C}}, \mathcal{K}_\Delta) < \infty$. Hence, from Lemma 3.14, $\phi_{\mathbf{C}}$ is transverse to $\Delta_{\mathbf{C}}$.

There is a canonical Whitney-stratification $\{\widetilde{\Delta}_i\}$ of $\widetilde{\Delta}_{\mathbf{C}}$. As $\widetilde{\Delta}_{\mathbf{C}}$ is defined by a Γ_L-invariant equation, the strata are Γ_L-invariant (Field [1976]). Following Damon & Mond [1991], $\tilde{\mu}$ is tangent to $\widetilde{\Delta}_i$ for all $\tilde{\mu} \in \Theta_{\widetilde{\Delta}_{\mathbf{C}}}$ and for all $\widetilde{\Delta}_i$. In particular, any $\mu \in \widetilde{\Theta}_{\Delta_{\mathbf{C}}}$ is tangent to every $\Delta_i = \widetilde{\Delta}_i \cap \mathrm{Fix}\,\Gamma_L$. From the definition of transversality, $\phi_{\mathbf{C}}$ is thus geometrically transverse to each Δ_i (in a punctured neighborhood of the origin), that is $\phi_{\mathbf{C}}(\lambda_0) + T_{\phi_{\mathbf{C}}(\lambda_0)}\Delta_i = \mathbf{C}^l$.

In the one-parameter setting, finite $\mathcal{K}_\lambda$-codimension is equivalent to the restricted tangent space $\widehat{\mathcal{T}}(\nabla_x F_{\mathbf{C}}(\cdot, \phi), \mathcal{K}_\lambda)$ being of finite codimension, and either imply finite $\mathcal{K}^\Gamma_\lambda$-codimension. By coherence of those modules, this is equivalent to $(0,0)$ being an isolated singularity in the zero-set of $\nabla_x F_{\mathbf{C}}$.

If $(0,0)$ is not isolated in $\tilde{\pi}_{\mathbf{C}}^{-1}((\mathrm{Im}\ \phi_{\mathbf{C}}, 0)) \cap C_{\tilde{\pi}} \overset{\mathrm{def}}{=} Y$, by the Curve Selection Lemma (Looijenga [1984, p.22]), there exists a nontrivial analytic curve r in Y. $\tilde{\pi}_{\mathbf{C}}$ being finite, it means we have an analytic curve in $\mathrm{Im}\ \phi_{\mathbf{C}} \cap \Delta_{\mathbf{C}}$. As $\dim(\mathrm{Im}\ \phi_{\mathbf{C}}) = 1$ and $\dim \Delta_{\mathbf{C}} = L - 1$, this contradicts the transversality of the intersection of $\phi_{\mathbf{C}}$ and $\Delta_{\mathbf{C}}$.

The converse is immediate :

$$\begin{aligned}\mathrm{cod}(\nabla_x F(\cdot, \phi), \mathcal{K}^\Gamma_\lambda) < \infty &\Rightarrow \dim_{\mathbf{K}} (\mathrm{Im}\ \Omega_{\phi,e}) < \infty \\ &\Rightarrow \dim \mathcal{N}_e(\phi, \mathcal{K}_\Delta) < \infty \\ &\Rightarrow \mathrm{cod}(\phi, \mathcal{K}_\Delta) < \infty .\end{aligned}$$

(b)(i) Suppose that $(\phi \sim \phi_1 ; \mathcal{K}_\Delta)$. Because $\mathcal{K}_\Delta$ is connected, there exists a path ϕ_t connecting $\phi_0 \overset{\mathrm{def}}{=} \phi$ and ϕ_1:

$$\phi_t = (\mu_t, \Lambda_t) \cdot \phi_0\ ,\ t \in [0,1],\ (\mu_t, \Lambda_t) \in \mathcal{K}_\Delta\ .$$

Let us define

$$\xi_t = \frac{\partial}{\partial s}\bigg|_{s=0} \mu_{t+s} \in \widetilde{\Theta}_{\Delta,\lambda}.$$

The subscript λ indicates that we are considering λ-parametrized vector fields. From Lemma 3.15 (λ-parametrized version) we have a lift $\tilde{\eta}_t$ of $\tilde{\xi}_t$, the extension of ξ_t in $\Theta^{\Gamma_L}_{\Delta,\lambda}$. Denote by $\tilde{\pi}_n$ the natural projection from $\mathbf{K}^{n+L}$ onto $\mathbf{K}^n$ and by $\tilde{\zeta}_t$ a Γ-equivariant extension to $\mathbf{K}^{n+L+1}$ of $\tilde{\pi}_n(\tilde{\eta}_t)$. We can define a λ-parametrized equivariant vector field on $\mathbf{K}^{n+L}$: $X_{t,\lambda} = (\tilde{\zeta}_t, \tilde{\xi}_t)$. From local integration of X, we are going to define changes of coordinates. Using the connectedness of $[0,1]$, we get the result.

Let $\psi_{t,\lambda}$ be the local equivariant flow near t_0 of $\frac{\partial}{\partial t}\ \psi_{t,\lambda} = X_{t,\lambda} \circ \psi_{t,\lambda}$ such that $\psi_{t_0,\lambda} = I$. We denote

$$\psi_{t,\lambda}(x,u) \stackrel{\text{def}}{=} (\rho_{t,\lambda}(x,u), \tilde{\mu}_{t,\lambda}(u))$$

and

$$\psi^{-1}_{t,\lambda}(x,u) \stackrel{\text{def}}{=} (\bar{\rho}_{t,\lambda}(x,u), \tilde{\mu}^{-1}_{t,\lambda}(u)).$$

Clearly $\rho_{t,\lambda}(\bar{\rho}_{t,\lambda}(x,u), \tilde{\mu}^{-1}_{t,\lambda}(u)) = x$. We denote by $\tilde{\phi}_t \stackrel{\text{def}}{=} (\phi_t, 0)$ the trivial extension of ϕ_t into $\mathbf{K}^L$. Define $\bar{u}_t(\lambda) = \tilde{\mu}_{t,\lambda} \circ \tilde{\phi}_{t_0}\ (\Lambda_t(\lambda))$ and so

$$\rho_{t,\lambda}\left(\bar{\rho}_{t,\lambda}(x, \bar{u}_t(\lambda)), \tilde{\phi}_{t_0}(\Lambda_t(\lambda))\right) = x.$$

Define

$$\Omega_t : \mathbf{K}^{n+1} \to \mathbf{K}^n$$

$$(x,\lambda) \mapsto \bar{\rho}_{t,\lambda}(x, \bar{u}_t(\lambda)).$$

We verify that $(x,\lambda) \mapsto (\Omega_t(x,\lambda), \Lambda_t(\lambda))$ is in $\mathcal{K}^\Gamma_\lambda$.

For t and λ fixed, $\psi_{t,\lambda}(Z(\widetilde{F})) \subset Z(\widetilde{F})$ and since $Z(\widetilde{F})$ is a manifold, it is known that we can find $\widetilde{T} : \mathbf{K}^{n+L+2} \to \mathrm{GL}(n,\mathbf{K})$ such that

$$\nabla_x \widetilde{F}(\psi_{t,\lambda}(x,u)) = \nabla_x \widetilde{F}(\rho_{t,\lambda}(x,u), \tilde{\mu}_{t,\lambda}(u)) = \widetilde{T}_{t,\lambda}(x,u)\ \nabla_x \widetilde{F}(x,u).$$

Finally

$$\begin{aligned}\nabla_x \widetilde{F}(x, \tilde{\phi}_t) &= \nabla_x \widetilde{F}\Big(\rho_{t,\lambda}(\Omega_{t,\lambda}(x,\lambda), \tilde{\phi}_{t_0}(\Lambda_t(\lambda))), \tilde{\mu}_{t,\lambda}(\tilde{\phi}_{t_0}(\Lambda_t(\lambda)))\Big)\\ &= \widetilde{T}_{t,\lambda}\Big(\Omega_t(x,\lambda), \tilde{\phi}_{t_0}(\Lambda_t(\lambda))\Big)\ \nabla_x \widetilde{F}\Big(\Omega_t(x,\lambda), \tilde{\phi}_{t_0}(\Lambda_t(\lambda))\Big)\end{aligned}$$

that is

$$(\nabla_x \widetilde{F}(\cdot, \phi_t) \sim \nabla_x \widetilde{F}(\cdot, \phi_{t_0});\ \mathcal{K}_\lambda).$$

Averaging $\widetilde{T}$, as Ω_t is already Γ-equivariant, we conclude that

$$(\nabla_x \widetilde{F}(\cdot, \phi_t) \sim \nabla_x \widetilde{F}(\cdot, \phi_{t_0});\ \mathcal{K}^{\Gamma_{n+L}}_\lambda).$$

Restricting to the fixed point subspace

$$(\nabla_x F(\cdot,\phi_t) \sim \nabla_x F(\cdot,\phi_{t_0});\, \mathcal{K}^\Gamma_\lambda).$$

(ii) For the reverse implication, we assume that $\nabla_x F(\cdot,\phi_0)$ and $\nabla_x F(\cdot,\phi_1)$ are in a connected component of the intersection of a $\mathcal{K}^\Gamma_\lambda$-orbit with $\mathcal{E}^\Gamma_{\nabla,\lambda}$. By hypothesis $\mathrm{cod}(\phi,\mathcal{K}_\Delta)$ and $\mathrm{cod}\big(\nabla_x F(\cdot,\phi_0),\mathcal{K}^\Gamma_\lambda\big)$ are finite, and so, for a large enough k we can work in the spaces of jets $\mathsf{M}' = J^k(\mathcal{P}_l)$, $\mathsf{M} = J^k(\mathcal{E}^\Gamma_\lambda)$ and $\mathsf{M}_\nabla = J^k(\mathcal{E}^\Gamma_{\nabla,\lambda})$. In these, $\mathcal{K}_\Delta$, respectively $\mathcal{K}^\Gamma_\lambda$, induce smooth algebraic actions. To simplify the notation we use the same symbols to denote the jets and the functions. From our hypotheses there is a path P connecting $\nabla_x F(\cdot,\phi_0)$ and $\nabla_x F(\cdot,\phi_1)$ in M_∇. Explicitly we assume that P is given by

$$\nabla_x f(\cdot,\cdot,t) = (T_t, X_t, \Lambda_t)\cdot \nabla_x F(\cdot,\phi_0)\,,\ t\in[0,1],\ (T_t,X_t,\Lambda_t)\in\mathcal{K}^\Gamma_\lambda\,.$$

Now, consider $\phi_t = \phi_0\circ\Lambda_t$, $t\in[0,1]$. This is the parametrization of a path $\mathsf{P}'\subset\mathsf{M}'$ connecting ϕ_0 to ϕ_1. As in classical singularity theory, we can use

Lemma 3.18 (Mather [1968]). *Let G be a Lie group acting smoothly on a smooth manifold M and $\mathsf{P}\subset\mathsf{M}$ a connected smooth submanifold. Then P is contained in a simple G-orbit if and only if the following conditions are fulfilled:*

(1) $\mathrm{T}_x(G\cdot x)\supset \mathrm{T}_x\mathsf{P}$, $\forall x\in\mathsf{P}$,

(2) $\dim \mathrm{T}_x(G\cdot x)$ *is constant*, $\forall x\in\mathsf{P}$.

Let us verify the conditions (1) and (2) of Mather's Lemma for the group $J^k(\mathcal{K}_\Delta)$ and the manifold $\mathsf{P}'\subset\mathsf{M}'$.

$$\begin{aligned}\mathrm{T}_{\phi_t}\mathsf{P}' &= \frac{\partial}{\partial s}\Big|_{s=0}\phi_{t+s}\\ &= \frac{\partial}{\partial s}\Big|_{s=0}\omega^{-1}_{\phi_t}(\nabla_x F(\cdot,\phi_{t+s})) \qquad (\text{since } \omega_{\phi_t} \text{ is injective})\\ &= \omega^{-1}_{\phi_t}\left(\frac{\partial}{\partial s}\Big|_{s=0}\nabla_x F(\cdot,\phi_{t+s})\right).\end{aligned}$$

Now

$$\frac{\partial}{\partial s}\Big|_{s=0}\nabla_x F(\cdot,\phi_{t+s}) = \frac{\partial}{\partial s}\Big|_{s=0}\Big((T^{-1}_{t+s}\nabla_x f_{t+s})\circ(X^{-1}_{t+s},I)\Big).$$

Differentiating $T_{t+s}\cdot T^{-1}_{t+s} = I$ and $X_{t+s}\circ X^{-1}_{t+s} = x$, we get

$$\frac{\partial}{\partial s}\Big|_{s=0}\nabla_x F(\cdot,\phi_{t+s}) \in \mathcal{T}\big(\nabla_x F(\cdot,\phi_t),\mathcal{K}^\Gamma_\lambda\big)\,.$$

And so,

$$\text{(1)}\quad \mathrm{T}_{\phi_t}\mathsf{P}' \in \omega_{\phi_t}^{-1}\Big(\mathcal{T}(\nabla_x F(\cdot,\phi_t),\mathcal{K}_\lambda^\Gamma)\cap \mathcal{E}_{\nabla,\lambda}^\Gamma\Big) = \omega_{\phi_t}^{-1}\Big(\mathcal{T}_\nabla(\nabla_x F(\cdot,\phi_t),\mathcal{K}_\lambda^\Gamma)\Big)$$
$$\subset \mathcal{T}(\phi_t,\mathcal{K}_\Delta)\quad (\Omega_{\phi_t} \text{ is injective}).$$

$$\text{(2)}\quad \dim \mathrm{T}_{\phi_t}(\mathcal{K}_\Delta\cdot\phi_t)\ (\text{in } \mathsf{M}') = \dim \mathsf{M}' - \operatorname{codim} \mathrm{T}_{\phi_t}(\mathcal{K}_\Delta\cdot\phi_t)$$
$$= \dim \mathsf{M}' - \dim_{\mathsf{R}}\big(\mathsf{M}'/\mathcal{T}(\phi_t,\mathcal{K}_\Delta)\big).$$

From part (c) of the Fundamental Lemma,

$$\dim_{\mathsf{R}}\big(\mathsf{M}'/\mathcal{T}(\phi_t,\mathcal{K}_\Delta)\big) = \dim_{\mathsf{R}}\big(\mathsf{M}'/\mathcal{T}_e(\phi_t,\mathcal{K}_\Delta)\big)+1$$
$$= \dim_{\mathsf{R}}\Big(\mathsf{M}_\nabla/\mathcal{T}_{\nabla,e}(\nabla_x F(\cdot,\phi_t),\mathcal{K}_\lambda^\Gamma)\Big)+1\,.$$

This is a constant (independent of t). Hence Mather's Lemma can be applied. P' is a $J^k(\mathcal{K}_\Delta)$-orbit, that is $\big(j^k(\phi_0)\sim j^k(\phi_1)\,;\,J^k(\mathcal{K}_\Delta)\big)$. From the finite determinacy of ϕ_0, we can conclude.

For the proof of part (c), assume that for all unfoldings Ψ of ϕ, $\nabla_x F(\cdot,\Psi)$ $\mathcal{K}_\lambda^\Gamma$-maps into $\nabla_x F(\cdot,\Phi)$. We need to prove that Φ is thus a $\mathcal{K}_\Delta$-versal unfolding of ϕ. In coordinates we can write $\Phi(\lambda,\delta)$, $\delta\in\mathsf{R}^p$ (for some p), and $\Phi(\lambda,0)=\phi(\lambda)$. $\{\frac{\partial}{\partial\delta_i}\big|_{\delta=0}\Phi\}_{i=1}^p$ generates the K-vector subspace of $\mathcal{P}_l$ we denote by $\mathcal{N}_\Phi$. It is then enough to prove that Φ is infinitesimally stable; that is, that $\mathcal{P}_l=\mathcal{T}_e(\phi,\mathcal{K}_\Delta)+\mathcal{N}_\Phi$.

For any $\xi\in\mathcal{P}_l$, consider the one parameter unfolding $\Psi(\cdot,t)=\phi+t\xi$. By hypothesis $\nabla_x F(\cdot,\Psi)$ $\mathcal{K}_\lambda^\Gamma$-maps into $\nabla_x F(\cdot,\Phi)$; that is,

$$\nabla_x F(x,\phi(\lambda)+t\xi(\lambda)) = T(x,\lambda,t)\,\nabla_x F\Big(X(x,\lambda,t),\Phi\big(\Lambda(\lambda,t),D(t)\big)\Big) \tag{3.14}$$

for $(T,X,\Lambda,D)\in\mathcal{K}_{\lambda,un}^\Gamma(1)$. Differentiating (3.14) with respect to t at $t=0$ results in

$$\omega_\phi(\xi)\in\mathcal{T}_e\big(\nabla_x F(\cdot,\phi),\mathcal{K}_\lambda^\Gamma\big)+\omega_\phi(\mathcal{N}_\Phi). \tag{3.15}$$

Because ξ is arbitrary in $\mathcal{P}_l$, through $\Omega_{\phi,e}$, it follows that (3.15) implies that

$$\Omega_{\phi,e}\big(\mathcal{P}_l/\mathcal{T}_e(\phi,\mathcal{K}_\Delta)\big)\subset\Omega_{\phi,e}(\mathcal{N}_\Phi).$$

From the Fundamental Lemma $\Omega_{\phi,e}$ is injective, which means that $\mathcal{P}_l/\mathcal{T}_e(\phi,\mathcal{K}_\Delta)\subset\mathcal{N}_\Phi$, and so Φ is a $\mathcal{K}_\Delta$-versal unfolding of ϕ.

For the reverse implication we follow the classical approach using the idea that infinitesimal stability implies stability (proof of Theorem XV.2.1, GSS II [1988, p.238]). We can verify that the proof goes through if we can establish the ODE (7.14); that is, the relation (7.13) in GSS II [1988, p.241], which we now elaborate on.

Let $\Phi(\cdot,\delta)$ be a $\mathcal{K}_\Delta$-versal unfolding of ϕ, $\delta \in \mathbb{R}^p$, and so $\mathcal{N}_\Phi + \mathcal{T}_e(\phi, \mathcal{K}_\Delta) = \mathcal{P}_l$. Via $\Omega_{\phi,e}$ it follows that

$$\mathcal{E}^\Gamma_{\nabla,\lambda} = \Omega_{\phi,e}(\mathcal{N}_\Phi) + \mathcal{T}_{\nabla,e}(\nabla_x F(\cdot,\phi), \mathcal{K}^\Gamma_\lambda). \tag{3.16}$$

For (7.13) (GSS II [1988, p.241]) to hold true it is sufficient that

$$\mathcal{E}^\Gamma_{\nabla,\lambda,un}(p) = \Omega_{\phi,e}(\mathcal{N}_\Phi(p)) + \mathcal{T}_{\nabla,e,un}(\nabla_x F(\cdot,\Phi), \mathcal{K}^\Gamma_{\lambda,un}(p)),$$

where $\mathcal{N}_\Phi(p)$ is the module generated over the functions in δ by $\left\{\frac{\partial}{\partial\delta_i}\Big|_{\delta=0}\Phi\right\}_{i=1}^p$. Define

$$\mathcal{W} = \mathcal{E}^\Gamma_{\nabla,\lambda,un}(p)/\mathcal{T}_{\nabla,e,un}(\nabla_x F(\cdot,\Phi), \mathcal{K}^\Gamma_{\lambda,un}(p)).$$

Clearly $\Omega_{\phi,e}(\mathcal{N}_\Phi) = \mathcal{W}/m_\delta\mathcal{W}$ is a finite dimensional vector space. From Lemma 3.17, $\mathcal{W}$ is finitely generated $\mathcal{O}_{\lambda,\delta}$-module. The proof now follows from the Equivariant Preparation Theorem as in Corollary 7.2 of GSS II [1988, p.234-36]. And so $\mathcal{W}$ is finitely generated as an $\mathcal{O}_\delta$-module by the generators of $\Omega_{\Phi,e}(\mathcal{N}_\Phi)$. ∎

4. Classification of Zq-Equivariant Gradient Bifurcation Problems

Suppose $\nabla_x f(x, \lambda, \alpha)$ is a $\mathbf{Z}_q$-equivariant gradient bifurcation problem with f a smooth potential (germ), $x \in \mathbf{R}^2$, $\lambda \in \mathbf{R}$ (the main bifurcation parameter), $\alpha \in \mathbf{R}^m$ (the unfolding parameters). We assume that $q \geq 3$.

To classify $\mathbf{Z}_q$-equivariant gradient bifurcation problems we follow the program detailed in Chapter 3. First, the singularities of the potential without distinguished parameter are classified up to codimension 3 using $\mathcal{A}^{\mathbf{Z}_q}$-equivalence. Then, using the path formulation and the Fundamental Theorem, the singularities of $\mathbf{Z}_q$-equivariant gradient maps on $\mathbf{R}^2$ with a distinguished parameter are classified up to topological codimension 2.

It is convenient to identify $\mathbf{R}^2$ with $\mathbf{C}$ through $z = x_1 + ix_2$. Then $\mathbf{Z}_q = <\delta_q>$ with $\delta_q \cdot z = e^{i\theta} z$ with $\theta = 2\pi p/q$ and any nonstandard action of $\mathbf{Z}_q$ can be recovered by a group homomorphism.

We begin by recording some basic results that are either well known or routine to verify.

(a) The ring of real $\mathbf{Z}_q$-invariant functions, $\mathcal{E}_o^{\mathbf{Z}_q}$, is generated by

$$u = z\bar{z}, \quad v = \tfrac{1}{2}(z^q + \bar{z}^q) \quad \text{and} \quad w = \tfrac{1}{2} i (\bar{z}^q - z^q). \tag{4.1}$$

We denote by $\mathcal{E}_o^{\mathbf{Z}_q}(\mathbf{C}) = \{\, \hat{p} + i\hat{q} \mid \hat{p}, \hat{q} \in \mathcal{E}_o^{\mathbf{Z}_q} \,\}$ the ring of complex valued $\mathbf{Z}_q$-invariant functions.

When $q = 4$ there is a more convenient (but equivalent) basis for the invariants. Let

$$\delta = -\tfrac{1}{2}(z^2 + \bar{z}^2), \; \mu = \tfrac{1}{2} i (z^2 - \bar{z}^2), \; \Delta = \delta^2 \quad \text{and} \quad w = 2\delta\mu.$$

Then $(u, \Delta, 2\delta\mu)$ is used instead of (u, v, w). This choice of coordinates is an extension of the invariants used in Golubitsky & Roberts [1987] for $\mathbf{D}_4$.

(b) The addition of the reflection $z \mapsto \bar{z}$ to $\mathbf{Z}_q$ results in the standard representation of $\mathbf{D}_q$ on $\mathbf{C}$. The ring of $\mathbf{D}_q$-invariant functions, $\mathcal{E}_o^{\mathbf{D}_q}$, is generated by u and v (respectively u and Δ when $q = 4$).

(c) From the relation $w^2 + v^2 = u^q$ (respectively $w^2 = 4\Delta(u^2 - \Delta)$ when $q = 4$), the ring $\mathcal{E}_o^{\mathbf{Z}_q}$ has the structure of a $\mathcal{E}_o^{\mathbf{D}_q}$-module, freely generated by 1 and w. And so any $\mathbf{Z}_q$-invariant potential can be uniquely written as

$$H(z) = \begin{cases} F(u, v) + w\, G(u, v), \\ F(u, \Delta) + 2\delta\mu\, G(u, \Delta), \quad q = 4. \end{cases} \tag{4.2}$$

(d) The $\mathbf{Z}_q$-equivariant gradient map associated with (4.2) is

$$\nabla_z H(z) = \begin{cases} [2F_u - i(2vG_u + qu^{q-1}G_v)]\, z + [qF_v + i(qG + 2uG_u + qvG_v)]\, \overline{z}^{q-1}, \\ [2F_u - 4i(uG + \Delta G_u + 2\Delta u G_\Delta)]\, z \\ \qquad -[4F_\Delta + 4i(2G + uG_u + 2\Delta G_\Delta)]\, \delta\overline{z}, \quad q = 4. \end{cases}$$

(e) The $\mathcal{E}_o^{\mathbf{Z}_q}$-module of equivariant maps from $\mathbf{R}^2 \to \mathbf{R}^2$, denoted by $\mathcal{E}^{\mathbf{Z}_q}$, is freely generated over $\mathcal{E}_o^{\mathbf{D}_q}$. The generators are z, iz, $\overline{z}^{q-1}$ and $i\overline{z}^{q-1}$ and if $q = 4$, $\overline{z}^{q-1}$ is replaced by $\delta\overline{z}$. Hence we can represent an element h of $\mathcal{E}^{\mathbf{Z}_q}$ by

$$h(z) = f(u,v)\, z + g(u,v)\, \overline{z}^{q-1} \quad \text{with} \quad f, g \in \mathcal{E}_o^{\mathbf{D}_q}(\mathbf{C}).$$

(f) Let $h = f\, z + g\, \overline{z}^{q-1}$ (respectively $f\, z + g\, \delta\overline{z}$ if $q = 4$). Then h is a gradient iff

$$\mathrm{Re}\,(qf_v - 2g_u) = \mathrm{Im}\,(2f + 2uf_u + qvf_v + 2vg_u + qu^{q-1}g_v) \equiv 0$$

or, if $q = 4$,

$$\mathrm{Re}\,(2f_\Delta + g_u) = \mathrm{Im}\,(f + uf_u + 2\Delta f_\Delta - ug - \Delta g_u - 2u\Delta g_\Delta) \equiv 0\,.$$

(g) The above results in (a)-(f) are unmodified by the introduction of (bifurcation and unfolding) parameters. When the bifurcation parameter λ is present, a λ-subscript is added to the symbol of the space under consideration; for example,

$$\mathcal{E}_{o,\lambda}^{\mathbf{Z}_q} = \{\, H(z,\lambda) \mid H : \mathbf{C} \times \mathbf{R} \to \mathbf{R} \quad \text{and} \quad H \text{ is } \mathbf{Z}_q\text{-invariant} \,\}.$$

4.1 $\mathcal{A}^{\mathbf{Z}_q}$-classification of potentials

We classify $\mathbf{Z}_q$-invariant potentials on $\mathbf{R}^2$ via $\mathcal{A}^{\mathbf{Z}_q}$-equivalence. The group of changes of coordinates corresponding to $\mathcal{A}^{\mathbf{Z}_q}$ on $\mathcal{E}_o^{\mathbf{Z}_q}$ is given by

$$\{\, (a, Z) \mid a : (\mathbf{R}, 0) \to \mathbf{R}\,,\ a^o \neq 0\,,\ Z = b\, z + c\, \overline{z}^{q-1} \text{ with } b, c \in \mathcal{E}_o^{\mathbf{D}_q}(\mathbf{C}),\ b^o \neq 0 \,\},$$

with the action $(a, Z) \cdot H = a(H(Z))$ for $H \in (m\mathcal{E}_o)^{\mathbf{Z}_q}$. Note that a^o can be any non-zero real number because $-\mathbf{I}$ belongs to the connected component of the identity in the space of $\mathbf{Z}_q$-commuting matrices.

Through (4.2), $(m\mathcal{E}_o)^{\mathbf{Z}_q}$ is identified with $\left((m\mathcal{E}_o)^{\mathbf{D}_q}\right)^2$. And so, a given potential H can be identified with the pair of functions (F, G) in $(m\mathcal{E}_o)^{\mathbf{D}_q}$. The generators $t_1 \ldots t_4$ for the different tangent spaces that we will use are given by

$$t_1(z) = \begin{cases} 2uF_u + qvF_v + w\,(qG + 2uG_u + qvG_v)\,, \\ uF_u + 2\Delta F_\Delta + w\,(2G + uG_u + 2\Delta G_\Delta)\,, \quad q = 4\,, \end{cases}$$

$$t_2(z) = \begin{cases} vG + v^2 G_v - u^q G_v - w\, F_v\,, \\ 2(\Delta - u^2)G + 4\Delta(\Delta - u^2)G_\Delta - w\, F_\Delta\,, \quad q = 4, \end{cases}$$

$$t_3(z) = \begin{cases} 2vF_u + qu^{q-1}F_v + w\,(2vG_u + qu^{q-1}G_v), \\ \Delta F_u + 2u\Delta F_\Delta + w\,(uG + \Delta G_u + 2u\Delta G_\Delta), \quad q = 4, \end{cases}$$

and

$$t_4(z) = \begin{cases} qu^{q-1}G - 2v^2G_u + 2u^qG_u + 2w\,F_u\,, \\ 4u\Delta G - 4\Delta(\Delta - u^2)\,G_u + w\,F_u\,, \qquad q = 4\,. \end{cases}$$

The "unipotent" tangent space is given by

$$\mathcal{T}(H,\mathcal{U}(\mathcal{A}^{\mathbf{Z}_q})) = <m^{\mathbf{D}_q}t_1, m^{\mathbf{D}_q}t_2, t_3, t_4> + \mathbf{R} <H^2, H^3 \ldots>$$

and the extended tangent space is

$$\mathcal{T}_e(H,\mathcal{A}^{\mathbf{Z}_q}) = \mathcal{T}(H,\mathcal{U}(\mathcal{A}^{\mathbf{Z}_q})) + \mathbf{R} <t_1, t_2, H>\,.$$

The codimension of H in $(m\mathcal{E}_o)^{\mathbf{Z}_q}$ is given by the real dimension, as a vector space, of

$$\mathcal{N}_e(H,\mathcal{A}^{\mathbf{Z}_q}) = (m\mathcal{E}_o)^{\mathbf{Z}_q}/\mathcal{T}_e(H,\mathcal{A}^{\mathbf{Z}_q})\,.$$

The $\mathbf{Z}_q$-intrinsic ideals are generated by $m^{\mathbf{Z}_q}$ or $<u^{q-1}, v, w>$ (respectively by $m^{\mathbf{Z}_4}$ or $<u^2, \Delta, \delta\mu>$ if $q = 4$). The ideal of higher order terms $\mathcal{P}(H,\mathcal{A}^{\mathbf{Z}_q})$ contains the intrinsic part of $\mathcal{T}(H,\mathcal{U}(\mathcal{A}^{\mathbf{Z}_q}))$.
Now suppose $\{h_i\}_{i=1}^k$ is a basis for $\mathcal{N}_e(H,\mathcal{A}^{\mathbf{Z}_q})$ then

$$\widehat{H}(z,\nu) = H(z) + \sum_{i=1}^{k} \nu_i\,h_i(z)$$

is a universal unfolding of H.

The last set of algebraic data necessary for the classification are formulae for explicit changes of coordinates. For a change of coordinates $Z(z) = b(z)\,z + c(z)\,\overline{z}^{q-1}$, $b, c \in \mathcal{E}_o^{\mathbf{D}_q}(\mathbf{C})$, the invariants $(\tilde{u},\tilde{v},\tilde{w})$ in the transformed coordinates for $q = 3, 4$ and $q \geq 5$ are recorded in Table 4.1.

The first elementary results about the $\mathcal{A}^{\mathbf{Z}_q}$-classification of potentials are as follows.

Lemma 4.1. *Let $H = F + w\,G$ be a potential in $(m\mathcal{E}_o)^{\mathbf{Z}_q}$.*

(a) *If $F_u^o \neq 0$ then $(H \sim u\,;\, \mathcal{A}^{\mathbf{Z}_q})$. Moreover, u is a potential of codimension zero without bifurcation.*

(b) *If $F_u^o = 0$ but $(F_v^o + iG^o) \neq 0$, then there exists $\widehat{F}(u)$, depending on u only, such that $(H \sim v + \widehat{F}(u)\,;\, \mathcal{A}^{\mathbf{Z}_q})$. Furthermore, the universal unfolding of an H of finite codimension is also of the form $v + \widehat{F}_e(u,\nu)$ where $\widehat{F}_e(u,0) = \widehat{F}(u)$. Moreover, when $q = 3$, $\widehat{F} = \widehat{F}_e \equiv 0$.*

Proof. For part (a), a first order change of coordinates will result in

$$(H \sim u + \widehat{H}(u,v,w)\,;\, \mathcal{A}^{\mathbf{Z}_q})$$

where $\widehat{H} \in (m^{\mathbf{Z}_q})^2$. It is then a simple calculation to show that

$$(m^{\mathbf{Z}_q})^2 \subset \mathrm{Intr}\Big\{\mathcal{T}\big(u, \mathcal{U}(\mathcal{A}^{\mathbf{Z}_q})\big)\Big\}$$

resulting in $(H \sim u\,;\, \mathcal{A}^{\mathbf{Z}_q})$. Moreover $\mathcal{T}_e(u, \mathcal{A}^{\mathbf{Z}_q}) = m^{\mathbf{Z}_q}$ and so $\mathrm{cod}(u, \mathcal{A}^{\mathbf{Z}_q}) = 0$.

To prove part (b), we assume that the first part of (b) is true; it is then a simple computation to show that

$$(m^{\mathbf{Z}_3})^2 \subset \mathrm{Intr}\Big\{\mathcal{T}\big(v + \widehat{F}, \mathcal{U}(\mathcal{A}^{\mathbf{Z}_3})\big)\Big\} \quad \text{and} \quad m^{\mathbf{Z}_3} \subset \mathcal{T}_e(v, \mathcal{A}^{\mathbf{Z}_3}).$$

To verify the first part of statement (b), consider an explicit change of coordinates $Z = b(z)\,z$ with $b \in \mathcal{E}_o^{\mathbf{D}_q}(\mathbf{C})$. We can then write $b(z) = b_0 + u\,b_1(u) + v\,b_2(u,v)$ with $b_0 \neq 0$. Now consider a similar splitting for H,

$$H = v\,F_v^o + u\,H_1(u) + v\,H_2(u,v) + w\,G^o + w\,G_1(u,v)\,,$$

where $F_v^o \in \mathbf{R}$, H_1, H_2 and $G_1 \in m^{\mathbf{D}_q}$. The idea is to remove H_2 and G_1 using b_1 and b_2 and to take care of (F_v^o, G^o) with b_0.

After some calculation, the problem can be reduced to the solvability of the following equations in $\mathbf{C} \approx \mathbf{R}^2$:

$$b_0^q \overline{a}_0 = 1\,,$$

$$b_1 \overline{a}_0 b_0^{q-1} + L(u, b_0, b_1) = 0\,, \tag{4.3}$$

$$b_2 \overline{a}_0 b_0^{q-1} + M(u, v, b_0, b_1, b_2) = 0\,, \tag{4.4}$$

where $a_0 = F_v^o + iG^o$, $L(u, b_0, b_1) = L(b_0) + u\,\hat{L}(u, b_0, b_1)$ and

$$M(u,v,b_0,b_1,b_2) = M(b_0) + u\,M_1(u,v,b_0,b_1,b_2) + v\,M_2(u,v,b_0,b_1,b_2)\,.$$

Clearly $b_0^q = 1/\overline{a}_0$ and the solvability of (4.3) and (4.4) is established using the Implicit Function Theorem.

In the cases when $q \neq 3$, we have no a priori control over the transformation of the u part of H, giving rise to some $\widehat{F}(u)$. Using t_1 and t_2, it is a simple verification to show that

$$< v, w > \subset \mathcal{T}_e(v + \widehat{F}(u), \mathcal{A}^{\mathbf{Z}_q})\,,$$

and so the universal unfolding of $v + \widehat{F}(u)$ can be chosen to depend only on u and some parameters α. This completes the proof. ■

It is important to note that if H satisfies either part (a) or (b) of Lemma 4.1 then, with an elementary change of coordinates, H can be transformed into a potential that is invariant with respect to the standard action of $\mathbf{D}_q$. It is not hard to see that the $\mathbf{Z}_q$-theory, in that case, reduces to the $\mathbf{D}_q$-theory, although there will be non-$\mathbf{D}_q$-symmetric potentials in the equivalence classes.

$$q=3 \quad \begin{cases} \tilde{u} = |b|^2 u + (\bar{b}c + b\bar{c})\, v + |c|^2 u^2 + i\,(b\bar{c} - \bar{b}c)\, w \\ \tilde{v} = \frac{1}{2}(b^3 + \bar{b}^3)\, v + \frac{3}{2}(b^2 c + \bar{b}^2 \bar{c})\, u^2 + \frac{3}{2}(bc^2 + \bar{b}\bar{c}^2)\, uv + (c^3 + \bar{c}^3)\, v^2 \\ \qquad + i\,\frac{1}{2}(b^3 - \bar{b}^3)\, w + i\,\frac{3}{2}(\bar{b}\bar{c}^2 - bc^2)\, uw + i\,(\bar{c}^3 - c^3)\, vw \\ \tilde{w} = i\,\frac{1}{2}(\bar{b}^3 - b^3)\, v - i\,\frac{3}{2}(b^2 c - \bar{b}^2 \bar{c})\, u^2 + i\,\frac{3}{2}(\bar{b}\bar{c}^2 - bc^2)\, uv + i\,(\bar{c}^3 - c^3)\, v^2 \\ \qquad + \frac{1}{2}(b^3 + \bar{b}^3)\, w - \frac{3}{2}(bc^2 + \bar{b}\bar{c}^2)\, uw - (c^3 + \bar{c}^3)\, vw \end{cases}$$

$$q=4 \quad \begin{cases} \tilde{u} = |b|^2 u - (\bar{b}c + b\bar{c})\, \Delta + i\,(\bar{b}c - b\bar{c})\, \delta\mu + \cdots \\ \tilde{\Delta} = \frac{1}{2}(b^4 + \bar{b}^4)\, \Delta - \frac{1}{4}(b^2 - \bar{b}^2)^2 u^2 - (b^2 + \bar{b}^2)(bc + \bar{b}\bar{c})\, \Delta u \\ \qquad + i\,\frac{1}{2}(b^4 - \bar{b}^4)\, \delta\mu - i\,(bc + \bar{b}\bar{c})(b^2 - \bar{b}^2)\, \delta\mu u + \cdots \\ 2\tilde{\delta}\tilde{\mu} = i\,(\bar{b}^4 - b^4)\, \Delta + i\,\frac{1}{2}(b^4 - \bar{b}^4)\, u^2 + (b^4 + \bar{b}^4)\, \delta\mu - 2(b^3 c + \bar{b}^3 \bar{c})\, \delta\mu u + \cdots \end{cases}$$

$$q \geq 5 \quad \begin{cases} \tilde{u} = |b|^2 u + (\bar{b}c + \bar{b}c)\, v + i\,(b\bar{c} - c\bar{b})\, w + \cdots \\ \tilde{v} = \frac{1}{2}(b^q + \bar{b}^q)\, v + i\,\frac{1}{2}(b^q - \bar{b}^q)\, w + \cdots \\ \tilde{w} = i\,\frac{1}{2}(\bar{b}^q - b^q)\, v + \frac{1}{2}(\bar{b}^q + b^q)\, w + \cdots \end{cases}$$

Table 4.1: Invariants in the transformed coordinates $Z = bz + c\bar{z}^{q-1}$

Up to topological codimension 3, the classification of $\mathbf{Z}_q$-invariant potentials is as follows (for nontrivial bifurcation problems we always assume that $F_u^o = 0$).

The classification of $\mathbf{Z}_3$-invariant potentials is given in Tables 4.2 and 4.3.

Normal Form	Defining Condition	Nondegeneracy Conditions	$\mathcal{A}^{\mathbf{Z}_3}$-cod	Universal Unfolding
$\frac{1}{3}v$	—	$F_v^{o2} + G^{o2} \neq 0$	1	$+\frac{1}{2}\nu u$
$\frac{1}{4}u^2 + \frac{1}{6}v^2$	$F_v^{o2} + G^{o2} = 0$	$F_{uu}^o(\hat{\alpha} + i\hat{\beta}) \neq 0$	3	$+\frac{1}{2}\nu_1 u + \frac{1}{3}\nu_2 v + \frac{1}{3}\nu_3 w$

$$\hat{\alpha} = F_{uu}^o F_{vv}^o - F_{uv}^{o2} + G_u^{o2} \qquad \hat{\beta} = F_{uu}^o G_v^o - F_{uv}^o G_u^o$$

Table 4.2: Classification of $\mathbf{Z}_3$-invariant potentials up to top-cod 3 $\quad (H = F + w\,G)$

$$
\begin{array}{ccc}
F_v^{o2} + G^{o2} & \xrightarrow{\neq 0} & \frac{1}{3}v \\
\Big\downarrow{=0} & & \\
F^o_{uu} \cdot (\widehat{\alpha} + i\,\widehat{\beta}) & \xrightarrow{\neq 0} & \frac{1}{4}u^2 + \frac{1}{6}v^2 \\
\Big\downarrow{=0} & & \\
\text{top-cod} \geq 4 & & (\widehat{\alpha},\ \widehat{\beta} \text{ are given above in Table 4.2})
\end{array}
$$

Table 4.3: Flowchart for $\mathbf{Z}_3$-invariant potentials up to top-cod 3

The algebraic data for the higher order terms for the two normal forms in Table 4.2 are:

$$(m^{\mathbf{Z}_3})^2 \subset P(\tfrac{1}{3}v, \mathcal{A}^{\mathbf{Z}_3})$$

and

$$(m^{\mathbf{Z}_3})^3 \subset P(\tfrac{1}{4}\epsilon u^2 + \tfrac{1}{6}v^2, \mathcal{A}^{\mathbf{Z}_3}).$$

The classification for $\mathbf{Z}_4$-potentials is given in Tables 4.4-6 with the algebraic data for the higher order terms recorded in Table 4.7.

For $q \geq 5$ the classification is given in Tables 4.8-10 with the algebraic data for higher order terms recorded in Table 4.11.

The last point we should make is that up to topological codimension 3, the classification under $\mathcal{A}^{\mathbf{Z}_q}$-equivalence of the potentials corresponds to the classification under $\mathcal{K}^{\mathbf{Z}_q}$-equivalence for the gradients (cf. Section 3.3).

Remarks. (a) When $F^o_{uu} = 0$ the normal forms are q-dependent with special cases for $q = 5, 6, 7, 8$ and $q \geq 9$.

(b) Note that the invariants v and w are q-dependent in Tables 4.9-11.

(c) When $q = 6$, if $h_6^2 = 1$ and $\hat{\epsilon} = 0$ then top-cod$(H, \mathcal{A}^{\mathbf{Z}_6}) \geq 4$.

(d) When $q = 8$ and if $F^o_{uuu} \cdot (h_8^2 - 1) = 0$ then top-cod$(H, \mathcal{A}^{\mathbf{Z}_8}) \geq 4$.

(e) When $F_u^{o2} + F_v^{o2} + G^{o2} \neq 0$, the $\mathbf{Z}_q$ and $\mathbf{D}_q$-codimension are the same (from Lemma 4.1). However, when $F_u^o \cdot F_v^o \cdot G^o = 0$ many of the resulting $\mathbf{Z}_q$-invariant normal forms can still be put into $\mathbf{D}_q$-invariant normal form *but their* $\mathbf{Z}_q$*-codimension is higher than the* $\mathbf{D}_q$*-codimension.*

4.2 Classification of $\mathbf{Z}_q$-equivariant bifurcation problems

Now we consider the subsequent parts of the classification of $\mathbf{Z}_q$-equivariant gradient maps. Here we consider paths through the unfolding of the $\mathbf{Z}_q$-potentials and then use path equivalence. Ultimately the result is a classification up to contact equivalence of gradient maps preserving the special role of the bifurcation parameter.

Given a $\mathbf{Z}_q$-invariant potential $h(z,\lambda)$, set $\lambda = 0$ and consider the $\mathcal{A}^{\mathbf{Z}_q}$-universal unfolding $H(z,\nu)$, $\nu \in \mathbf{R}^k$, of $h_o \stackrel{\text{def}}{=} h(\cdot, 0)$.

Let $\phi : \mathbf{R} \to \mathbf{R}^k$ be a path, then $\nabla_z h(z,\lambda)$ can be represented by $\nabla_z H\big(z, \phi(\lambda)\big)$ with a $\mathcal{K}_\lambda^{\mathbf{Z}_q}$-change of coordinates.

With the exception of the potential $\frac{1}{4}\epsilon u^2 + \frac{1}{2q} v^2$ all $\mathbf{Z}_q$-invariant potential germs up to topological codimension 3 and their universal enfoldings are identical to $\mathbf{D}_q$-invariant potentials with their universal unfoldings (see the discussion in previous section).

Therefore we can apply $\mathbf{D}_q$-theory to $\mathbf{Z}_q$-invariant potentials in an equivalence class with a $\mathbf{D}_q$-invariant potential. Classifications of the corresponding $\mathbf{D}_q$-equivariant bifurcation diagrams up to topological codimension 2 can be found in Furter [1990], Golubitsky & Roberts [1987] and Furter [1991] for $\mathbf{D}_3$, $\mathbf{D}_4$ and $\mathbf{D}_q$, $q \geq 5$, respectively. For these cases the $\mathbf{Z}_q$-classification will follow the $\mathbf{D}_q$-results with minor adaptation; in particular, the values of the coefficients in the normal forms will depend on the G-part of the potential $H = F + w\,G$.

For $\mathbf{D}_4$ it is necessary to modify the normal forms of Golubitsky & Roberts to adapt the coefficients to the gradient setting (see Chapter 6).

The fundamental result is stated in Theorem 4.2 for which we introduce the following quantities (when defined). Let $H\big(z,\phi(\lambda)\big) = F\big(u,v,\phi(\lambda)\big) + w\,G\big(u,v,\phi(\lambda)\big)$ and define:

$$\text{if } q \neq 4 \quad \begin{cases} \xi = (F^o_{v\lambda} + i\,G^o_\lambda) - \frac{F^o_{u\lambda}}{F^o_{uu}} (F^o_{uv} + i\,G^o_u), \\ \zeta = -\frac{1}{F^o_{uu}} (F^o_{uv} + i\,G^o_u)^2 + (F^o_{vv} + i\,2G^o_v), \\ \mu^2 = \frac{F^o_{uu}}{\zeta} \quad (\mu \in \mathbf{C}), \\ k = \frac{\operatorname{sign} F^o_{uu}}{|F^o_{u\lambda}|} \left(\frac{q}{2}\right)^{\frac{1}{2}} \xi\mu \end{cases} \tag{4.7a}$$

and

$$\text{if } q = 4 \quad \begin{cases} \xi = -4\,(F^o_{\Delta\lambda} + i\,2G^o_\lambda) + 4\,\frac{F^o_{u\lambda}}{F^o_{uu}} (F^o_{u\Delta} + i\,2G^o_u), \\ \zeta = \frac{4}{F^o_{uu}} (F^o_{u\Delta} + i\,2G^o_u)^2 - 4(F^o_{\Delta\Delta} + i\,4G^o_\Delta), \\ \mu^2 = 2\,\frac{F^o_{uu}}{\zeta} \quad (\mu \in \mathbf{C}), \\ k = \frac{1}{2} (\operatorname{sign} F^o_{uu})\, \frac{\xi\mu}{|F^o_{u\lambda}|}. \end{cases} \tag{4.7b}$$

Theorem 4.2. *Up to topological codimension 2, the classification of $\mathbf{Z}_q$-equivariant gradient bifurcation problems is as follows:*

(a) *if the associated potential satisfies $F_v^{o2} + G^{o2} \neq 0$ then the normal forms and universal unfoldings are equivalent to the results for $\mathbf{D}_q$-equivariant gradient bifurcation problems (up to topological codimension 2),*

(b) *when $q \neq 4$ and $F_v^{o2} + G^{o2} = 0$ but $F_{uu}^o \cdot F_{u\lambda}^o \cdot \zeta \neq 0$ then the normal form is*

$$(u + \epsilon\lambda)\, z + (v + k\lambda)\, \overline{z}^{q-1} \tag{4.8}$$

where $\epsilon = \operatorname{sign}(F_{uu}^o F_{u\lambda}^o)$ and $k \in \mathbf{C}^*$ is a modal parameter as defined in (4.7a). The universal unfolding of (4.8) is

$$(u + \epsilon\lambda)\, z + (v + k\lambda + \alpha)\, \overline{z}^{q-1} \tag{4.9}$$

where $\alpha \in \mathbf{C}$ is the unfolding parameter. The normal form (4.8) is of codimension 4 but of topological codimension 2.

(c) *When $q = 4$ and $F_v^{o2} + G^{o2} = 0$ but $F_{uu}^o \cdot F_{u\lambda}^o \cdot \zeta \neq 0$ then the normal form is*

$$(u + \epsilon\lambda + i\,\tfrac{1}{2}(\operatorname{Im} k)\,\lambda u)\, z + (\Delta + k\lambda)\, \delta\overline{z}$$

where $\epsilon = \operatorname{sign}(F_{uu}^o F_{u\lambda}^o)$ and $k \in \mathbf{C}^*$ is a modal parameter as defined in (4.7b). The universal unfolding is

$$(u + \epsilon\lambda + i\,\tfrac{1}{2}(\operatorname{Im} k)\,\lambda u)\, z + (\Delta + k\lambda + \alpha)\, \delta\overline{z}\,.$$

where $\alpha \in \mathbf{C}$ is the unfolding parameter.

The normal forms that are also $\mathbf{D}_q$-equivariant do not need as many sign constants when considered as $\mathbf{Z}_q$-equivariant problems; this is because $-I$ belongs to the connected component of the identity in the set of $\mathbf{Z}_q$-commuting matrices.

Proof. Because the group $\mathcal{K}_\Delta$, of equivalence for paths, is a subgroup of the usual group of contact equivalence for functions $\phi : \mathbf{R} \to \mathbf{R}^k$ and because $\phi(0) = 0$, the topological $\mathcal{K}_\Delta$-codimension of ϕ is at least $k-1$. As we are interested in bifurcation diagrams of topological codimension at most 2, this implies we should only consider potentials of topological $\mathcal{A}^{\mathbf{Z}_q}$-codimension at most 3. In other words, the only cases which are not analyzable through the $\mathbf{D}_q$-theory are given by paths through the potential

$$H(z,\nu) = \begin{cases} \tfrac{1}{4}\epsilon u^2 + \tfrac{1}{2q}v^2 + \tfrac{1}{2}\nu_1 u + \tfrac{1}{q}\nu_2 v + \tfrac{1}{q}\nu_3 w\,, \\ \tfrac{1}{4}\epsilon u^2 - \tfrac{1}{8}\Delta^2 + \tfrac{1}{2}\nu_1 u - \tfrac{1}{4}\nu_2 \Delta - \tfrac{1}{2}\nu_3\, \delta\mu\,, \quad q = 4\,. \end{cases}$$

Consider a $\mathbf{Z}_q$-equivariant gradient bifurcation problem $\nabla_z h(z,\lambda)$ and assume that $h(z,0)$ is $\mathcal{A}^{\mathbf{Z}_q}$-equivalent to $\frac{1}{4}\epsilon u^2 + \frac{1}{2q}v^2$. The first step is to introduce a change of coordinates in $\mathcal{K}_\lambda^{\mathbf{Z}_q}$ that brings $\nabla_z h(z,\lambda)$ into the normal form (4.8) modulo higher order terms. This is a straightforward albeit lengthy computation that proceeds as follows.

First note that we can identify $\mathcal{E}_\lambda^{\mathbb{Z}_q}$ and $\left(\mathcal{E}_{o,\lambda}^{\mathbb{Z}_q}(\mathbb{C})\right)^2$ via

$$\nabla_z h(z,\lambda) = f(u,v,\lambda)\, z + g(u,v,\lambda)\, \overline{z}^{q-1}\ , \quad f,g \in \mathcal{E}_{o,\lambda}^{\mathbb{Z}_q}(\mathbb{C}).$$

In our particular case all the second order terms in u, v, λ can be neglected. The change of coordinates $(T, Z, \Lambda) \in \mathcal{K}_\lambda^{\mathbb{Z}_q}$ can then be written as

$$T(z,\lambda)\cdot\omega = \beta\,\omega + \gamma\overline{z}^{q-2}\,\overline{\omega}\ , \quad \beta,\gamma \in \mathcal{E}_{o,\lambda}^{\mathbb{Z}_q}(\mathbb{C}),$$

and

$$Z(z,\lambda) = b\,z + c\,\overline{z}^{q-1}\ , \quad b,c \in \mathcal{E}_{o,\lambda}^{\mathbb{Z}_q}(\mathbb{C}).$$

The object is to find β, γ, b, c such that $(T, Z, \Lambda)\cdot\nabla_z h(z,\lambda)$ is of the form (4.8) modulo higher order terms.

To simplify the formulas we consider the change of coordinates in two steps: first $(I, Z, \Lambda)\cdot(f,g) = (f', g')$ and then $(T, I, I)\cdot(f', g') = (f'', g'')$ producing the final result. We find that

$$\begin{aligned}
&\tilde u = z\overline{z} = |b|^2 u + (\overline{b}c + b\overline{c})\, v + i\,(b\overline{c} - \overline{b}c)\, w + m^2,\\
&\tilde v = v(Z) = \tfrac12 (b^q + \overline{b}^q)\, v + i\,\tfrac12 (b^q - \overline{b}^q)\, w + m^2,\\
&\tilde f = f(\tilde u, \tilde v, \Lambda)\ , \quad \tilde g = g(\tilde u, \tilde v, \Lambda)\ ,\\
&f' = b\,\tilde f\ , \quad g' = c\,\tilde f + \overline{b}^{q-1}\,\tilde g\ , \quad f'' = \beta\, f' \quad \text{and} \quad g'' = \beta\, g' + \gamma\,\overline{f'}.
\end{aligned}$$

Explicitly, if we begin with $h(z,\lambda) = F(u,v,\lambda) + w\,G(u,v,\lambda)$, where

$$\begin{aligned}
F &= A\,u^2 + B\,uv + C\,v^2 + D\,u\lambda + E\,v\lambda + \cdots\ ,\\
G &= P\,u + Q\,v + R\lambda + \cdots\ ,
\end{aligned}$$

then

$$\begin{aligned}
f &= 4A\,u + 2(B - iP)\, v + 2D\,\lambda + \cdots\ ,\\
g &= \big(qB + i(q+2)P\big)\, u + 2q\,(C + iQ)\, v + q\,(E + iR)\,\lambda + \cdots\ ,
\end{aligned}$$

from which we obtain,

$$\begin{aligned}
&f'_u = 4A\,b|b|^2\ , \qquad f'_v = 8A\,b^2\overline{c} + 2\,(B - iP)\, b^{q+1}\ , \qquad f'_\lambda = 2D\,b\Lambda_\lambda\ ,\\
&g'_u = 8A\,|b|^2 c - 4A\,b^2\overline{c} + (q+1)(B + iP)\, b\overline{b}^q + (-B + iP)\, b^{q+1}\ ,\\
&g'_v = 8A\,\overline{b}c^2 + 2(q+1)(B + iP)\,\overline{b}^q c + 2q\,(C + iQ)\,\overline{b}^{2q-1}\ ,\\
&g'_\lambda = 2D\,c\Lambda_\lambda + q\,(E + iR)\,\overline{b}^{q-1}\,\Lambda_\lambda\ ,\\
&f''_u = \beta\, 4A\,b|b|^2\ , \qquad f''_v = \beta\, f'_v\ , \qquad f''_\lambda = \beta\, 2D\,b\Lambda_\lambda\ ,\\
&g''_u = \beta\, g'_u + \gamma\overline{f'_u}\ , \qquad g''_v = \beta\, g'_v + \gamma\overline{f'_v}\ , \qquad g''_\lambda = \beta\, g'_\lambda + \gamma\,\overline{f'_\lambda}\,.
\end{aligned}$$

Since (f'', g'') represents a gradient we want to take $f''_v = g''_u = 0$. This gives us the values of $c_0 = -\frac{1}{4A}(B + iP)\overline{b}_o^{q-1}$ and $\gamma_o = -\frac{(q-1)}{4A}(B + iP)\beta_o \overline{b}_o^{q-2}$. Substituting these coefficients into the remaining equations results in the normal form (4.8) provided that $A \cdot D \cdot \zeta \neq 0$. In the case $q = 4$, a similar computation gives the analogous result.

Now we have a path

$$\phi(\lambda) = \left(\epsilon\lambda + m^2, (\mathrm{Re}\, k)\lambda + m^2, (\mathrm{Im}\, k)\lambda + m^2\right), \quad \forall q \geq 3.$$

The next step is to determine if the set of higher order terms for ϕ, denoted $\mathcal{P}(\phi, \mathcal{K}_\Delta)$, contains any second order terms and if so a change of coordinates is introduced to remove them.

$\mathcal{P}(\phi, \mathcal{K}_\Delta)$ contains the intrinsic part of $\mathcal{T}(\phi, \mathcal{U}(\mathcal{K}_\Delta))$. In the Fundamental Lemma of Chapter 3, it is proved that

$$\xi \in \mathcal{T}(\phi, \mathcal{U}(\mathcal{K}_\Delta)) \Leftrightarrow \frac{\partial}{\partial \tau}\bigg|_{\tau=0} (\nabla_z H(z, \phi + t\xi)) \in \mathcal{T}\left(\nabla_z H(z, \phi(\lambda)), \mathcal{U}(\mathcal{K}_\lambda^{\mathbf{Z}_q})\right).$$

It is straightforward to compute the generators of the $\mathcal{K}_\lambda^{\mathbf{Z}_q}$-tangent spaces for a given $\mathbf{Z}_q$-equivariant germ.

The $\mathcal{E}_{o,\lambda}^{\mathbf{D}_q}$-free module of $\mathbf{Z}_q$-equivariant maps, $\mathcal{E}_\lambda^{\mathbf{Z}_q}$, is generated by $z, iz, \overline{z}^{q-1}$ and $i\overline{z}^{q-1}$ and can be represented by $(\hat{p}, \hat{q}, \hat{r}, \hat{s}) \in (\mathcal{E}_{o,\lambda}^{\mathbf{D}_q})^4$.

Similarly, the $\mathcal{E}_{o,\lambda}^{\mathbf{D}_q}$-free module, $\mathcal{M}_\lambda^{\mathbf{Z}_q}$, of $\mathbf{Z}_q$-commuting matrices is generated by (in complex notation, for all $q \geq 3$):

$$S_1 : \omega \mapsto \omega, \quad S_2 : \omega \mapsto z^2\overline{\omega}, \quad S_3 : \omega \mapsto \overline{z}^{q-2}\overline{\omega}, \quad S_4 : \omega \mapsto z^q\omega,$$

as well as iS_1, iS_2, iS_3, iS_4 and the set of generators for the tangent spaces is

$$(\hat{p}, \hat{q}, \hat{r}, \hat{s}), \quad (-\hat{q}, \hat{p}, -\hat{s}, \hat{r}), \tag{4.10}$$

$$\left.\begin{array}{ll} (u\hat{p} + v\hat{r}, -v\hat{s}, 0, u\hat{s}), & (v\hat{s}, u\hat{p} + v\hat{r}, -u\hat{s}, 0), \\ (u^{q-2}\hat{r}, -u^{q-2}\hat{s}, \hat{p}, -\hat{q}), & (u^{q-2}\hat{s}, u^{q-2}\hat{r}, \hat{q}, \hat{p}), \\ (v\hat{p}, u^{q-1}\hat{s} + v\hat{q}, -u\hat{p}, 0), & (-v\hat{q}, u^{q-1}\hat{r} + v\hat{p}, u\hat{q}, 0), \end{array}\right\} \tag{4.11}$$

$$\left.\begin{array}{l} (2u\hat{p}_u + qv\hat{p}_v, 2u\hat{q}_u + qv\hat{q}_v, (q-2)\hat{r} + 2u\hat{r}_u + qv\hat{r}_v, (q-2)\hat{s} + 2u\hat{s}_u + qv\hat{s}_v), \\ (-v\hat{q}_v - u^{q-1}\hat{s}_v, v\hat{p}_v + u^{q-1}\hat{r}_v, u\hat{q}_v + \hat{s} + v\hat{s}_v, -u\hat{p}_v - \hat{r} - v\hat{r}_v), \end{array}\right\} \tag{4.12}$$

$$\left.\begin{array}{l} \big(2v\hat{p}_u + qu^{q-1}\hat{p}_v + qu^{q-2}\hat{r}, 2v\hat{q}_u + qu^{q-1}\hat{q}_v + (q-2)u^{q-2}\hat{s}, \\ \qquad 2\hat{p} + 2v\hat{r}_u + qu^{q-1}\hat{r}_v, 2v\hat{s}_u + qu^{q-1}\hat{s}_v\big), \\ \big(2v\hat{q}_u + qu^{q-2}\hat{s} + 2u^{q-1}\hat{s}_u, -2v\hat{p}_u - (q-2)u^{q-2}\hat{r} - 2u^{q-1}\hat{r}_u, \\ \qquad -2u\hat{q}_u - 2v\hat{s}_u, 2\hat{p} + 2u\hat{p}_u + 2v\hat{r}_u\big). \end{array}\right\} \tag{4.13}$$

In the present case: $\hat{p} = u + \epsilon\lambda$, $\hat{q} = 0$, $\hat{r} = v + (\mathrm{Re}\, k)\lambda$ and $\hat{s} = (\mathrm{Im}\, k)\lambda$.

Now $\mathcal{T}(\nabla_z H(z,\phi), \mathcal{U}(\mathcal{K}_\lambda^{\mathbf{Z}_q}))$ is generated by $m_\lambda^{\mathbf{D}_q}$ times the generators (4.10), (4.12) and $\mathcal{E}_{o,\lambda}^{\mathbf{D}_q} \cdot \{(4.11),(4.12)\}$. The linear map $\frac{\partial}{\partial\tau}\Big|_{\tau=0} (\nabla_z H(z, \phi + \tau\xi))$ is given by

$$\xi_1(\lambda)\, z + \big(\xi_2(\lambda) + i\xi_3(\lambda)\big)\, \bar{z}^{q-1},$$

where the ξ_i correspond to the components of $\xi \in \mathcal{P}_3$.

We obtain the desired result for $\mathcal{P}(\phi, \mathcal{K}_\Delta)$ if we can prove that

$$\Big((m_\lambda^{\mathbf{D}_q})^2,\ 0\ ,(m_\lambda^{\mathbf{D}_q})^2, (m_\lambda^{\mathbf{D}_q})^2\Big) \subset \mathcal{T}\big(\nabla_z H(z,\phi), \mathcal{U}(\mathcal{K}_\lambda^{\mathbf{Z}_q})\big). \qquad (4.14)$$

It is important to note that the zero in the second component comes from the gradient structure of the problems we are investigating. This is the essential difference with non-gradient $\mathbf{Z}_q$-equivariant problems. A calculation shows that

$$(m_\lambda^{\mathbf{D}_q})^3 \mathcal{E}_\lambda^{\mathbf{Z}_q} \subset \mathcal{T}\big(\nabla_z H(z,\phi), \mathcal{U}(\mathcal{K}_\lambda^{\mathbf{Z}_q})\big).$$

From Nakayama's Lemma it is enough to verify (4.14) modulo third order terms. The generators (4.9)-(4.13) can be rearranged so that their second component is zero. After some simplification (modulo third order terms), the result is,

(i) to be multiplied by $m^{\mathbf{Z}_q}$

$$(\hat{p}, 0, \hat{r}, \hat{s})\,, \quad (v, 0, 0, 0)\,, \quad (0, 0, \hat{p}, 0)\,, \quad (0, 0, 0, \hat{p})\,,$$
$$(0, 0, \hat{s}, -v - \hat{r})\,, \quad (0, 0, qv + (q-2)\hat{r}, (q-2)\hat{s})\,,$$

(ii) already of second order

$$(u\hat{p}, 0, 0, 0)\,, \quad (0, 0, v\hat{s}, -v\hat{r})\,, \quad (0, 0, 0, u\hat{r}).$$

If $a \neq 0$, it is now a simple verification to show that (4.14) is satisfied.

The final computation is to construct a basis for the extended tangent space of ϕ, $\mathcal{T}_e(\phi, \mathcal{K}_\Delta)$. Here again the relation

$$\xi \in \mathcal{T}_e(\phi, \mathcal{K}_\Delta) \ \Leftrightarrow\ \frac{\partial}{\partial\tau}\Big|_{\tau=0} (\nabla_z H(z, \phi + t\xi)) \in \mathcal{T}_e\big(\nabla_z H(z,\phi), \mathcal{K}_\lambda^{\mathbf{Z}_q}\big)$$

is used, from which we obtain that $(0, \lambda, 0)$, $(0, 1, 0)$, $(0, 0, \lambda)$ and $(0, 0, 1)$ generate $\mathcal{P}_3 / \mathcal{T}_e(\phi, \mathcal{K}_\Delta)$, completing the proof. ■

Corollary 4.3. *The $\mathbf{Z}_q$-equivariant bifurcation diagrams of topological codimension 0 are:*

(a) *when $q = 3$ and $F_{u\lambda}^o \cdot (F_v^{o2} + G^{o2}) \neq 0$*

$$\epsilon\lambda\, z + \bar{z}^2\,,$$

(b) *when $q = 4$ and $F_{u\lambda}^o \cdot m \cdot (m-1) \neq 0$*

$$(m\, u + \epsilon\lambda)\, z + \delta \bar{z}\,,$$

(c) *when* $q \geq 5$ *and* $F^o_{u\lambda} \cdot F^o_{uu} \cdot (F^{o2}_v + G^{o2}) \neq 0$

$$(\epsilon_1 u + \epsilon\lambda) z + \bar{z}^{q-1},$$

where $m = \frac{1}{2}\left(1 + \frac{F^o_\Delta + F^o_{uu}}{|F^{o2}_\Delta + 4G^{o2}|^{\frac{1}{2}}}\right)$, $\epsilon = \operatorname{sign} F^o_{u\lambda}$ and $\epsilon_1 = \operatorname{sign} F^o_{uu}$.

Note that ϵ_1 can actually be eliminated by incorporating the sign of F^o_{uu} in ϵ. There is also a small discrepancy between the definition of m for $\mathbf{Z}_4$-equivariant and $\mathbf{D}_4$-equivariant problems, that is, when $G \equiv 0$ and $F^o_\Delta > 0$ we do not recover exactly the formula of Chapter 6. On a more general level, this is the price to pay to be able to impose some coefficients to be 1 in the $\mathbf{Z}_q$-theory instead of being equal to the sign of some quantities.

In Chapter 8, for the study of the collision of multipliers we shall need the following additional normal forms of topological codimension 1.

Corollary 4.4. *Let* $H(z,\lambda) = F(u,v,\lambda) + w\,G(u,v,\lambda)$ *be a* $\mathbf{Z}_q$*-invariant potential. Suppose that* $F^{o2}_v + G^{o2} \neq 0$ *but* $F^o_{u\lambda} = 0$. *Then the diagrams of topological codimension 1,* $\nabla_z H(z,\lambda)$, *are* $\mathcal{K}^{\mathbf{Z}_q}_\lambda$*-equivalent to the following normal forms (*α *is the unfolding parameter):*

(a) *when* $q = 3$ *and* $\epsilon \neq 0$

$$(\epsilon\lambda^2 + \alpha) z + \bar{z}^2,$$

(b) *when* $q = 4$ *and* $\epsilon \cdot \epsilon_2 \cdot m \cdot (m-1) \neq 0$

$$(m u + \epsilon_2 \lambda u + \epsilon\lambda^2 + \alpha) z + \delta \bar{z},$$

(c) *When* $q \geq 5$ *and* $\epsilon \cdot F^o_{uu} \neq 0$

$$(\epsilon_1 u + \epsilon\lambda^2 + \alpha) z + \bar{z}^{q-1},$$

where $m = \frac{1}{2}\left(1 + \frac{F^o_\Delta + F^o_{uu}}{|F^{o2}_\Delta + 4G^{o2}|^{\frac{1}{2}}}\right)$, $\epsilon = \operatorname{sign} F^o_{u\lambda\lambda}$, $\epsilon_1 = \operatorname{sign} F^o_{uu}$ *and*

$$\epsilon_2 = \operatorname{sign}\left(F^o_\Delta \,(F^o_\Delta F^o_{uu\lambda} - F^o_{uu} F^o_{\Delta\lambda}) + 4G^{o2}(F^o_{\Delta\lambda} + F^o_{uu\lambda}) - 8G^o G^o_\lambda (F^o_\Delta + F^o_{uu})\right).$$

4.3 Bifurcation diagrams for the unfolding (4.9)

Normal forms for $\mathbf{Z}_q$-equivariant gradient bifurcation problems fall into two categories: those with $F^{o2}_v + G^{o2} \neq 0$ and those with $F^{o2}_v + G^{o2} = 0$.

Bifurcation diagrams for the former category can be obtained from the analogous $\mathbf{D}_q$-normal forms by a suitable coordinate transformation.

When $F_v^{o2} + G^{o2} = 0$ the resulting normal form was given in equation (4.8) with unfolding

$$(u + \epsilon\lambda)\, z + (v + k\lambda + \alpha)\, \bar{z}^{q-1} \tag{4.15}$$

where $k, \alpha \in \mathbf{C}$. When $k, \alpha \in \mathbf{R}$ the normal form (4.15) reduces to the $\mathbf{D}_q$-equivariant case for which the bifurcation diagrams are well-known (cf. GSS II [1988, p.200]) and this will be our point of reference when analyzing the bifurcation diagrams for (4.15).

With $k, \alpha \in \mathbf{C}$, non-trivial differences (compared with $k, \alpha \in \mathbf{R}$) arise in the bifurcation diagrams; in particular, the secondary bifurcation points break-up into limit points.

The analysis of (4.15) is simplified by using polar coordinates rather than cartesian coordinates for the complex variables z, k and α. Technically the resulting equations in polar coordinates do not satisfy the unfolding conditions (difficulties with the mapping $(\alpha_r, \alpha_i) \mapsto (|\alpha|, \operatorname{Arg}\alpha)$ in the unfolding space) so we must verify that the diagrams are preserved under the inverse mapping $(|\alpha|, \operatorname{Arg}\alpha) \mapsto (\alpha_r, \alpha_i)$.

Let $z = r\,e^{i\theta}$, $\alpha = R\,e^{i(\Phi+\Psi)}$ and $k = \hat{k}\,e^{i\Psi}$. As k is a modal parameter, the parameters $\hat{k}$ and Ψ can be treated as any admissible fixed numbers. However, α is the unfolding parameter which requires that $|\alpha| = R$ be small and although Φ can be any angle, it represents the deviation from $\Psi = \operatorname{Arg} a$.

Given $(\hat{k}, \Psi)$ the idea is to determine solutions of (4.15) with (R, Φ) variable and R small but fixed relative to z. Keeping in mind that (4.15) is an unfolding germ, admissible solutions are those that persist in the limit $R \to 0$.

Equating the real and imaginary parts of (4.15) to zero results in the following system for the non-trivial solutions of (4.15)

$$\begin{aligned}\epsilon\lambda\big(1 + \epsilon\hat{k}\, r^{q-2} \cos(\Psi - q\theta)\big)& \\ +R\, r^{q-2} \cos(\Psi + \Phi - q\theta) + r^2 + r^{2q-2} \cos^2 q\theta &= 0\,,\end{aligned} \tag{4.16}$$

$$\hat{k}\lambda \sin(\Psi - q\theta) + R \sin(\Psi + \Phi - q\theta) - r^q \cos q\theta \sin q\theta = 0\,. \tag{4.17}$$

It is clear that λ can be eliminated – for r sufficiently small – using (4.16), resulting in the single function of (r, θ, R, Φ):

$$\begin{aligned}0 = H(r, \theta, R, \Phi) = R \sin(\Psi + \Phi - q\theta) + \epsilon\hat{k}\, R\, r^{q-2} \sin\Phi - \epsilon\hat{k}\, r^2 \sin(\Psi - q\theta)& \\ - \tfrac{1}{2} r^q \sin 2q\theta - \epsilon\hat{k}\, r^{2q-2} \cos q\theta \sin\Psi\,.&\end{aligned} \tag{4.18}$$

Given (r, θ, R, Φ) as a solution of (4.18), back substitution into (4.16) provides the corresponding value of λ. Recall that $k, \alpha \in \mathbf{R}$ ($\Phi = \Psi = 0 \bmod \pi$) corresponds to the $\mathbf{D}_q$-normal form. In the polar coordinates it is $\Psi = 0, \pi$ that recovers the real modal parameter and $\Phi = 0, \pi$ that recovers the real unfolding parameter.

In the absence of the unfolding term – setting $R = 0$ – the Implicit Function Theorem is easily applied to $r^{-2}H(r,\theta,0,\Phi)$ to show that there exists $2q$ branches $\theta_i(r)$, $1 \le i \le 2q$, with θ_i^o satisfying $\sin(\Psi - q\theta_i^o) = 0$ and furthermore

$$\theta_i(r) = \theta_i^o + \tfrac{\epsilon\delta_i}{2qk}(\sin 2\Psi)\, r^{q-2} + \cdots \quad \text{with} \quad \delta_i = \cos(\Psi - q\theta_i^o) \quad (\delta_i^2 = 1).$$

The initial direction of the bifurcating branches is determined by Ψ, the argument of the modal parameter. There are two special situations to consider.

When $\sin 2\Psi = 0$, returning to (4.18), we see that the bifurcating branches lie in the planes $\sin q\theta = 0$ or $\cos q\theta = 0$. The first case ($\Psi = 0, \pi$) is the known D_q-situation where $k \in \mathbf{R}$. The second ($\Psi = \frac{1}{2}\pi, \frac{3}{2}\pi$) is when a is purely imaginary. In that case the diagram is a rotation of the D_q-diagram with the modal parameter $|k|$. In both cases, the stability on branch i is given by the sign of $\epsilon\delta_i$. And so, we can consider the role of Ψ as to introduce a "twist" in the bifurcation diagram relative to the special situation when $\sin 2\Psi = 0$.

When $R \neq 0$ the implicit function theorem can again be used to show that there exist $2q$ branches $\theta_i(r)$, $1 \le i \le 2q$, bifurcating from the origin with θ_i^o given by $\sin(\Phi + \Psi - q\theta_i^o) = 0$ and

$$\theta_i(r) = \theta_i^o + \tfrac{\epsilon\delta_i}{q}\,\hat{k}\,(\sin\Phi)\, r + \cdots, \quad q = 3$$
$$\theta_i(r) = \theta_i^o + \tfrac{\epsilon}{q}\,\hat{k}\,(\sin\Phi)\, r^2 + \ldots, \quad q \ge 4$$

with $\delta_i = \cos(\Psi + \Phi - 3\theta_i^o)$ $(\delta_i^2 = 1)$. The argument Φ of the unfolding parameter introduces an additional "twist". Figures 4.1(a) and (b) show the (r,θ) plane for small twist when $q = 3$ and 4.

Unless we are in the situation where k is real or purely imaginary with α to match , that is $\sin 2\Psi = \sin\Phi = 0$, it is also easily established that there is at most one solution of (4.18) along any given ray $\{\, (r,\theta) \mid r \text{ small}, \ \theta \text{ fixed} \,\}$ and exactly one if

$$\epsilon\ \sin(\Psi + \Phi - q\theta)\ \sin(\Psi - q\theta) > 0\,. \tag{4.19}$$

The important value of Φ is π where the two sinus functions of (4.19) change their respective position. Hence the only remaining possible "secondary bifurcations" are turning points given by $H = H_\theta = 0$. In particular, when (4.19) is satisfied, in contrast to the real unfolding of the D_q-case, there are no symmetry breaking secondary bifurcations: all the bifurcating branches in the Z_q-case are non-symmetric.

The transition variety for the unfolding corresponds to the collapse of two of the turning points and is given by

$$\{\, (R,\Phi) \mid \exists\, (r,\theta) \text{ with } H(r,\theta,R,\Phi) = H_\theta(r,\theta,R,\Phi) = H_{\theta\theta}(r,\theta,R,\Phi) = 0 \,\}\,. \tag{4.20}$$

A careful, but routine, analysis of this system shows that when (4.19) is satisfied, the solution set of (4.20) is empty; there is no change in the behavior of the turning points modulo periodicity.

The branches are in the cones defined by (4.19), with or without turning points. As ϕ varies, these cones shrink or expand without changing the behavior of the branch inside them (cf. Figure 4.1).

(a) $\mathbb{D}_3 \to \mathbb{Z}_3$ (b) $\mathbb{D}_4 \to \mathbb{Z}_4$

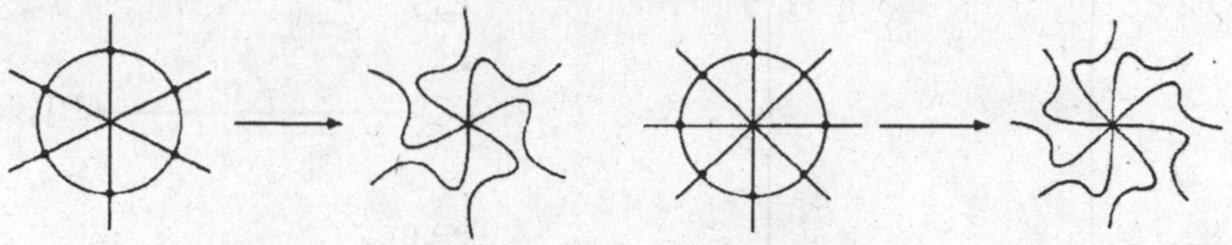

Figure 4.1: Contrast between $\mathbb{D}_q$ and $\mathbb{Z}_q$-bifurcation diagrams in (r, θ)-space for equation (4.9) when: (a) $q = 3$ and (b) $q = 4$

Stability in the unfolding (4.8), as given by the sign of the Jacobian, results in a complicated formula which is given in Appendix H. The best way to understand it is to continue the stability assignments of the branches of the $\mathbb{D}_q$-unfolding as the "screws", that is Φ and Ψ, are applied to it. Note that stability does not follow the turning points of H.

A bifurcation diagram in $\mathbb{R}^3 = (\lambda, r, \theta)$ for the case $q = 3$ is shown below.

(a) $\mathbb{D}_3$ (b) $\mathbb{Z}_3$

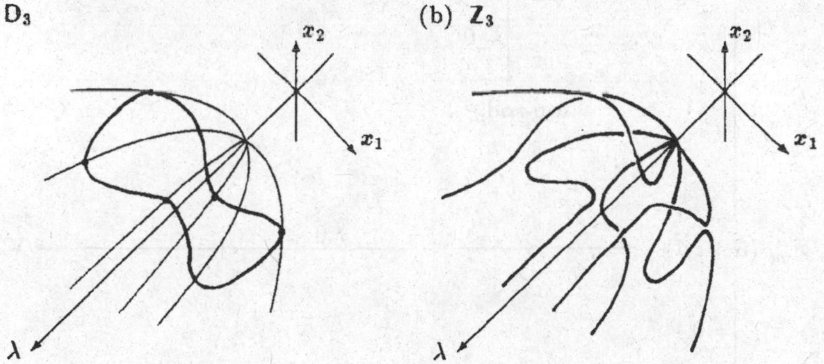

Figure 4.2: Bifurcation diagram in (λ, r, θ)-space for normal form (4.8) when $q = 3$: (a) $(a, \alpha) \in \mathbb{R}^2$ ($\mathbb{D}_3$-normal form) and (b) $(a, \alpha) \in \mathbb{C}^2$ ($\mathbb{Z}_3$-normal form)

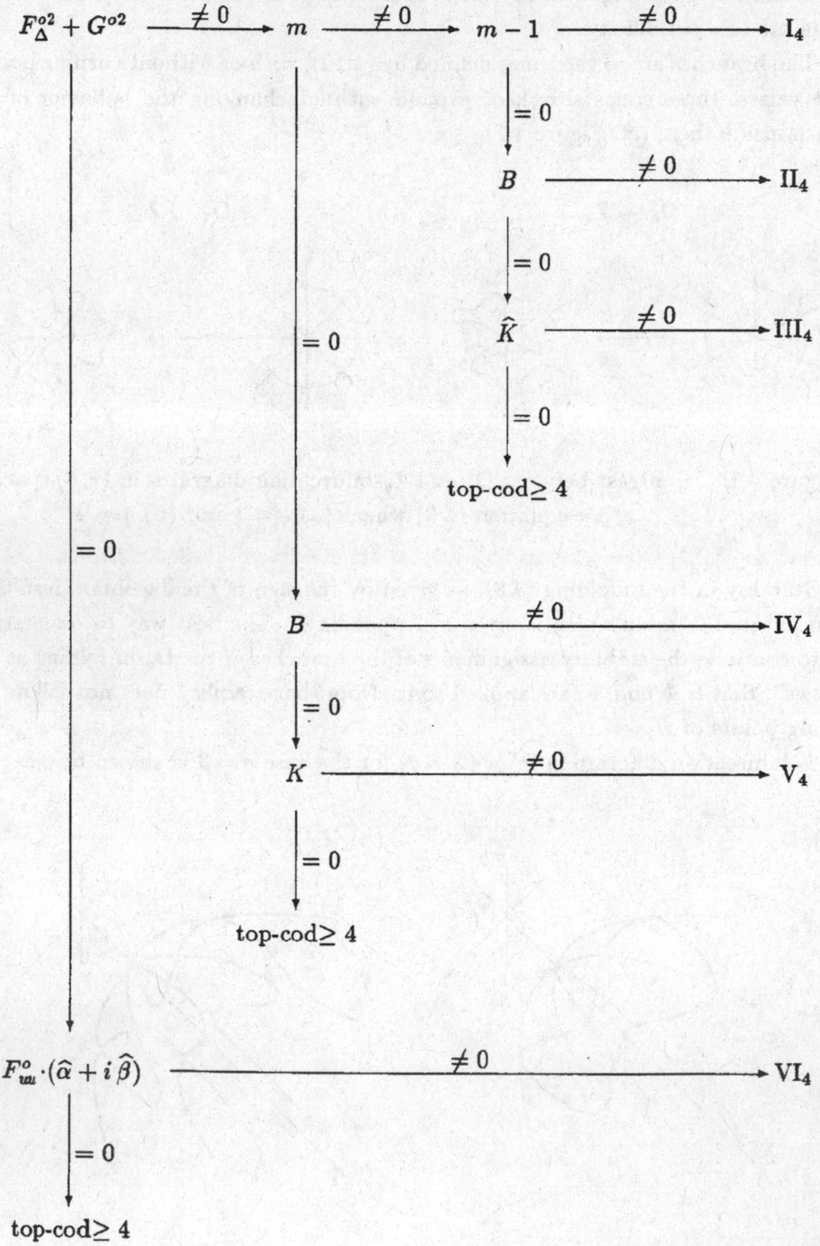

Table 4.4: Flowchart for $\mathbf{Z}_4$-invariant potentials up to top-cod 3

Case	Normal Form	Defining Condition	$\mathcal{A}^{\mathbf{Z}_4}$-cod (top-cod)	Universal Unfolding
I_4	$\frac{1}{4}\, m\, u^2 - \frac{1}{4}\,\Delta$	—	2(1)	$+\frac{1}{2}\,\nu_1\, u$
II_4	$\frac{1}{6}\epsilon u^3 + \frac{1}{4}u^2 - \frac{1}{4}\Delta$	$m = 1$	2	$+\frac{1}{2}\,\nu_1\, u + \frac{1}{4}\,\nu_2\, u^2$
III_4	$\frac{1}{8}\bar{\epsilon} u^4 + \frac{1}{4}u^2 - \frac{1}{4}\Delta$	$m = 1$ $B = 0$	3	$+\frac{1}{2}\,\nu_1 u + \frac{1}{4}\,\nu_2 u^2 + \frac{1}{6}\,\nu_3 u^3$
IV_4	$\frac{1}{6}\, u^3 - \frac{1}{4}\,\Delta$	$m = 0$	2	$+\frac{1}{2}\,\nu_1\, u + \frac{1}{4}\,\nu_2\, u^2$
V_4	$\frac{1}{8}\, u^4 - \frac{1}{4}\,\Delta$	$m = 0$ $B = 0$	3	$+\frac{1}{2}\,\nu_1 u + \frac{1}{4}\,\nu_2 u^2 + \frac{1}{6}\,\nu_3 u^3$
VI_4	$\frac{1}{4}\, u^2 - \frac{1}{8}\,\Delta^2$	$F_\Delta^{o2} + G^{o2} = 0$	3	$+\frac{1}{2}\,\nu_1 u - \frac{1}{4}\,\nu_2 \Delta - \frac{1}{4}\,\nu_3 w$

(m and B are defined in Table 4.6)

Table 4.5: Classification of $\mathbf{Z}_4$-invariant potentials up to top-cod 3 $\quad (H = F + w\,G)$

Case	Nondegeneracy Conditions	Coefficients
I_4	$m \cdot (m-1) \neq 0$ $F_\Delta^{o2} + G^{o2} \neq 0$	$m = \frac{1}{2}\left(1 + \frac{F_\Delta^o + F_{uu}^o}{\sqrt{F_\Delta^{o2} + 4\,G^{o2}}}\right)$
II_4	$m \cdot B \neq 0$ $F_\Delta^{o2} + G^{o2} \neq 0$	$B = F_{uuu}^o\,(F_\Delta^o + F_{uu}^o) + 6\,F_\Delta^o\,F_{u\Delta}^o - 12\,G_o\,G_u^o$ $\epsilon = \operatorname{sign} B$
III_4	$m \cdot \widehat{K} \neq 0$ $F_\Delta^{o2} + G^{o2} \neq 0$	$\bar{\epsilon} = \operatorname{sign} \widehat{K},\ \widehat{K} \in \mathbf{R}$
IV_4	$B \neq 0$ $F_\Delta^{o2} + G^{o2} \neq 0$	—
V_4	$K \neq 0$ $F_\Delta^{o2} + G^{o2} \neq 0$	$K \in \mathbf{R}$
VI_4	$F_{uu}^o(\widehat{\alpha} + i\,\widehat{\beta}) \neq 0$	$\widehat{\alpha} = F_{uu}^o\,F_{\Delta\Delta}^o - F_{u\Delta}^{o2} + 3\,G_u^{o2}$ $\widehat{\beta} = F_{uu}^o\,G_\Delta^o - F_{u\Delta}^o\,G_u^o$

Table 4.6: Data for $\mathbf{Z}_4$-invariant potentials up to top-cod 3

Normal Form	$\mathcal{P}(\cdot) \supset$
$\frac{1}{4}\, m\, u^2 - \frac{1}{4}\,\Delta$	$(m^{\mathbf{Z}_4})^3 + (m^{\mathbf{Z}_4})(\Delta, w)$
$\frac{1}{6}\,\epsilon\, u^3 + \frac{1}{4}\, u^2 - \frac{1}{4}\,\Delta$	$(m^{\mathbf{Z}_4})^4 + (m^{\mathbf{Z}_4})^2\,(u^2, \Delta, w) + (u^2, \Delta, w)^2$
$\frac{1}{8}\,\bar{\epsilon}\, u^4 + \frac{1}{4}\, u^2 - \frac{1}{4}\,\Delta$	$(m^{\mathbf{Z}_4})^5 + (m^{\mathbf{Z}_4})^3\,(u^2, \Delta, w) + (m^{\mathbf{Z}_4})\,(u^2, \Delta, w)^2$
$\frac{1}{6}\, u^3 - \frac{1}{4}\,\Delta$	$(m^{\mathbf{Z}_4})^4 + (m^{\mathbf{Z}_4})^2\,(u^2, \Delta, w) + (u^2, \Delta, w)^2$
$\frac{1}{8}\, u^4 - \frac{1}{4}\,\Delta$	$(m^{\mathbf{Z}_4})^5 + (m^{\mathbf{Z}_4})^3\,(u^2, \Delta, w) + (m^{\mathbf{Z}_4})\,(u^2, \Delta, w)^2$
$\frac{1}{4}\, u^2 - \frac{1}{8}\,\Delta^2$	$(m^{\mathbf{Z}_4})^3$

Table 4.7: Algebraic data for $\mathbf{Z}_4$-invariant potentials

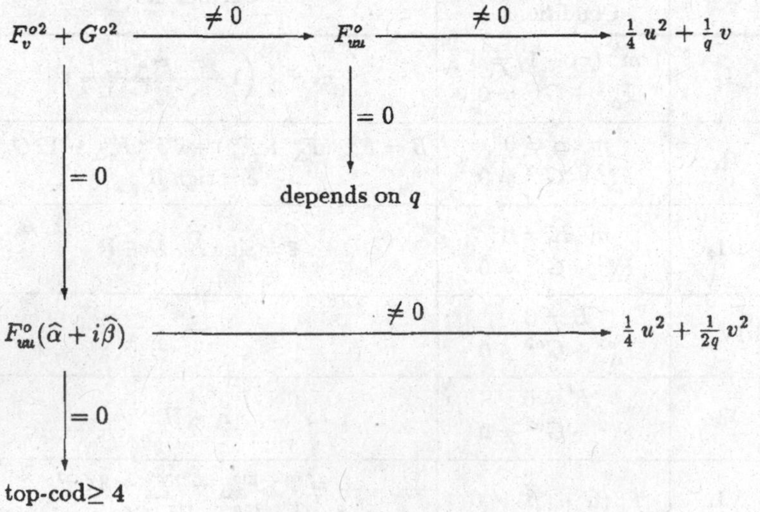

Table 4.8: Flowchart for $\mathbf{Z}_q$-invariant potentials up to top-cod 3, $q \geq 5$

q	Normal Form	Defining Condition	$\mathcal{A}^{\mathbb{Z}_q}$-cod	Universal Unfolding
≥ 5	$\frac{1}{4}u^2 + \frac{1}{q}v$	—	1	$+\frac{1}{2}\nu_1 u$
≥ 5	$\frac{1}{4}u^2 + \frac{1}{2q}v^2$	$F^o_v = 0$ $G^o = 0$	3	$+\frac{1}{2}\nu_1 u + \frac{1}{q}\nu_2 v$ $+\frac{1}{q}\nu_3 w$
5	$\frac{1}{6}\hat{\epsilon}u^3 + \frac{1}{5}v$	$F^o_{uu} = 0$	2/3(2)	$+\frac{1}{2}\nu_1 u + \frac{1}{4}\nu_2 u^2$
6	$\frac{1}{8}\hat{\epsilon}u^4 + \frac{1}{6}h_6 u^3 + \frac{1}{6}v$	$F^o_{uu} = 0$	3/4(2)	$+\frac{1}{2}\nu_1 u + \frac{1}{4}\nu_2 u^2$
6	$\frac{1}{8}\hat{\epsilon}k(u)u^4 + \frac{1}{6}h_6 u^3 + \frac{1}{6}v$	$F^o_{uu} = 0$ $h_6^2 = 1$	4(3)	$+\frac{1}{2}\nu_1 u + \frac{1}{4}\nu_2 u^2$ $+\frac{1}{6}\nu_3 u^3$
7	$\frac{1}{8}ku^4 + \frac{1}{6}u^3 + \frac{1}{7}v$	$F^o_{uu} = 0$	3(2)	$+\frac{1}{2}\nu_1 u + \frac{1}{4}\nu_2 u^2$
7	$\frac{1}{8}k(u)u^4 + \frac{1}{7}v$	$F^o_{uu} = 0$ $F^o_{uuu} = 0$	4/5(3)	$+\frac{1}{2}\nu_1 u + \frac{1}{4}\nu_2 u^2$ $+\frac{1}{6}\nu_3 u^3$
8	$\frac{1}{8}ku^4 + \frac{1}{6}\hat{\epsilon}u^3 + \frac{1}{8}v$	$F_{uu} = 0$	3(2)	$+\frac{1}{2}\nu_1 u + \frac{1}{4}\nu_2 u^2$
8	$\frac{1}{10}k(u)u^5 + \frac{1}{8}h_8 u^4 + \frac{1}{8}v$	$F^o_{uu} = 0$ $F^o_{uuu} = 0$	5/6(3)	$+\frac{1}{2}\nu_1 u + \frac{1}{4}\nu_2 u^2$ $+\frac{1}{6}\nu_3 u^3$
≥ 9	$\frac{1}{8}ku^4 + \frac{1}{6}\hat{\epsilon}u^3 + \frac{1}{q}v$	$F_{uu} = 0$	3(2)	$+\frac{1}{2}\nu_1 u + \frac{1}{4}\nu_2 u^2$
≥ 9	$\frac{1}{10}k(u)u^5 + \frac{1}{8}\tilde{\epsilon}u^4 + \frac{1}{q}v$	$F^o_{uu} = 0$ $F^o_{uuu} = 0$	5(3)	$+\frac{1}{2}\nu_1 u + \frac{1}{4}\nu_2 u^2$ $+\frac{1}{6}\nu_3 u^3$

Table 4.9: Classification of $\mathbb{Z}_q$-invariant potentials up to top-cod 3, $q \geq 5$, $(H = F + w\,G)$

q	Normal Form	Nondegeneracy Condition	Coefficients
≥ 5	$\frac{1}{4}u^2 + \frac{1}{q}v$	$F^o_{uu}(F^{o2}_v + G^{o2}) \neq 0$	—
≥ 5	$\frac{1}{4}u^2 + \frac{1}{2q}v^2$	$F^o_{uu}(\widehat{\alpha} + i\widehat{\beta}) \neq 0$	$\widehat{\alpha} = F^o_{uu}F^o_{vv} - F^{o2}_{uv} + G^{o2}_u$ $\widehat{\beta} = F^o_{uu}G^o_v - F^o_{uv}G^o_u$
5	$\frac{1}{6}\hat{\epsilon}u^3 + \frac{1}{5}v$	$(F^{o2}_v + G^{o2}) \neq 0$	$\hat{\epsilon}^* = \lvert \mathrm{sign} F^o_{uuu} \rvert$
6	$\frac{1}{8}\hat{\epsilon}u^4 + \frac{1}{6}h_6u^3 + \frac{1}{6}v$	$h_6^2 \neq 1$ $(F^{o2}_v + G^{o2}) \neq 0$	$h_6 = \frac{1}{6}F^o_{uuu}/\sqrt{F^{o2}_v + G^{o2}}$ $\hat{\epsilon}^* = \mathrm{sign}\big(\frac{1}{24}F^o_{uuuu} - \frac{F^o_{uv}F^o_v - G^o_uG^o}{F^{o2}_v + G^{o2}}\big)$
6	$\frac{1}{8}\hat{\epsilon}k(u)u^4 + \frac{1}{6}h_6u^3 + \frac{1}{6}v$	$\hat{\epsilon}(F^{o2}_v + G^{o2}) \neq 0$	$\hat{\epsilon}$ as above $k(u) = 1 + k_1^* u$
7	$\frac{1}{8}ku^4 + \frac{1}{6}u^3 + \frac{1}{7}v$	$F^o_{uuu}(F^{o2}_v + G^{o2}) \neq 0$	$k^* \in \mathbf{R}$
7	$\frac{1}{8}ku^4 + \frac{1}{7}v$	$(F^{o2}_v + G^{o2}) \neq 0$	$k(u) = \lvert \mathrm{sign}\, F^o_{uuuu} \rvert^* + k_1^* u$
8	$\frac{1}{8}ku^4 + \frac{1}{6}\hat{\epsilon}u^3 + \frac{1}{8}v$	$F^o_{uuu}(F^{o2}_v + G^{o2}) \neq 0$	$\hat{\epsilon} = \mathrm{sign}\, F^o_{uuu}$, $k^* \in \mathbf{R}$
8	$\frac{1}{10}k(u)u^5 + \frac{1}{8}h_8u^4 + \frac{1}{8}v$	$h_8^2 \neq 1$ $(F^{o2}_v + G^{o2}) \neq 0$	$h_8 = \frac{1}{24}\frac{F^o_{uuuu}}{\sqrt{F^{o2}_v + G^{o2}}}$ $k(u)^* = k_0 + k_1 u$
≥ 9	$\frac{1}{8}ku^4 + \frac{1}{6}\hat{\epsilon}u^3 + \frac{1}{q}v$	$F^o_{uuu}(F^{o2}_v + G^{o2}) \neq 0$	$\hat{\epsilon} = (\mathrm{sign}\, F^o_{uuu})^{q-1}$, $k^* \in \mathbf{R}$
≥ 9	$\frac{1}{10}k(u)u^5 + \frac{1}{8}\tilde{\epsilon}u^4 + \frac{1}{q}v$	$F^o_{uuu}(F^{o2}_v + G^{o2}) \neq 0$	$\hat{\epsilon} = (\mathrm{sign}\, F_{uuuu})^{q-1}$ $k(u)^* = k_0 + k_1 u$

* can be zero   (k, k_0, k_1 are modal parameters)

Table 4.10: Data for $\mathbf{Z}_q$-invariant potentials up to top-cod 3, $q \geq 5$

q	Normal Form	$\mathcal{P}(\cdot) \supset$
≥ 5	$\frac{1}{4}u^2 + \frac{1}{q}v$	$(m^{\mathbb{Z}_q})^3 + (m^{\mathbb{Z}_q})(u^{q-1}, v, w)$
≥ 5	$\frac{1}{4}u^2 + \frac{1}{2q}v^2$	$(m^{\mathbb{Z}_q})^3$
5	$\frac{1}{6}\hat{\epsilon}u^3 + \frac{1}{5}v$	$(m^{\mathbb{Z}_5})^4 + (m^{\mathbb{Z}_5})(u^4, v, w)$
6	$\frac{1}{8}\hat{\epsilon}u^4 + \frac{1}{6}h_6 u^3 + \frac{1}{6}v$	$(m^{\mathbb{Z}_6})^5 + (m^{\mathbb{Z}_6})^2(u^5, v, w) + (u^5, v, w)^2$
6	$\frac{1}{8}\hat{\epsilon}k(u)u^4 + \frac{1}{6}h_6 u^3 + \frac{1}{6}v$	$(m^{\mathbb{Z}_6})^6 + (m^{\mathbb{Z}_6})^3(u^5, v, w) + (u^5, v, w)^2$
7	$\frac{1}{8}k u^4 + \frac{1}{6}u^3 + \frac{1}{7}v$	$(m^{\mathbb{Z}_7})^5 + (m^{\mathbb{Z}_7})^2(u^6, v, w) + (u^6, v, w)^2$
7	$\frac{1}{8}k(u)u^4 + \frac{1}{7}v$	$(m^{\mathbb{Z}_7})^6 + (m^{\mathbb{Z}_7})^2(u^6, v, w) + (u^6, v, w)^2$
8	$\frac{1}{8}k u^4 + \frac{1}{6}\hat{\epsilon}u^3 + \frac{1}{8}v$	$(m^{\mathbb{Z}_8})^5 + (m^{\mathbb{Z}_8})^2(u^7, v, w) + (u^7, v, w)^2$
8	$\frac{1}{10}k(u)u^5 + \frac{1}{8}h_8 u^4 + \frac{1}{8}v$	$(m^{\mathbb{Z}_8})^7 + (m^{\mathbb{Z}_8})^3(u^7, v, w) + (u^7, v, w)^2$
≥ 9	$\frac{1}{8}k u^4 + \frac{1}{6}\hat{\epsilon}u^3 + \frac{1}{q}v$	$(m^{\mathbb{Z}_q})^5 + (m^{\mathbb{Z}_q})^2(u^{q-1}, v, w)$ $+(u^{q-1}, v, w)^2$
≥ 9	$\frac{1}{10}k(u)u^5 + \frac{1}{8}\tilde{\epsilon}u^4 + \frac{1}{q}v$	$(m^{\mathbb{Z}_q})^7 + (m^{\mathbb{Z}_q})^3(u^{q-1}, v, w)$ $+(u^{q-1}, v, w)^2$

Table 4.11: Algebraic data for $\mathbb{Z}_q$-invariant potentials, $q \geq 5$

5. Period-3 points of the generalized standard map

As an example to illustrate the theory of Chapters 3 and 4, we are going to look at the bifurcation of period-3 points in the generalized standard map, denoted by **S**. This map is area-preserving and takes the form

$$\begin{aligned} x' &= x + u(y', \Lambda) \,, \\ y' &= y - v(x, \Lambda) \,, \end{aligned} \tag{5.1}$$

where $u, v : \mathsf{R} \times \mathsf{R}^k \to \mathsf{R}$ are smooth functions depending on the (multi)parameter Λ. We assume that $x = y = 0$ is a fixed point of the map for some value Λ_o. Without loss of generality take $\Lambda_o = 0$; then, at such a fixed point,

$$u^\circ = v^\circ = 0 \,.$$

Associated with the map (5.1) is the generating function

$$h(x, x', \Lambda) = U(x - x', \Lambda) - V(x, \Lambda) \,, \tag{5.2}$$

where $U_t(t, \Lambda) = -u^{-1}(-t, \Lambda)$ and $V_t(t, \Lambda) = v(t, \Lambda)$.

If U is an even function note that (5.1) is reversible on configuration space (cf. Appendix F) with reversor κ. We refer to the beginning of Chapter 7 for further discussions of the reversibility of (5.1). An example of non κ-reversible map is the inverse of the Rannou map (Rannou [1974]) defined by:

$$u(t) = \cos t - t - 1 \quad \text{and} \quad v(t, \lambda) = (\lambda - 3)(\cos t - \sin t - 1) \,. \tag{5.3}$$

Without limiting the generality, we can normalize the twist u°_t to 1 by considering positive scalings and $\mathbf{S}^{-1}$ instead of **S** (to change the sign of the twist). And so, setting $a \stackrel{\text{def}}{=} V^\circ_{xx}$, the linearization of (5.1) about the trivial solution is

$$\mathbf{DS}^\circ = \begin{pmatrix} 1 - a & 1 \\ -a & 1 \end{pmatrix} .$$

If the origin is an elliptic fixed point, the multipliers lie at $\mathrm{e}^{\pm i\theta}$, $\theta \in \mathsf{R}$, and a satisfies

$$a = 2(1 - \cos\theta) \,.$$

In this chapter we take $\theta = \frac{2\pi}{3}$ and compute normal forms for the bifurcation equations for the period-3 points as a function of U and V; that is, as a function of the Taylor coefficients of U and V at the origin:

$$\begin{aligned} U(t,0) &= \tfrac{1}{2}t^2 + \tfrac{1}{3}Bt^3 + \tfrac{1}{4}Ct^4 + \tfrac{1}{5}Dt^5 + \tfrac{1}{6}Et^6 + \cdots, \\ V(t,0) &= \tfrac{1}{2}at^2 + \tfrac{1}{3}bt^3 + \tfrac{1}{4}ct^4 + \tfrac{1}{5}dt^5 + \tfrac{1}{6}et^6 + \cdots. \end{aligned} \tag{5.4}$$

In practice U and V may have only finitely many terms (cf. the quadratic map) but based on the singularity theory of Chapter 3 we include all terms that may contribute to finite determinacy. In the simplest cases (the generic theory for example) only the first few Taylor coefficients contribute to the finite determinacy question, but for the higher codimension singularities all 9 Taylor coefficients $a, (b, B) \dots (e, E)$ contribute to the normal form.

5.1 Computations of the bifurcation equation

Without limiting the generality, as in Chapter 2, we can assume that $V_{xx}(0, \Lambda)$ depends on only one bifurcation parameter, $\lambda \in \mathsf{R}$, and is of the form $3 - \lambda$. When $\lambda = 0$ the multipliers are at third roots of unity ($\cos\theta = -\frac{1}{2}$) and on the space $\mathbf{X}_3^1$ of period-3 scalar sequences the action is

$$\begin{aligned} \mathrm{W}_3(\mathbf{x}, \Lambda) &= \sum_{j=1}^{3} h(x_j, x_{j+1}, \Lambda) \\ &= \tfrac{1}{2}\langle \mathbf{x}, \mathbf{L}^o\mathbf{x} \rangle + \tfrac{1}{2}\lambda \|x\|^2 \\ &\qquad + \widehat{U}(x_1 - x_2, \Lambda) + \widehat{U}(x_2 - x_3, \Lambda) + \widehat{U}(x_3 - x_1, \Lambda) - \sum_{j=1}^{3} \widehat{V}(x_j, \Lambda), \end{aligned}$$

where

$$\begin{aligned} \widehat{U}(t,\Lambda) &\stackrel{\text{def}}{=} U(t,\Lambda) - \tfrac{1}{2}t^2 = \tfrac{1}{3}\widetilde{B}(\Lambda)\left(\tfrac{t}{\sqrt{3}}\right)^3 + \tfrac{1}{4}\widetilde{C}(\Lambda)\left(\tfrac{t}{\sqrt{3}}\right)^4 + \cdots, \\ \widehat{V}(t,\Lambda) &\stackrel{\text{def}}{=} V(t,\Lambda) - \tfrac{1}{2}at^2 = \tfrac{1}{3}b(\Lambda)t^3 + \cdots. \end{aligned}$$

Moreover,

$$\mathbf{L}^o = -(\Gamma_3 + \Gamma_3^T + \mathbf{I}_3) = \begin{pmatrix} -1 & -1 & -1 \\ -1 & -1 & -1 \\ -1 & -1 & -1 \end{pmatrix}.$$

From Proposition 2.3, $\sigma(\mathbf{L}^o) = \{0, -3\}$ (with 0 of multiplicity 2) and the eigenvectors are

$$\xi_1 = \tfrac{1}{\sqrt{6}}\begin{pmatrix} 2 \\ -1 \\ -1 \end{pmatrix}, \quad \xi_2 = \tfrac{1}{\sqrt{2}}\begin{pmatrix} 0 \\ 1 \\ -1 \end{pmatrix}, \quad \xi_3 = \tfrac{1}{\sqrt{3}}\begin{pmatrix} 1 \\ 1 \\ 1 \end{pmatrix}.$$

Any $\mathbf{x} \in \mathbf{X}_3^1$ can be expressed as $\mathbf{x} = \chi_1\xi_1 + \chi_2\xi_2 + \Upsilon\,\xi_3$ with $\chi \in \mathsf{R}^2$ and $\Upsilon \in \mathsf{R}$.

The idea is to construct the reduced Z_3-invariant functional

$$\widehat{\mathrm{W}}_3(\chi, \Lambda) = \mathrm{W}_3(\chi_1\xi_1 + \chi_2\xi_2 + \Upsilon(\chi, \Lambda)\,\xi_3, \Lambda) \tag{5.5}$$

given by the Splitting Lemma and apply the singularity theory results for $\mathbf{Z}_3$-equivariant gradient maps.

The function $\Upsilon(\chi, \Lambda)$ is determined by solving the complementary equation

$$\langle \nabla_{\mathbf{x}} W_3(\chi_1\xi_1 + \chi_2\xi_2 + \Upsilon\,\xi_3, \Lambda), \xi_3 \rangle = 0 \tag{5.6}$$

for Υ with $\|\chi\|$ and $|\Lambda|$ sufficiently small. Writing out (5.6) explicitly, we find that

$$(3-\lambda)\Upsilon + \tfrac{1}{\sqrt{3}} \sum_{j=1}^{3} \widehat{V}_t(x_j, \Lambda) = 0. \tag{5.7}$$

In particular, note that (5.7) has the same form when (5.1) is reversible, and so Υ is actually $\mathbf{D}_3$-invariant and a short computation results in

$$\Upsilon = \Upsilon_1 u + \Upsilon_2 v + \Upsilon_3 u^2 + \cdots$$

where $u = \chi_1^2 + \chi_2^2$, $v = \chi_1(\chi_1^2 - 3\chi_2^2)$ are the $\mathbf{D}_3$-invariants and

$$\Upsilon_1 = -\tfrac{1}{3\sqrt{3}}\, b\,, \quad \Upsilon_2 = -\tfrac{1}{9\sqrt{2}}\, c \quad \text{and} \quad \Upsilon_3 = -\tfrac{\sqrt{3}}{9}\left(\tfrac{1}{27}b^3 - \tfrac{1}{3}bc + \tfrac{1}{2}d\right).$$

Substitution of Υ into (5.5) results in

$$\begin{aligned}\widehat{W}_3(\chi, \Lambda) = {} & \tfrac{1}{2}\lambda u + \tfrac{1}{3}w_2\, v + \tfrac{1}{4}w_3\, u^2 + w_4\, uv + \tfrac{1}{6}w_5\, v^2 + \tfrac{1}{6}w_6\, u^3 + \cdots \\ & + w\left(\tfrac{1}{3}\widetilde{w}_2 + w_4\, u + \cdots\right)\end{aligned} \tag{5.8}$$

where

$$\left.\begin{aligned} & w_2 = -\tfrac{1}{\sqrt{6}}\, b\,, \quad w_3 = \tfrac{1}{18}\left(4b^2 - 9c + 9\widetilde{C}\right), \\ & \widetilde{w}_2 = -\tfrac{1}{\sqrt{6}}\, \widetilde{B}\,, \quad \widetilde{w}_4 = -\tfrac{1}{6\sqrt{6}}\, \widetilde{D}\,, \\ & w_4 = \tfrac{1}{18\sqrt{6}}\,(2bc - 3d)\,, \quad w_5 = \tfrac{1}{18}\left(c^2 - e + \widetilde{E}\right) \\ \text{and} \quad & w_6 = \tfrac{2}{243}b^4 - \tfrac{1}{9}b^2c + \tfrac{1}{3}bd - \tfrac{1}{4}e + \tfrac{1}{4}\widetilde{E}\,. \end{aligned}\right\} \tag{5.9}$$

We can now apply the classification results of Chapter 4 to (5.8).

5.2 Analysis of the bifurcation equations

The basic result is the generic case; that is, if $w_2 + i\,\widetilde{w}_2 \neq 0$ (or equivalently if $b^2 + B^2 \neq 0$) then the gradient of (5.8) is equivalent to (see Corollary 4.3(a))

$$g(z, \lambda) = \lambda z + \overline{z}^2 \quad \text{with} \quad z = \chi_1 + i\chi_2\,,$$

which recovers the classic result on period-3 points in Meyer [1970].

From the Stability Lemma I 2.10, the stability is determined from

$$J = \operatorname{sign}\left|\operatorname{Hess}_{\chi} \widehat{W}_3(\chi, \lambda)\right|.$$

For $\lambda z + \bar{z}^2$ a simple calculation shows that $J = -\text{sign}\, u = -1$, recovering the well-known result that generic period 3 points are unstable (Meyer [1971]). In particular the Rannou map in (5.3) satisfies the generic non-degeneracy condition.

Note that when $b + iB \neq 0$, the only further degeneracies that can occur for $\mathbb{Z}_3$-equivariant problems are related to the number of λ-derivatives of $V_{xx}(0, \lambda)$ vanishing at $\lambda = 0$; that is, they are of normal form $\lambda^l z + \bar{z}^2$ for some $l \geq 2$.

A new interesting feature appears when the coefficients w_2 and $\tilde{w}_2$ (or equivalently b and B) go through 0: the period-3 points can stabilize through a turn in the bifurcation branch.

Using Theorem 4.2(b), the first normal form for (5.8) in that direction is

$$(u + \epsilon\lambda)\, z + (v + k\lambda)\, \bar{z}^2. \tag{5.10}$$

The coefficients in the normal form can be expressed in terms of the coefficients in the generating function; we find,

$$\epsilon = \text{sign}\,(\widetilde{C} - c) \quad \text{and} \quad k = \epsilon\sqrt{\tfrac{3}{2}}\,(d + i\,\widetilde{D})\,\psi^{-1}$$

where $\psi \in \mathbb{C}$ is defined by $\psi^2 = (\widetilde{C} - c)\,(c^2 + \widetilde{E} - e) - (d + i\,\widetilde{D})^2$.

The nondegeneracy condition is $\epsilon\cdot\psi\cdot(d + i\,\widetilde{D}) \neq 0$. Stability, given by the sign of $|\text{Hess}_\chi \widehat{W}_3|$, is a rather complicated expression. The bifurcation diagrams that are obtained have been studied in the last chapter (see Figures 4.1, 4.2).

We end this chapter with some remarks on the area of the polygon formed by the bifurcating period-3 points. The area function takes a particularly simple form in the coordinates $\mathbf{x} = \chi_1\xi_1 + \chi_2\xi_2 + \Upsilon\,\xi_3$. Given the ordered coordinates (x_j, y_j) of the group orbit of a family of period-3 points, $A = \frac{1}{2}\sum_{j=1}^{3}(x_{j+1}y_j - x_j y_{j+1})$ gives the oriented area. In the generalized standard map $y_{j+1} = u^{-1}(x_{j+1} - x_j, \Lambda)$, therefore

$$A = \tfrac{1}{2}\sum_{j=1}^{3}(x_{j+1} - x_{j-1})\, u^{-1}(x_j - x_{j-1}, \Lambda).$$

When $q = 3$ we can compute the first terms in the Taylor expansion of A in function of the coefficients of the expansion of $\widehat{U}$:

$$A = \tfrac{3}{4}u + \tfrac{1}{4\sqrt{2}}\widetilde{B}(\Lambda)\, v + \tfrac{1}{8}\widetilde{C}(\Lambda)\, u^2 - \tfrac{1}{4\sqrt{6}}\widetilde{B}(\Lambda)\, w + \cdots.$$

In principle, the parameters $\widetilde{B}$, $\widetilde{C}$ ... are functions of the parameters Λ. Other geometric properties such as the center of gravity of the polygon formed by the orbit of a period-3 point can be computed. Although we don't pursue it here the geometry of period-q points is an interesting area (see Davis [1979] and Schoenberg [1950]).

6. Classification of Dq-Equivariant Gradient Bifurcation Problems

In this chapter the theory of Chapter 3 is used to give a classification of $\mathbf{D}_q$-equivariant gradient bifurcation problems on $\mathbf{R}^2$. $\mathbf{D}_q$-equivariant gradient bifurcation problems arise, in the theory of bifurcating period-q points, when the symplectic map is also reversible (for example see Theorem 2.9 in Chapter 2).

The details of the classification up to topological $\mathcal{K}_\lambda^{\mathbf{D}_q}$-codimension two (that is when there is no gradient structure) is given in Furter [1990,1991] for $q \geq 3$ ($q \neq 4$). Therefore we need only a sketch of the results, with emphasis on the restrictions imposed by the gradient structure.

In Section 6.1 we record a list of the singularities and their unfoldings which we use to study the bifurcation of period-q points in two-parameter families of reversible-symplectic maps in Chapters 7 and 10.

A different basis is used for the set of invariants when $q = 4$ and therefore – in Section 6.2 – we give a separate treatment of this case. The classification for $\mathbf{D}_4$-equivariant germs up to topological codimension 2 – without a gradient structure – can be found in Golubitsky & Roberts [1987], and in Section 6.2 we introduce the modifications required by the gradient structure.

6.1 $\mathbf{D}_q$-normal forms when $q \neq 4$

In Proposition 2.6 it was shown that every parametrized family H of $\mathbf{D}_q$-invariant potentials on $\mathbf{R}^2$ is a function of u, v, the $\mathbf{D}_q$-invariants on $\mathbf{R}^2$, and the parameters Λ: $H(u, v, \Lambda)$. Here we record the details necessary to complete the proof of Theorem 2.9 with one dimensional distinguished parameter λ.

Let $H(u, v, \lambda)$ be a $\mathbf{D}_q$-invariant potential ($q \neq 4$). If $H_v^o \cdot H_{u\lambda}^o \neq 0$ (and $H_{uu}^o \neq 0$ when $q \geq 5$), then $\nabla_z H$ is $\mathcal{K}_\lambda^{\mathbf{D}_q}$-contact equivalent to the gradient of

$$\tfrac{1}{2}\epsilon_1 \lambda u + \tfrac{1}{3}\delta_1 v\,, \qquad q = 3\,,$$

$$\tfrac{1}{2}\epsilon_1 \lambda u + \tfrac{1}{4}\epsilon_2 u^2 + \tfrac{1}{q}\delta_1 v\,, \qquad q \geq 5\,,$$

with $\epsilon_1 = \operatorname{sign} H_{u\lambda}^o$, $\epsilon_2 = \operatorname{sign} H_{uu}^o$ and $\delta_1 = \operatorname{sign} H_v^o$.

These two potentials correspond to the generic situation (codimension 0) and are therefore their own universal unfolding.

In Chapter 10 we consider the collision of multipliers at rational points in reversible-symplectic maps. After application of the Splitting Lemma we reduce the analysis of bifurcating period-q points to a $\mathbf{D}_q$-equivariant gradient map on $\mathbf{R}^2$ with the collision singularity. For the $\mathbf{D}_q$-invariant potential H, the collision singularity corresponds to the generic case satisfying $H^o_{u\lambda} = 0$.

Suppose that $H^o_{u\lambda} = 0$ but that $H^o_v \cdot H^o_{u\lambda\lambda} \neq 0$ (and $H^o_{uu} \neq 0$ when $q \geq 5$), then $\nabla_z H$ is $\mathcal{K}_\lambda^{\mathbf{D}_q}$-contact equivalent to the gradient of

$$\tfrac{1}{2}\epsilon_1 \lambda^2 u + \tfrac{1}{3}\delta_1 v\,, \qquad q = 3$$

$$\tfrac{1}{2}\epsilon_1 \lambda^2 u + \tfrac{1}{4}\epsilon_2 u^2 + \tfrac{1}{q}\delta_1 v\,, \qquad q \geq 5$$

with $\epsilon_1 = \operatorname{sign} H^o_{u\lambda\lambda}$, $\epsilon_2 = \operatorname{sign} H^o_{uu}$ and $\delta_1 = \operatorname{sign} H^o_v$.

A $\mathbf{D}_q$-equivariant gradient map with collision singularity is of $\mathcal{K}_\lambda^{\mathbf{D}_q}$-codimension 1 and its universal unfolding is given by

$$\widehat{H}(u,v,\lambda,\alpha) = H(u,v,\lambda) + \tfrac{1}{2}\alpha\, u\,.$$

In Chapter 7 we also discuss the effect of various other singularities that arise in the bifurcation of period-q points in two-parameter reversible maps. These singularities can be found in the classification up to topological codimension 2 (cf. Furter [1990,1991]).

6.2 $\mathbf{D}_4$-invariant potentials with a distinguished parameter path

In this section we use the path formulation of Section 3.3 to give a classification of $\mathbf{D}_4$-equivariant gradient bifurcation problems.

Let $z = x + iy$ then the $\mathbf{D}_4$-invariants on $\mathbf{R}^2$ are

$$N = z\bar{z} \quad \text{and} \quad \Delta = \delta^2 \quad \text{where} \quad \delta = -\operatorname{Re} z^2.$$

Given a $\mathbf{D}_4$-invariant potential dependent on a parameter λ: H, the corresponding gradient takes the form

$$g(z) = 2\, H_N\, z - 4\, H_\Delta\, \delta \bar{z}\,. \tag{6.1}$$

Note that a general $\mathbf{D}_4$-equivariant map has the form

$$\hat{p}(N,\Delta,\lambda)\, z + \hat{q}(N,\Delta,\lambda)\, \delta\bar{z}\,,$$

where $\hat{p}$ and $\hat{q}$ are real $\mathbf{D}_4$-invariant functions, and the gradient structure forces the following relation

$$2\,\hat{p}_\Delta + \hat{q}_N \equiv 0\,.$$

For the general class of $\mathbf{D}_4$-equivariant maps on $\mathbf{R}^2$, Golubitsky & Roberts obtain the classification up to topological codimension 2 using $\mathcal{K}_\lambda^{\mathbf{D}_4}$-equivalence. They find 15 normal forms families (some with moduli).

Proposition 6.1. *For each of the* 15 *normal forms families (excluding the case X) of topological codimension less than or equal to 2 in the Golubitsky & Roberts* *classification, the $\mathcal{K}_\lambda^{\mathbf{D}_4}$-orbit of each normal form contains a gradient. Moreover, their gradient $\mathcal{K}_\lambda^{\mathbf{D}_4}$-codimension is equal to the $\mathcal{K}_\lambda^{\mathbf{D}_4}$-codimension.*

For normal form X, the potential is of $\mathcal{R}^{\mathbf{D}_4}$-codimension 3 (including the modulus) with unfolding

$$H = \tfrac{1}{6}mN^3 - \tfrac{1}{4}\epsilon_1 N\Delta - \tfrac{1}{8}\epsilon_2\Delta^2 + \tfrac{1}{2}\gamma N + \tfrac{1}{4}\mu N^2 - \tfrac{1}{4}\kappa\Delta, \quad \epsilon_1 m \in \mathbf{R}\setminus\{0,1,\tfrac{3}{2}\}. \tag{6.2}$$

where $m \in \mathbf{R}$ is a modulus, $\epsilon_1, \epsilon_2 = \pm 1$ and $(\gamma, \mu, \kappa) \in \mathbf{R}^3$ are the unfolding parameters. Moreover, there exists a two-parameter unfolding path

$$\Phi(\lambda, \alpha, \beta) = (\epsilon_0\lambda, \alpha, \beta) \in (\gamma, \mu, \kappa).$$

This result means that the classification for $\mathbf{D}_4$-equivariant gradient maps is roughly identical to the $\mathcal{K}_\lambda^{\mathbf{D}_4}$-classification. On a practical level, the only difference is that in the case X the modulus n (in the $\mathcal{K}_\lambda^{\mathbf{D}_4}$-classification) is forced by the gradient structure to take the value $-\frac{1}{2}\epsilon_1$. However, to be precise there are changes that must be made to the normal forms (scalings and different unfoldings) to account for the gradient structure.

Using the path formulation we list the potentials and paths corresponding to the universal unfoldings up to topological codimension 2 (with parameters m (modal), λ, (α, β, γ)). The codimensions in the list refer to the $\mathcal{A}^{\mathbf{D}_4}$-codimension of the associated potentials (with parameters $\tilde{m}$ (modal), (ν, μ, κ)). The case numbers correspond to the $\mathcal{K}_\lambda^{\mathbf{D}_4}$-classification of Golubitsky & Roberts (a * indicates that the $\mathcal{K}_\lambda^{\mathbf{D}_4}$ normal form had to be altered to fit the gradient structure and a + indicates that the normal form contains moduli parameters, that is, that the C^∞-codimension is strictly bigger).

cod 1_+ $\quad \frac{1}{4}\tilde{m}N^2 - \frac{1}{4}\epsilon_1\Delta + \frac{1}{2}\nu N$

$$\mathrm{I}_+ : (m, \epsilon_o\lambda)$$
$$\mathrm{V}_+^* : (m + \epsilon_2\lambda, \epsilon_o\lambda^2 + \alpha)$$
$$\mathrm{XIV}_+^* : (m + \epsilon_2\lambda, \epsilon_o\lambda^2 + \alpha + \beta\lambda)$$
$$\mathrm{XV}_+ : (m + \beta\lambda, \epsilon_o\lambda^2 + \alpha)$$

cod 2 $\quad \frac{1}{4}\epsilon_1 N^2 - \frac{1}{8}\epsilon_2\Delta^2 + \frac{1}{2}\nu N - \frac{1}{4}\mu\Delta$

$$\mathrm{IV}_+^* : (\epsilon_o\lambda, m\lambda + \alpha)$$
$$\mathrm{IX}_+ : (\epsilon_o\lambda, m\lambda^2 + \alpha + \beta\lambda)$$
$$\mathrm{XIII}_+^* : (m\lambda^2 + \alpha, \epsilon_o\lambda + \beta)$$

$$\tfrac{1}{6}\,\epsilon_1 N^3 - \tfrac{1}{4}\,\epsilon_2\,\Delta + \tfrac{1}{2}\,\nu N + \tfrac{1}{4}\,\mu N^2$$

II $: (\,\epsilon_o\lambda\,,\,\alpha\,)$

XI$_+$ $: (\,\epsilon_o\,\lambda^2 + \alpha\,,\,m\,\lambda + \beta\,)$

$$\tfrac{1}{6}\,\epsilon_1 N^3 + \tfrac{1}{4}\epsilon_1 N^2 - \tfrac{1}{4}\,\epsilon_1\,\Delta + \tfrac{1}{2}\,\nu N + \tfrac{1}{4}\,\mu N^2$$

III $: (\,\epsilon_o\lambda\,,\,\alpha\,)$

XII$_+$ $: (\,\epsilon_o\,\lambda^2 + \alpha\,,\,m\,\lambda + \beta\,)$

cod 3/3$_+$ $\quad \tfrac{1}{4}\,\epsilon_1 N^2 - \tfrac{1}{12}\,\epsilon_2\,\Delta^3 + \tfrac{1}{2}\,\nu N - \tfrac{1}{4}\,\mu\Delta - \tfrac{1}{8}\,\kappa\,\Delta^2$

VIII$^*_+$ $: (\,m\,\lambda\,,\,\epsilon_o\lambda + \alpha\,,\,\beta\,)$

$$\tfrac{1}{8}\,\epsilon_1 N^4 - \tfrac{1}{4}\,\epsilon_2\,\Delta + \tfrac{1}{2}\,\nu N + \tfrac{1}{4}\,\mu N^2 + \tfrac{1}{6}\,\kappa N^3$$

VI $: (\,\epsilon_o\lambda\,,\,\alpha\,,\,\beta\,)$

$$\tfrac{1}{8}\,\epsilon_2 N^4 + \tfrac{1}{4}\,\epsilon_1 N^2 - \tfrac{1}{4}\,\epsilon_1\,\Delta + \tfrac{1}{2}\,\nu N + \tfrac{1}{4}\,\mu N^2 + \tfrac{1}{6}\,\kappa N^3$$

VII $: (\,\epsilon_o\lambda\,,\,\alpha\,,\,\beta\,)$

$$\tfrac{1}{6}\,\tilde{m}\,N^3 - \tfrac{1}{4}\,\epsilon_1 N\Delta - \tfrac{1}{8}\,\epsilon_2\,\Delta^2 + \tfrac{1}{2}\,\nu N + \tfrac{1}{4}\,\mu N^2 - \tfrac{1}{4}\,\kappa\,\Delta$$

X$^*_+$ $: (\,m\,,\,\epsilon_o\lambda\,,\,\alpha\,,\,\beta\,)$

More detailed results on the problems of topological codimension 1 and case X are given in Table 6.1, with the defining conditions given in Table 6.2. These are the most important normal forms for studying bifurcation of period-4 points in reversible-symplectic maps.

The normal form X is a two moduli (m, n)-family in the non-gradient setting. The gradient structure fixes the second modulus:

$$n = -\tfrac{1}{2}\operatorname{sign} H^o_{N\Delta}\,.$$

With the reduction to one modulus in the normal form a complete classification of the bifurcation diagrams is now much simpler to achieve. Differentiating (6.2) we

get the normal form X whose branches of solutions, along with the eigenvalues of the Hessian of H, are given by

$$\begin{cases} \text{R branch}: & -\epsilon_0\lambda = (\alpha-\beta)\,x^2 + (m - \frac{3}{2}\epsilon_1)\,x^4 - \epsilon_2\,x^6, \quad y=0 \\ \text{Eigenvalues}: & (\beta + \epsilon_1\,x^2 + \cdots,\ (\alpha-\beta) + (2m - 3\epsilon_1)\,x^2 + \cdots) \end{cases}$$

$$\begin{cases} \text{S branch}: & -\epsilon_0\lambda = 2\alpha\,x^2 + 4m\,x^4, \quad x = y \\ \text{Eigenvalues}: & (-\beta - 2\epsilon_1 x^2 + \cdots,\ \alpha + 4m\,x^2) \end{cases}$$

$$\begin{cases} \text{T branch}: & -\epsilon_0\lambda = \alpha N + mN^2 - \frac{1}{2}\epsilon_1\Delta, \quad \beta + \epsilon_1 N + \epsilon_2\Delta = 0 \\ \text{Eigenvalues}: & |\text{Hess}_z H| < 0 \quad (\text{always}). \end{cases}$$

For the T-branch, note that the gradient structure forces the eigenvalues of the Hessian (the Jacobian of the normal form) to be of opposite sign,

$$\text{sign}\,|\text{Hess}_z H(z,\Lambda)| = -\tfrac{1}{2} - \epsilon_2\,\alpha - 2\epsilon_2 m\,N < 0$$

for $(N, \lambda, \alpha, \beta)$ sufficiently small. The T-branch corresponds to the secondary branch of solutions with trivial isotropy subgroup (orbits without symmetry). These branches are always unstable in the normal form X for the gradient case.

The transition varieties in the unfolding space (α, β) for the normal form X are given in Table 6.3 (using notation of Golubitsky & Roberts). Note that the transition varieties are functions of $\epsilon_1 m$, that is, the sign of ϵ_1 does not enter independently. And so we fix $\epsilon_1 = +1$.

The respective positions of the transition varieties are important to describe the diagrams of (6.2) and these positions are functions of m. There are 11 regions in the m-space and the properties of the bifurcation diagrams in each region are topologically equivalent.

The regions are separated by the following values of m:

$$\left\{ -\tfrac{1}{2}, 0, \tfrac{1}{4}(3-2\sqrt{2}), \tfrac{1}{2}(3-\sqrt{5}), \tfrac{1}{2}, 1, \tfrac{1}{4}(3+\sqrt{5}), \tfrac{1}{4}(3+2\sqrt{2}), \tfrac{3}{2}, 2 \right\}.$$

The points 0, 1 and $\frac{3}{2}$ are the only values of m affecting the *recognition* problem, all the other distinguished values of m are associated with the *universal* unfolding.

There is a "symmetry" with respect to m in the behavior of the bifurcation diagrams. For example, the difference in the diagrams for $m < -\frac{1}{2}$ and $m > 2$ is simply the exchange of label between the R and S branches. Similarly for $-\frac{1}{2} < m < 0$ and $\frac{3}{2} < m < 2$ and so on.

Although the example of normal form X treated in Golubitsky & Roberts is not a gradient, its diagrams can be modified for the region $0 < m < \frac{3}{2}$; that is, only the unfolding must be reorganized. Hence we only give here the bifurcation diagrams for $m < \frac{1}{2}$ (Figure 6.1) and $-\frac{1}{2} < m < 0$ (Figure 6.2).

It is the normal form V that is important for the collision of multipliers at $\pm i$ in reversible-symplectic maps. The classification in Tables 6.1 and 6.2 justifies the result stated in Chapter 10.

The normal forms I-IV are of interest for the bifurcation of generic period-4 points (normal form I) and for the simple codimension one singularities of bifurcating period-4 points (normal forms II-IV) in reversible-symplectic maps.

The normal form X corresponds to a singularity of codimension 2 in the bifurcation of period-4 points. Although it requires families of reversible-symplectic maps with three (at least) parameters (including the bifurcation parameter), it is not too difficult to construct examples of such a family. The unfolding of the codimension 2 singularity provides interesting insight into *global* behavior in the bifurcation of periodic points as it explains how some secondary bifurcations occur.

Case	Potential	Path	$C^0(C^\infty)$ cod_∇
I	$\frac{1}{4}\widehat{m}\,N^2 - \frac{1}{4}\epsilon_1\Delta + \frac{1}{2}\nu N$	$(m\,,\epsilon_0\lambda)$	0(1)
II	$\frac{1}{6}\epsilon_1 N^3 - \frac{1}{4}\epsilon_2\Delta + \frac{1}{2}\nu N + \frac{1}{q}\mu N^2$	$(\epsilon_0\lambda\,,\alpha)$	1(1)
III	$\frac{1}{6}\epsilon_2 N^3 + \frac{1}{4}\epsilon_1 N^2 - \frac{1}{4}\epsilon_1\Delta$ $+\frac{1}{2}\nu N + \frac{1}{4}\mu N^2$	$(\epsilon_0\lambda\,,\alpha)$	1(1)
IV	$\frac{1}{4}\epsilon_1 N^2 - \frac{1}{8}\epsilon_2\Delta^2 + \frac{1}{2}\nu N - \frac{1}{4}\mu\Delta$	$(\epsilon_0\lambda\,,m\lambda+\alpha)$	1(2)
V	$\frac{1}{4}\widehat{m}\,N^2 - \frac{1}{4}\epsilon_1\Delta + \frac{1}{2}\nu N$	$(m+\epsilon_2\lambda\,,\epsilon_0\lambda^2+\alpha)$	1(2)
X	$\frac{1}{6}\widehat{m}\,N^3 - \frac{1}{4}\epsilon_1 N\Delta - \frac{1}{8}\epsilon_2\Delta^2$ $+\frac{1}{2}\nu N + \frac{1}{4}\mu N^2 - \frac{1}{4}\kappa\Delta$	$(m\,,\epsilon_0\lambda\,,\alpha\,,\beta)$	2(3)

Table 6.1: D_4-invariant potentials and unfoldings up to top-cod 1 and case X

Case	Defining Condition	Nondegeneracy Conditions	Coefficients in the Normal Form
I	$\emptyset$	$H^o_{NN}H^o_{\Delta}H^o_{N\lambda} \neq 0$ $H^o_{NN} + 2\,H^o_{\Delta} \neq 0$	$\epsilon_0 = \mathrm{sign}\,H^o_{N\lambda}\ ,\ \epsilon_1 = -\mathrm{sign}\,H^o_{\Delta}$ $m = \frac{1}{2}\,H^o_{NN}\,\lvert H^o_{\Delta}\rvert^{-1}$
II	$H^o_{NN} = 0$	$H^o_{NNN}H^o_{\Delta}H^o_{N\lambda} \neq 0$	$\epsilon_0 = \mathrm{sign}\,H^o_{N\lambda}\ ,\ \epsilon_1 = \mathrm{sign}\,H^o_{NNN}$ $\epsilon_2 = -\mathrm{sign}\,H^o_{\Delta}$
III	$H^o_{NN}+2H^o_{\Delta} = 0$	$H^o_{NN}H^o_{N\lambda} \neq 0$ $H^o_{NNN} + 6\,H^o_{N\Delta} \neq 0$	$\epsilon_0 = \mathrm{sign}\,H^o_{N\lambda}\ ,\ \epsilon_1 = \mathrm{sign}\,H^o_{NN}$ $\epsilon_2 = \mathrm{sign}\,(H^o_{NNN} + 6\,H^o_{N\Delta})$
IV	$H^o_{\Delta} = 0$	$H^o_{NN}H^o_{N\Delta} \neq 0$ $H^{o2}_{N\Delta} - H^o_{NN}H^o_{\Delta\Delta} \neq 0$ $H^o_{NN}H^o_{\Delta\lambda} - H^o_{N\Delta}H^o_{N\lambda} \neq 0$	$\epsilon_0 = \mathrm{sign}\,H^o_{N\lambda}\ ,\ \epsilon_1 = \mathrm{sign}\,H^o_{NN}$ $\epsilon_2 = \epsilon_1\,\mathrm{sign}(H^{o2}_{N\Delta} - H^o_{NN}H^o_{\Delta\Delta})$ $m = \frac{\sqrt{2}\,\epsilon_1}{\lvert H^o_{N\lambda}\rvert}\,\frac{(H^o_{NN}H^o_{\Delta\lambda} - H^o_{N\Delta}H^o_{N\lambda})}{\lvert H^{o2}_{N\Delta} - H^o_{NN}H^o_{\Delta\Delta}\rvert^{\frac{1}{2}}}$
V	$H^o_{N\lambda} = 0$	$H^o_{N\lambda\lambda}H^o_{NN}H^o_{\Delta} \neq 0$ $H^o_{NN} + 2\,H^o_{\Delta} \neq 0$ $H^o_{NN}H^o_{\Delta\lambda} - H^o_{\Delta}H^o_{NN\lambda} \neq 0$	$\epsilon_0 = \mathrm{sign}\,H^o_{N\lambda\lambda}\ ,\ \epsilon_1 = -\mathrm{sign}\,H^o_{\Delta}$ $\epsilon_2 = \epsilon_1\,\mathrm{sign}(H^o_{NN}H^o_{\Delta\lambda} - H^o_{\Delta}H^o_{NN\lambda})$ $m = \frac{1}{2}\,H^o_{NN}\,\lvert H^o_{\Delta}\rvert^{-1}$
X	$H^o_{NN} = 0$ $H^o_{\Delta} = 0$	$H^o_{NNN}H^o_{N\Delta}H^o_{N\lambda}\xi^* \neq 0$ $H^o_{NNN} + 6\,H^o_{N\Delta} \neq 0$ $H^o_{NNN} + 4\,H^o_{N\Delta} \neq 0$	$\epsilon_0 = \mathrm{sign}\,H^o_{N\lambda}\ ,\ \epsilon_1 = -\mathrm{sign}\,H^o_{N\Delta}$ $\epsilon_2 = \epsilon_0\,\mathrm{sign}\,\xi$ $m = \frac{1}{4}\,H^o_{NNN}\,\lvert H^o_{N\Delta}\rvert^{-1}$

$$^*\xi = H^o_{N\Delta}H^o_{\Delta\lambda} - H^o_{N\lambda}H^o_{\Delta\Delta} + \frac{2\,H^o_{N\Delta}}{H^o_{NNN}\,(H^o_{NNN} + 4\,H^o_{N\Delta})} \cdot \left\{H^o_{NNNN}H^o_{N\Delta}H^o_{N\lambda} + H^{o2}_{NNN}H^o_{\Delta\lambda} - 2H^o_{NNN}H^o_{NN\Delta}H^o_{N\lambda} + 4\,H^o_{NNN}H^o_{N\Delta}H^o_{\Delta\lambda}\right\}$$

Table 6.2: Solution of the recognition problem for the normal forms of Table 6.1

Type	Equation (Conditions)	Description
S_{II}	$\alpha = 0$	Case II at the origin
S_{III}	$\alpha = \beta$	Case III at the origin
S_{IV}	$\beta = 0$	Case IV at the origin
Q_R	$\alpha = 2(\epsilon_1 m - 1)\beta + \ldots$ with $\epsilon_1\beta < 0$	Coalescence of a fold and a symmetry breaking point on a R branch
Q_S	$\alpha = 2\epsilon_1 m\beta$ with $\epsilon_1\beta < 0$	Coalescence of a fold and a symmetry breaking point on a S branch
$\mathcal{D}(S_I, \mathcal{P}_R)$	$\alpha = (\epsilon_1 m - \frac{1}{2})\beta + \ldots$ with $\epsilon_1\beta < 0$	Symmetry breaking bifurcation from a R branch at $\lambda = 0$
$\mathcal{D}(S_I, \mathcal{P}_S)$	$\alpha = \epsilon_1 m\beta$ with $\epsilon_1\beta < 0$	Symmetry breaking bifurcation from a S branch at $\lambda = 0$
$\mathcal{D}(\mathcal{F}_R, \mathcal{F}_S)$	$\alpha = \frac{1}{12}[\epsilon_1 m \pm \sqrt{2\epsilon_1 m(2\epsilon_1 m - 3)}]\beta + \ldots$ with $\alpha m < 0$, $(\alpha - \beta)m < 0$ $\epsilon_1 m(2\epsilon_1 m - 3) > 0$	Fold points on R and S branches at the same value of λ
$\mathcal{D}(\mathcal{F}_R, \mathcal{P}_S)$	$\alpha = [2(\epsilon_1 m - 1) \pm \sqrt{3 - 2\epsilon_1 m}]\beta + \ldots$ with $\epsilon_1\beta < 0$, $\epsilon_1(\alpha - \beta) > 0$ $3 - 2\epsilon_1 m > 0$	Fold on a R branch and symmetry breaking bifurcation from a S branch at the same value of λ
$\mathcal{D}(\mathcal{F}_S, \mathcal{P}_R)$	$\alpha = [2\epsilon_1 m \pm \sqrt{2\epsilon_1 m}]\beta + \ldots$ with $\epsilon_1\beta < 0$, $\epsilon_1\alpha < 0$ $\epsilon_1 m > 0$	Fold on a S branch and symmetry breaking bifurcation from a R branch at the same value of λ

Table 6.3: Transition varieties for normal form X

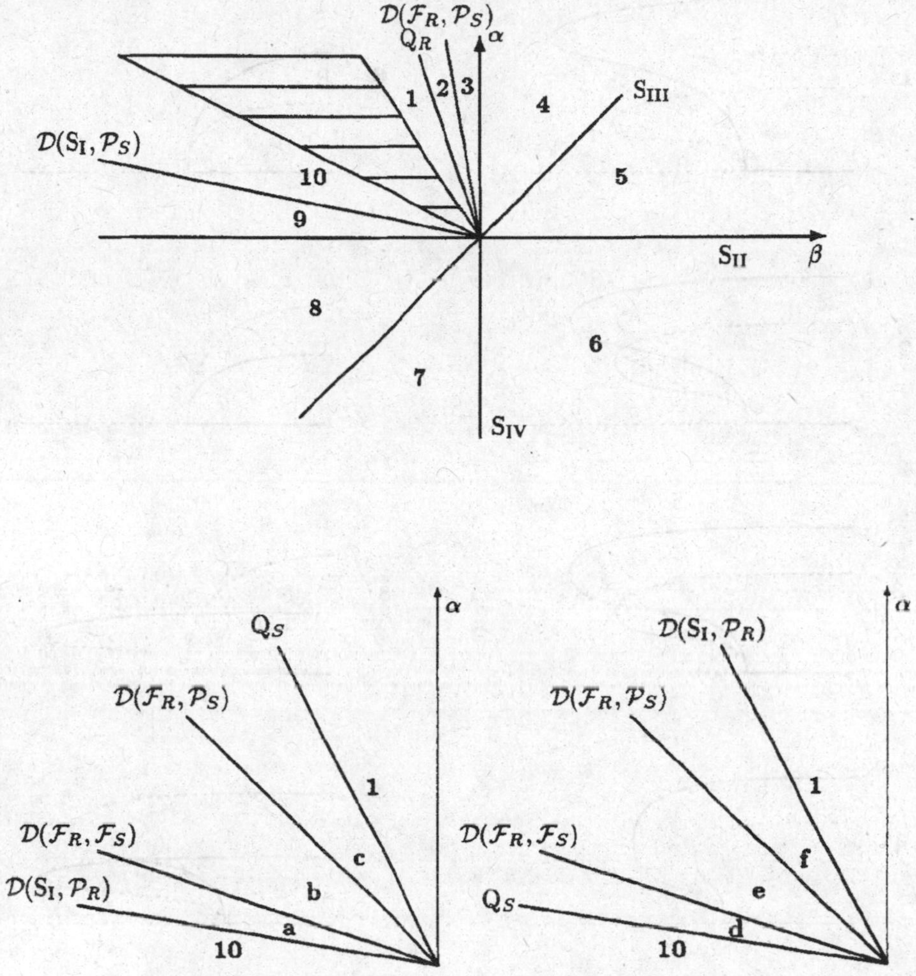

Shaded region for $m < -\frac{1}{2}$ Shaded region for $-\frac{1}{2} < m < 0$

Figure 6.1: Transition varieties for the case X

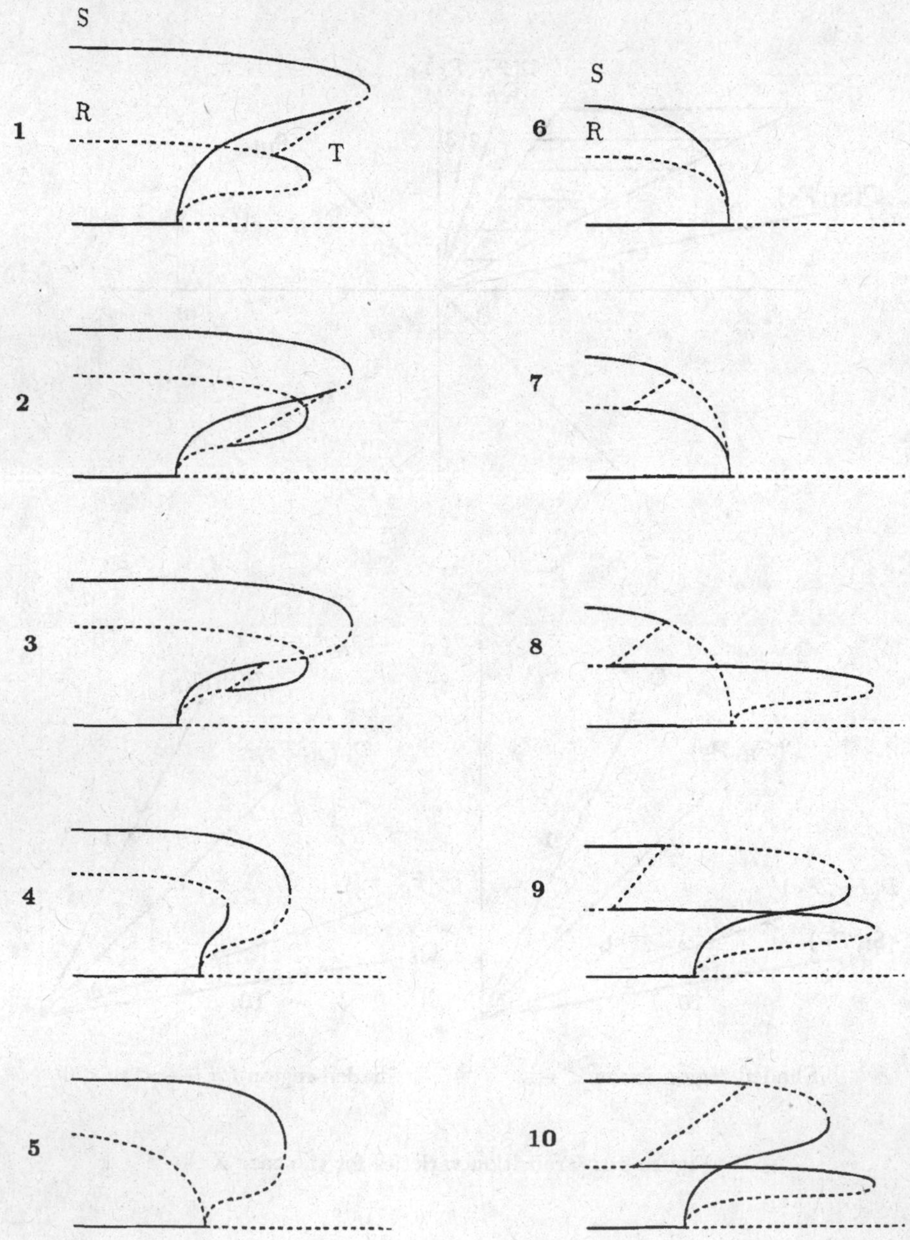

Figure 6.2: Common bifurcation diagrams for Figure 6.1

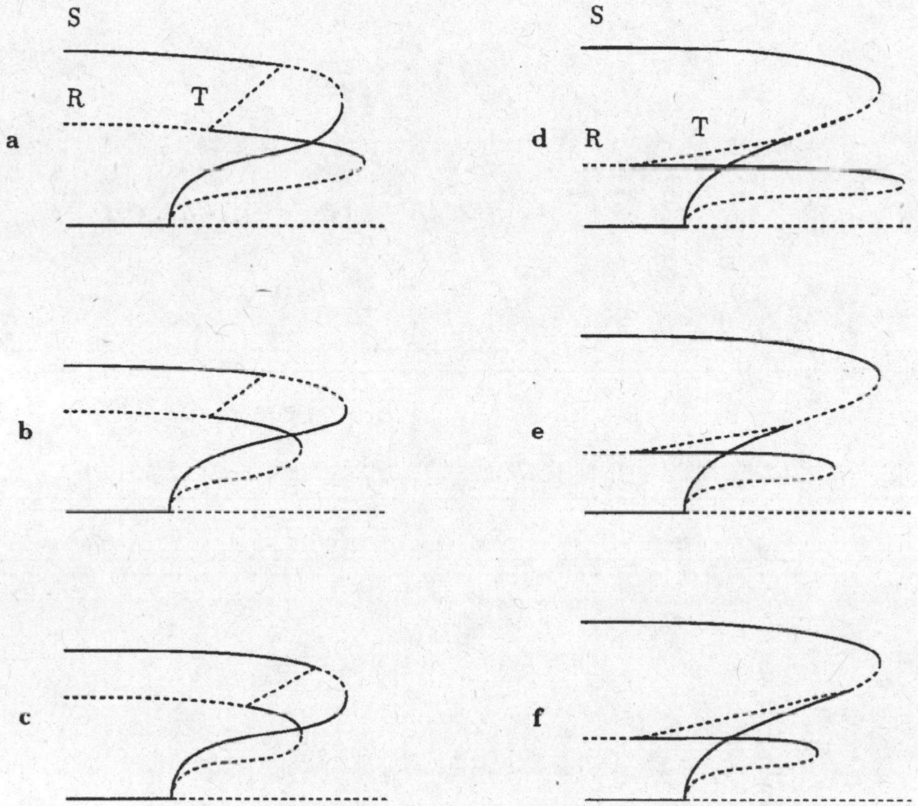

Figure 6.3: Bifurcation diagrams for Figure 6.1

7. Reversibility and degenerate bifurcation of period-q points of multiparameter maps

The theory of Chapters 3 and 6 is used to classify the singular bifurcations of period-q points ($q = 3, 4, 5$) for a class of maps based on the generalized standard map. These bifurcations will also occur in higher dimensional maps – to which the theory also applies – but the simplest context is the area-preserving case. Singularities that require a higher dimensional phase space are considered in Chapters 9 and 10.

Actually, all that is necessary to apply in a straightforward manner our $\mathbf{D}_q$-theory (that is in addition to the natural $\mathbf{Z}_q$-symmetry there is a non-trivial involution of the required type resulting in a $\mathbf{D}_q$-symmetry) are CS-reversible maps. CS-reversible maps are symplectic maps that are reversible in configuration space only (hence their name, see Appendix F). However, we show in Proposition F.4 that CS-reversibility is in fact *equivalent* to the standard definition of reversibility (that is anti-symplectic involution on the phase space in standard form: $\text{diag}(\mathbf{I}_n, -\mathbf{I}_n) = \kappa \otimes \mathbf{I}_n$, cf. (2.23)) modulo symplectic changes of coordinates. Therefore, without loss of generality, we consider here a family of maps (7.1) that are *not* κ-reversible but easily CS-reversible with respect to our classical action of $\mathbf{D}_q$.

In Section 7.1, the generic bifurcation of period-3 points in CS-reversible symplectic maps is considered. From the theory in Chapter 6 we know that the normal form is in 1-1 correspondence with the potential $H(u, v, \lambda) = \frac{1}{2}\epsilon_1 \lambda u + \frac{1}{3}\delta_1 v$ (see Section 6.1) *provided* $H_v^o \cdot H_{u\lambda}^o \neq 0$ (in Chapter 9 we consider the singularity $H_{u\lambda}^o = 0$).

In Sections 7.1 and 7.2 we study the effect of the singularity $H_v^o = 0$ on the bifurcation of period-3 and period-4 points, respectively. In Section 7.3 we carry out the computations for the generic bifurcation of period-5 points.

The theory is applied to a class of maps that is a special case of the generalized standard map (cf. Chapter 5):

$$\begin{aligned} x' &= x + y', \\ y' &= y - V_x(x, \Lambda), \end{aligned} \tag{7.1}$$

where $V : \mathbf{R} \times \mathbf{R}^k \to \mathbf{R}$ is a smooth function depending on parameters $\Lambda \in \mathbf{R}^k$. The origin is a fixed point of (7.1) if we assume that for some value $\Lambda = 0$

$$V^o = V_x^o = 0.$$

Associated with the map (7.1) is the generating function

$$h(x, x', \Lambda) = \tfrac{1}{2}(x - x')^2 - V(x, \Lambda). \tag{7.2}$$

Let $a = V^o_{xx}$ then the linearization of (7.1) about the trivial solution is

$$\mathbf{DT}^o = \begin{pmatrix} 1-a & 1 \\ -a & 1 \end{pmatrix}.$$

If the origin is an elliptic fixed point the multipliers lie at $e^{\pm i\theta}$, $\theta \in \mathbf{R}$, and a satisfies

$$a = 2(1 - \cos\theta).$$

The map (7.1) is reversible with the anti-symplectic reversor

$$\mathbf{R} = \begin{pmatrix} 1 & -1 \\ 0 & -1 \end{pmatrix} \qquad (\text{that is} \quad \mathbf{R}^T\mathbf{J}\mathbf{R} = -\mathbf{J}).$$

It is also CS-reversible as the action W_q is $\mathbf{D}_q$-invariant on $\mathbf{X}^1_q$ when h is given by (7.2). The group $\mathbf{D}_q = \langle \Gamma_q, \mathcal{K}_q \rangle$ with Γ_q the fundamental circulant matrix and $\mathcal{K}_q$ the involution defined in (2.21) in Chapter 2.

With h given by (7.2) the action is

$$\begin{aligned} \mathrm{W}_q(\mathbf{x}, \Lambda) &= \sum_{i=1}^{q} \left[\tfrac{1}{2}(x^i - x^{i+1})^2 - V(x^i, \Lambda)\right] = \sum_{i=1}^{q} \left[x^i x^i - x^i x^{i+1} - V(x^i, \Lambda)\right] \\ &= \|\mathbf{x}\|^2 - \tfrac{1}{2}\langle \mathbf{x}, (\Gamma_q + \Gamma_q^T)\mathbf{x} \rangle - \sum_{i=1}^{q} V(x^i, \Lambda). \end{aligned}$$

The nonlinear term $\sum_{i=1}^{q} V(x^i, \Lambda)$ is invariant under any permutation of $\mathbf{x} \in \mathbf{X}^1_q$ and the norm $\|\mathbf{x}\|$ is invariant with respect to the orthogonal group with contains Γ_q and $\mathcal{K}_q$. It remains to verify that the term $\langle \mathbf{x}, (\Gamma_q + \Gamma_q^T)\mathbf{x} \rangle$ is $\mathbf{Z}^{\kappa}_2$-invariant. But this follows from the identity $\mathcal{K}_q \Gamma_q \mathcal{K}_q = \Gamma_q^T$. By the Equivariant Splitting Lemma the $\mathbf{D}_q$-invariance of W_q is inherited by the reduced functional $\widehat{\mathrm{W}}_q$ on $\mathbf{R}^2$, in particular, the normal form for the bifurcating period-q points of the map (7.1) is a $\mathbf{D}_q$-equivariant gradient map on $\mathbf{R}^2$.

In this chapter we take $\theta = \frac{2\pi}{q}$ with $q = 3, 4$ and 5 and compute normal forms for the bifurcation equations for period-q points as a function of V; that is, as a function of the Taylor coefficients $a, b, c \ldots$ of V at the origin:

$$V(x, 0) = \tfrac{1}{2}a x^2 + \tfrac{1}{3}b x^3 + \tfrac{1}{4}c x^4 + \tfrac{1}{5}d x^5 + \tfrac{1}{6}e x^6 + \tfrac{1}{7}f x^7 + \tfrac{1}{8}g x^8 + \cdots. \tag{7.3}$$

Based on the singularity theory of Chapter 6 we include all terms that may contribute to finite determinacy. In the simplest cases (the generic theory for example) only the first few Taylor coefficients contribute to the finite determinacy question but for the higher codimension singularities (cod-1 bifurcation of period-4 points for example) all the Taylor coefficients $a \ldots g$ may contribute to the normal form.

7.1 Period-3 points with reversibility in multiparameter maps

As we are in a special case of (5.1), we can use the results for period-3 points we derived in Chapter 4.

In that case, with $V_{xx}(0,\lambda) = 3-\lambda$, we get the following reduced functional:

$$\widehat{W}_3(\chi,\lambda) = \tfrac{1}{2}\lambda u + \tfrac{1}{3}w_2 v + \tfrac{1}{4}w_3 u^2 + w_4 uv + \tfrac{1}{6}w_5 v^2 + \tfrac{1}{6}w_6 u^3 + \cdots$$

where

$$\left.\begin{aligned} w_2 &= -\tfrac{1}{\sqrt{6}}b, \quad w_3 = \tfrac{1}{18}(4b^2-9c), \\ w_4 &= \tfrac{1}{18\sqrt{6}}(2bc-3d), \quad w_5 = \tfrac{1}{18}(c^2-e) \\ \text{and} \quad w_6 &= \tfrac{2}{243}b^4 - \tfrac{1}{9}b^2c + \tfrac{1}{3}bd - \tfrac{1}{4}e. \end{aligned}\right\} \tag{7.4}$$

Using complex notation ($z = \chi_1 + i\chi_2$) the $\mathbf{D}_3$-equivariant gradient map associated with $\widehat{W}_3$ is

$$g(z,\lambda) = (\lambda + w_3 u + 2w_4 v + w_6 u^2 + \cdots)\,z + (w_2 + 3w_4 u + w_5 v + \cdots)\,\bar{z}^2 \tag{7.5}$$

to which the $\mathbf{D}_3$-classification of Section 6.1 is applicable.

The basic result is the generic case; that is, if $w_2 \neq 0$ (or equivalently $b \neq 0$) then (7.5) is equivalent to

$$g(z,\lambda) = \lambda z + \epsilon\,\bar{z}^2,$$

where $\epsilon = -\operatorname{sign} b$, which recovers the classic result on period-3 points (Meyer [1970]).

Using the Stability Lemma I 2.10, the stability is determined from

$$J = \operatorname{sign}|\mathrm{Hess}_\chi \widehat{W}_3(\chi,\lambda)|.$$

For the normal form $\lambda z + \epsilon\bar{z}^2$ a simple calculation shows that $J = -\operatorname{sign} u = -1$ recovering the well known result that generic bifurcating period-3 points are also unstable in the reversible context (Meyer [1971]).

The isotropy subgroup of the period-3 points is $\mathbf{Z}_2^\kappa$ which is generated by the reflection element in $\mathbf{D}_3$. This corresponds to the invariance under $z \mapsto \bar{z}$ in the normal form. In term of sequences, the elements of the subspace $\mathbf{X}_3^{\mathbf{Z}_2^\kappa} \subset \mathbf{X}_3^1$ are invariant under the rearrangement $\kappa \cdot \mathbf{x} = (x_1, x_3, x_2)$. They are sequences of the form $\ldots xooxooxoo\ldots$ with $x, o \in \mathbf{R}$. In particular generic period-3 points lie in a *two-dimensional* subspace of $\mathbf{X}_3^1$. But note that in phase space they lie in a 2-manifold of a subspace of dimension 3 of $\mathbf{X}_3^2$ as sequences $\{\cdots \binom{x}{0}\binom{o}{o_2}\binom{o}{-o_2}\cdots\}$ with $o_2 = h_{x'}(x,o)$.

Two new interesting features appear when the coefficient w_2 (or equivalently, the coefficient b) goes through 0: the period-3 points in $\mathbf{X}_3^{\mathbf{Z}_2^\kappa}$ can stabilize through a turn in the bifurcation branch and symmetry-breaking bifurcations $\mathbf{X}_3^{\mathbf{Z}_2^\kappa} \to \mathbf{X}_3^1$ can take place resulting in new branches of period-3 points.

Using the classification in Section 6.1 and Furter [1990], the normal form for (7.5) when $w_2 = 0$ is $\mathcal{K}_\lambda^{\mathbf{D}_3}$-equivalent to

$$g(z,\lambda) = (\epsilon_1 u + \lambda)\, z + (\epsilon_2 v + m\,\lambda)\,\overline{z}^2. \tag{7.6}$$

The coefficients in the normal form can be expressed in terms of the coefficients $b, c, \ldots$ in the generating function.

Let $\hat{p}(u,v,\lambda) = 2\,\widehat{\mathrm{W}}_u$, $\hat{q}(u,v,\lambda) = 3\,\widehat{\mathrm{W}}_v$, $\Delta_{u,v} = \hat{p}_u\hat{q}_v - \hat{p}_v\hat{q}_u$ and $\Delta_{u,\lambda} = \hat{p}_u\hat{q}_\lambda - \hat{p}_\lambda\hat{q}_u$. Then $\epsilon_1 = \operatorname{sign}\hat{p}_u^o = \operatorname{sign} w_3 = -\operatorname{sign} c$,

$$\epsilon_2 = \operatorname{sign}(\hat{p}_u^o \cdot \Delta_{u,v}^o) = \epsilon_1 \operatorname{sign}(w_3 w_5 - 6w_4^2) = -\epsilon_1 \operatorname{sign}(d^2 + c^3 - ec)$$

and the modal parameter is

$$m = \frac{\epsilon_1 \Delta_{u,\lambda}^o}{|\hat{p}_\lambda^o||\Delta_{u,v}^o|^{1/2}} = -3\epsilon_1 \frac{w_4}{|w_3 w_5 - 6w_4^2|^{1/2}} = \sqrt{\tfrac{3}{2}}\,\epsilon_1 \frac{d}{|d^2 + c^3 - ec|^{1/2}}.$$

The normal form (7.6) is of $\mathcal{K}_\lambda^{\mathbf{D}_3}$-codimension 2 and topological codimension 1. The universal unfolding is obtained by adding $(\beta + \widehat{m}\,\lambda)\,\overline{z}^2$ to (7.6) with $(\beta,\widehat{m}) \in \mathbf{R}^2$ in a neighborhood of 0. The non-degeneracy conditions are $c \neq 0$, $d \neq 0$ and $d^2 + c^3 - ec \neq 0$.

In addition to the trivial solution there are two classes of solutions of (7.6) and they are listed in Table 7.1.

Isotropy Subgroup	Branching Equations	Sign of the Hessian
$\mathbf{D}_3$	$u = 0$	$+1$
$\mathbf{Z}_2^\kappa$	$\lambda = -\alpha\,\chi_1 - \epsilon_1\chi_1^2 + \cdots$	$\operatorname{sign}\left[(\epsilon_2 v + m\,\lambda + \alpha)\,\frac{d\lambda}{d\chi_1}\right]$
$\mathbf{I}$	$\lambda + \epsilon_1 u = 0$ $\epsilon_2 v + m\,\lambda + \alpha = 0$	$\epsilon_1\epsilon_2$

Table 7.1: Solutions of $(\epsilon_1 u + \lambda)\, z + (\epsilon_2 v + m\,\lambda + \alpha)\,\overline{z}^2 = 0$

A complete set of bifurcation diagrams for the unfolding of (7.6) is easily obtained. Instead values of the parameters are chosen that illustrate the new phenomena; in particular stable $\mathbf{Z}_2^\kappa$-symmetric branches and the symmetry breaking bifurcation to non-symmetric period-3 points.

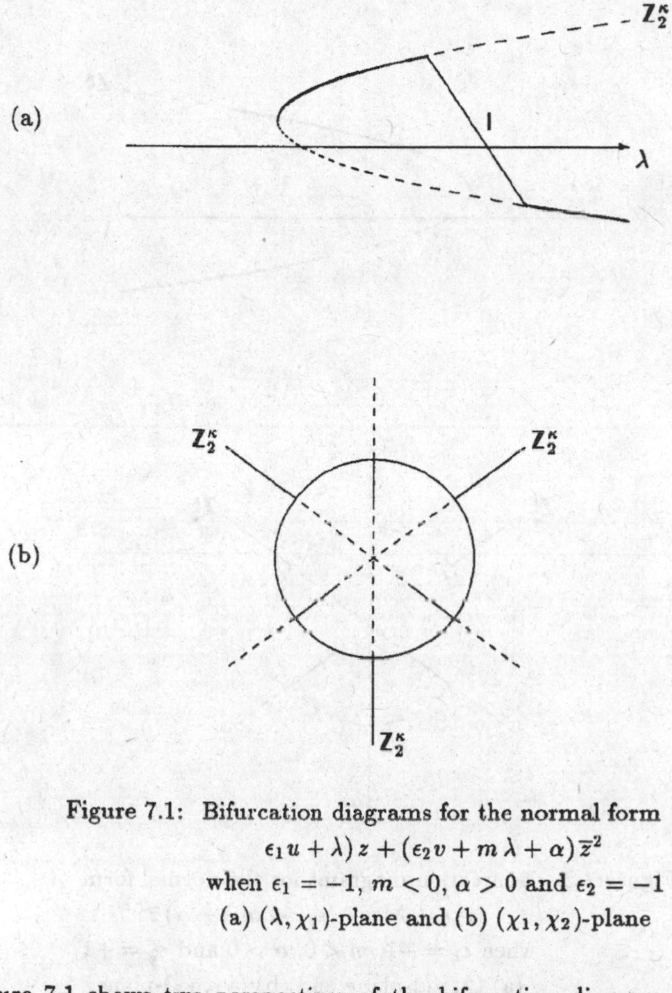

Figure 7.1: Bifurcation diagrams for the normal form
$\epsilon_1 u + \lambda) z + (\epsilon_2 v + m\,\lambda + \alpha)\,\overline{z}^2$
when $\epsilon_1 = -1$, $m < 0$, $\alpha > 0$ and $\epsilon_2 = -1$
(a) (λ, χ_1)-plane and (b) (χ_1, χ_2)-plane

Figure 7.1 shows two perspectives of the bifurcation diagram for $\epsilon_1 = -1$, $m < 0$, $\alpha > 0$ and $\epsilon_2 = -1$ showing the turn (and stabilization) of the $\mathbf{Z}_2^\kappa$-branches and the stable connecting non-symmetric branch. The projection onto the (λ, χ_1)-plane is misleading however as the non-symmetric branch actually connects the group orbit of $\mathbf{Z}_2^\kappa$-branches rather than exist as an individual branch as suggested in Figure 7.1(b). The associated bifurcation diagram in $\mathbf{R}^3$ can be seen in Figure 4.2.

Figure 7.2 is similar to 7.1 but with $\epsilon_2 = +1$ resulting in destabilization of the non-symmetric branch.

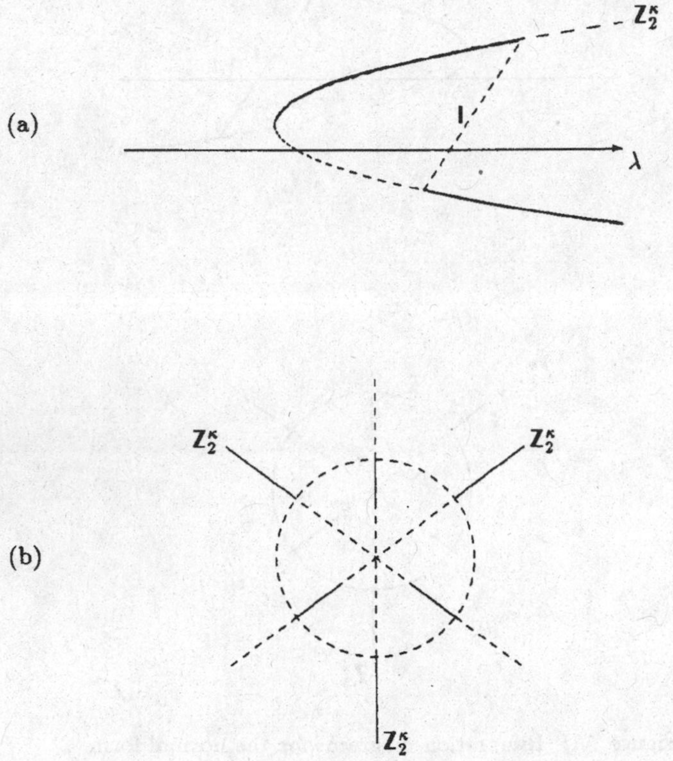

Figure 7.2: Bifurcation diagrams for the normal form
$$(\epsilon_1 u + \lambda) z + (\epsilon_2 v + m\lambda + \alpha)\bar{z}^2$$
when $\epsilon_1 = -1$, $m < 0$, $\alpha > 0$ and $\epsilon_2 = +1$
(a) (λ, χ_1)-plane and (b) (χ_1, χ_2)-plane

It is interesting to look at the affect of the symmetry breaking bifurcation of period-3 points on the phase portrait of the iterated map. A schematic of the bifurcation sequence that the phase portrait undergoes is shown in Figures 7.3(a) (corresponds to the parameters in Figure 7.1) and 7.3(b) (corresponds to the parameters in Figure 7.2) (note that our theory does not provide this result).

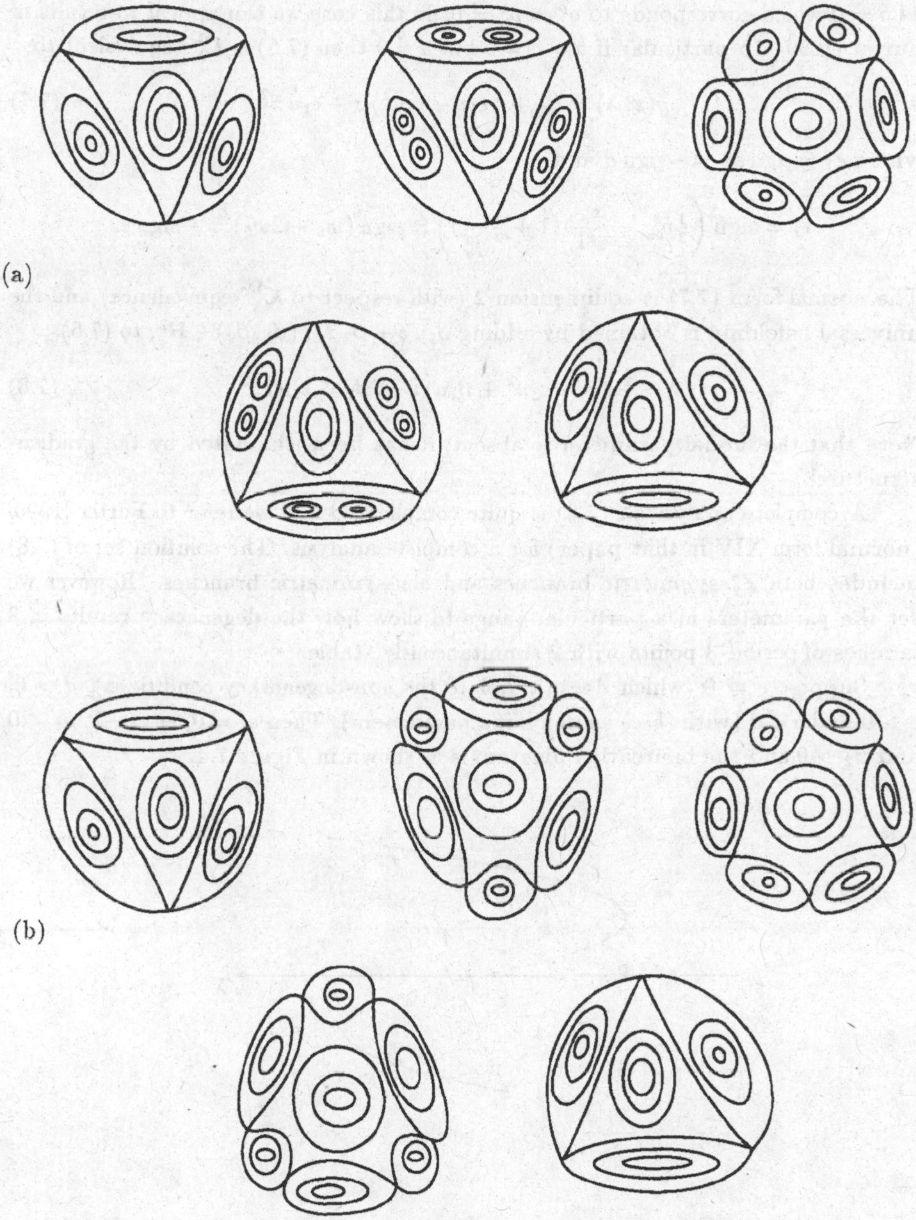

Figure 7.3: Schematic of the phase portrait surrounding the symmetry-breaking bifurcation of period-3 points when $\epsilon_1 = -1$, $m < 0$, $\alpha > 0$ and (a) $\epsilon_2 = -1$ or (b) $\epsilon_2 = +1$

There is an additional degeneracy in the normal form (7.5) when $c = 0$ as well as $b = 0$ which corresponds to $\hat{q}^o = \hat{p}^o_u = 0$. In this case we can appeal to results of Furter [1990]. In particular if $b = c = 0$ but $d \neq 0$ then (7.5) is $\mathcal{K}^{\mathbf{D}_3}_\lambda$-equivalent to

$$g(z,\lambda) = (\lambda + \tfrac{2}{3}\epsilon_1 v + \epsilon_2 u^2)\, z + \epsilon_1 u\, \overline{z}^2 \tag{7.7}$$

where $\epsilon_1 = \operatorname{sign} \hat{q}^o_u = -\operatorname{sign} d$ and

$$\epsilon_2 = \operatorname{sign}\left(\tfrac{1}{2}\hat{p}^o_{uu} - \frac{\Delta^o_{v,\lambda}}{\hat{p}^o_\lambda}(1 + \tfrac{3}{2}\frac{\hat{p}^o_v}{\hat{q}^o_u})\right) = \operatorname{sign}(w_6 + 2w_5) = -\operatorname{sign} e.$$

The normal form (7.7) is codimension 2 (with respect to $\mathcal{K}^{\mathbf{D}_3}_\lambda$-equivalence) and the universal unfolding is obtained by adding $\beta_1 u\, z + \beta_2\, \overline{z}^2$, $(\beta_1, \beta_2) \in \mathbf{R}^2$, to (7.6):

$$(\lambda + \tfrac{2}{3}\epsilon_1 v + \epsilon_2 u^2 + \beta_1 u)\, z + (\beta_2 + \epsilon_1 u)\, \overline{z}^2. \tag{7.8}$$

Note that the modal parameter is absent; it has been eliminated by the gradient structure!

A complete analysis of (7.8) is quite complicated and we refer to Furter [1990] (normal form XIV in that paper) for a complete analysis. The solution set of (7.8) includes both $\mathbf{Z}_2^\kappa$-symmetric branches and non-symmetric branches. However we set the parameters in a particular range to show how the degeneracy results in 3 families of period-3 points with 2 simultaneously stable.

Suppose $e = 0$ (which doesn't violate the non-degeneracy conditions), $d > 0$, $c > 0$ and $b < 0$ (with (b, c) as unfolding parameters). Then $\epsilon_2 = 0$, $\epsilon_1 = -1$, $\beta_1 < 0$ and $\beta_2 > 0$ and the bifurcation diagram is as shown in Figure 7.4.

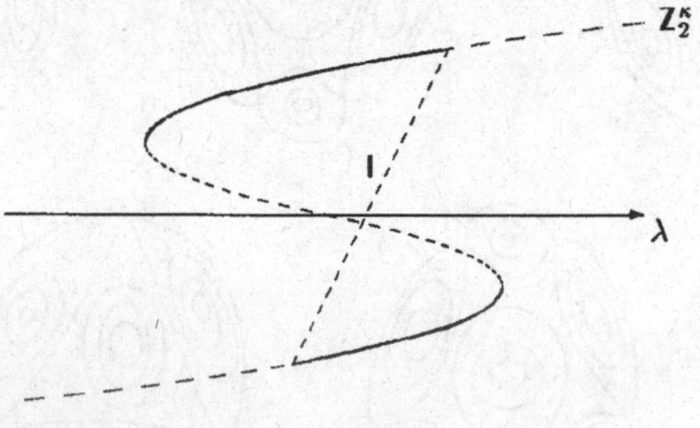

Figure 7.4: Bifurcation diagram in (λ, χ_1)-plane for normal form (7.8) when $\epsilon_1 = -1$, $\epsilon_2 = 0$, $\alpha_1 < 0$ and $\alpha_2 > 0$

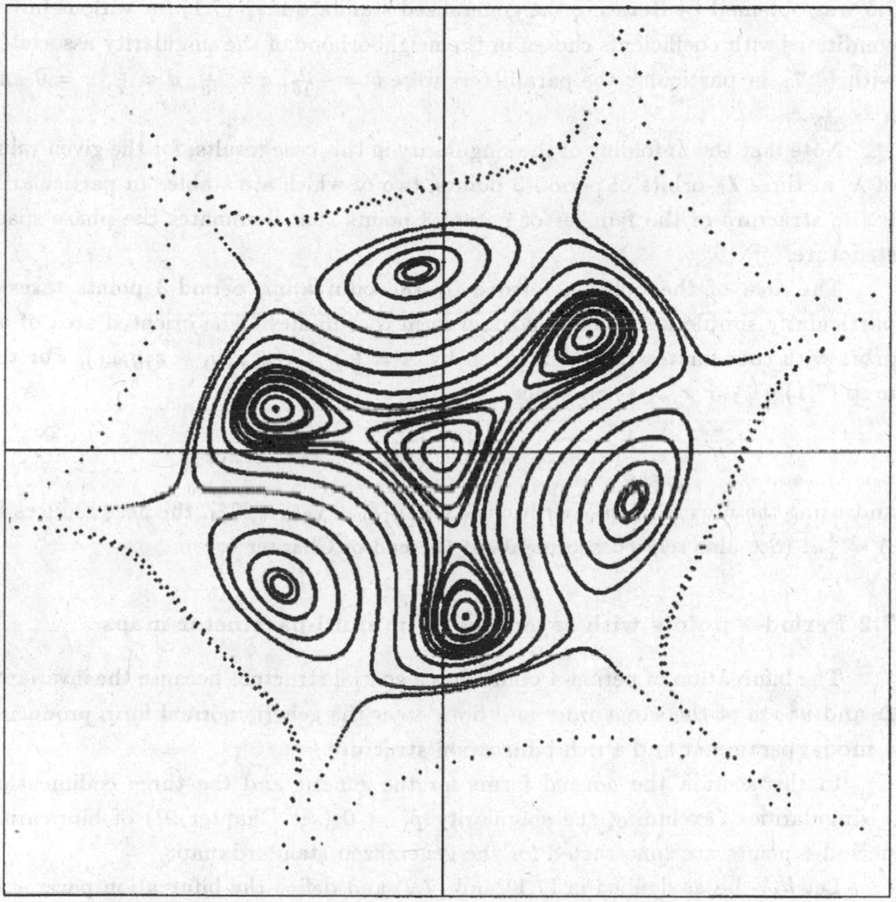

Figure 7.5 : Phase portrait in the neighborhood of singular bifurcation of period-3 points for the generalized standard map with parameters: $\lambda = \frac{1}{200}$, $b = -\frac{1}{10}$, $c = \frac{1}{10}$, $d = \frac{3}{2}$, and $e = 0$ (the phase space is rotated and stretched vertically for clarity).

Note that the group orbit of $\mathbf{Z}_2^{\kappa}$-branches is connected by the non-symmetric branch. The non-symmetric branch is uniformly unstable ($J < 0$ independent of parameters) but there exists a range of λ in which both the (upper and lower) branches of $\mathbf{Z}_2^{\kappa}$-symmetric period-3 points are stable.

In Figure 7.5 we show how the bifurcation structure affects the phase portrait; in particular the three families of period-3 points dominate the situation. Figure 7.5 was obtained by iterating the generalized standard map (7.1) for various initial conditions with coefficients chosen in the neighborhood of the singularity associated with (7.7); in particular the parameters were $b = -\frac{1}{10}$, $c = \frac{1}{10}$, $d = \frac{3}{2}$, $e = 0$ and $\lambda = \frac{1}{200}$.

Note that the unfolding of the singularity in this case results, for the given value of λ, in three $\mathbf{Z}_3$-orbits of period-3 points, two of which are stable. In particular it is the structure of the families of period-3 points that dominates the phase space structure.

The area of the polygon formed by the bifurcating period-3 points takes a particularly simple form in the normal form coordinates. The oriented area of an orbit with coordinates (x_j, y_j) is given by $A = \frac{1}{2}\sum_{j=1}^{3}(x_{j+1}y_j - x_j y_{j+1})$. For the map (7.1) $y_{j+1} = x_{j+1} - x_j$, hence

$$A = \frac{1}{2}(x_1^2 + x_2^2 + x_3^2) - \frac{1}{2}(x_1x_3 + x_2x_3 + x_1x_2)$$

and using the normal form coordinates $\mathbf{x} = \chi_1\xi_1 + \chi_2\xi_2 + \Upsilon\xi_3$, the area reduces to $A = \frac{3}{4}u$! (See also related comments at the end of Chapter 5.)

7.2 Period-4 points with reversibility in multi-parameter maps

The bifurcation of period-4 points has a special structure because the invariants Δ and u^2 are of the same order and both enter the generic normal form producing a modal parameter and a rich bifurcation structure.

In this section the normal forms for the generic and the three codimension 1 singularities (excluding the singularity $\hat{p}_X^o = 0$ (see Chapter 9)) of bifurcating period-4 points are constructed for the generalized standard map.

Let h, V be as defined in (7.1) and (7.2) and define the bifurcation parameter λ by $V_{xx}(0,\lambda) = 2 - \lambda$. Then when $\lambda = 0$ the multipliers of the linear map $\mathbf{DT}^o$ lie at $\pm i$.

On the space $\mathbf{X}_4^1$ the action takes the form

$$\mathrm{W}_4(\mathbf{x}, \Lambda) = \sum_{j=1}^{4} h(x_j, x_{j+1}, \Lambda) = \tfrac{1}{2}\langle \mathbf{x}, \mathbf{L}^o\mathbf{x}\rangle + \tfrac{1}{2}\lambda\|\mathbf{x}\|^2 - \sum_{j=1}^{4}\widehat{V}(x_j, \Lambda)$$

with

$$\mathbf{L}^o = -(\Gamma_4 + \Gamma_4^T) = \begin{pmatrix} 0 & -1 & 0 & -1 \\ -1 & 0 & -1 & 0 \\ 0 & -1 & 0 & -1 \\ -1 & 0 & -1 & 0 \end{pmatrix}.$$

From Proposition 2.3, $\sigma(\mathbf{L}^o) = \{0, -2, 2\}$ (with 0 of multiplicity 2) and the eigenvectors are

$$\xi_1 = \frac{1}{\sqrt{2}}\begin{pmatrix} 1 \\ 0 \\ -1 \\ 0 \end{pmatrix}, \quad \xi_2 = \frac{1}{\sqrt{2}}\begin{pmatrix} 0 \\ 1 \\ 0 \\ -1 \end{pmatrix}, \quad \eta_1 = \frac{1}{2}\begin{pmatrix} 1 \\ 1 \\ 1 \\ 1 \end{pmatrix} \quad \text{and} \quad \eta_2 = \frac{1}{2}\begin{pmatrix} -1 \\ 1 \\ -1 \\ 1 \end{pmatrix}.$$

Let $\mathbf{X}_4^1 = \mathbf{A} \oplus \mathbf{B}$ with $\mathbf{A}$ spanned by ξ_1 and ξ_2 and $\mathbf{B}$ spanned by η_1 and η_2. Then any $\mathbf{x} \in \mathbf{X}_4^1$ can be expressed as $\mathbf{x} = \chi_1\xi_1 + \chi_2\xi_2 + \Upsilon_1\eta_1 + \Upsilon_2\eta_2$.

An application of the Splitting Lemma results in the reduced functional

$$\widehat{\mathrm{W}}_4(\chi_1, \chi_2, \Lambda) = \mathrm{W}_4(\chi_1\xi_1 + \chi_2\xi_2 + \Upsilon_1(\chi, \Lambda)\eta_1 + \Upsilon_2(\chi, \Lambda)\eta_2, \Lambda). \tag{7.9}$$

It follows from the Splitting Lemma that the complementary functions Υ_1 and Υ_2 are $\mathbf{D}_4$-equivariant; in particular, using the reduced actions of $\mathbf{D}_4$ on the subspace $\mathbf{A}$ and $\mathbf{B}$ we find

$$\begin{aligned} \Upsilon_1(\chi_1, \chi_2, \Lambda) &= \Upsilon_1(\chi_2, -\chi_1, \Lambda) \\ -\Upsilon_2(\chi_1, \chi_2, \Lambda) &= \Upsilon_2(\chi_2, -\chi_1, \Lambda) \end{aligned} \quad \text{and} \quad \begin{aligned} \Upsilon_1(\chi_1, \chi_2, \Lambda) &= \Upsilon_1(\chi_1, -\chi_2, \Lambda) \\ \Upsilon_2(\chi_1, \chi_2, \Lambda) &= \Upsilon_2(\chi_1, -\chi_2, \Lambda). \end{aligned}$$

Therefore,

$$\begin{aligned} \Upsilon_1 &= \hat{\Upsilon}_1(u, \Delta) = \Upsilon_{11} u + \Upsilon_{12}\Delta + \Upsilon_{13} u^2 + \cdots \\ \Upsilon_2 &= \delta\, \hat{\Upsilon}_2(u, \Delta) = \delta\,(\Upsilon_{21} + \Upsilon_{22} u + \cdots) \end{aligned} \tag{7.10}$$

and a straightforward calculation results in

$$\begin{aligned} \Upsilon_{11} &= -\tfrac{1}{4}b - \tfrac{1}{8}b\,\lambda + \cdots, \quad \Upsilon_{21} = \tfrac{1}{4}b - \tfrac{1}{8}b\,\lambda + \cdots, \\ \Upsilon_{12} &= -\tfrac{1}{64}(b^3 + 6bc + 4d) + \tfrac{1}{128}(b^3 - 4d)\,\lambda + \cdots, \\ \Upsilon_{13} &= -\tfrac{1}{64}(b^3 - 6bc + 4d) + \tfrac{1}{128}(-3b^3 + 12bc - 4d)\,\lambda + \cdots, \\ \Upsilon_{22} &= -\tfrac{1}{32}(b^3 - 4d) + \tfrac{1}{64}(b^3 - 6bc - 4d)\,\lambda + \cdots, \end{aligned}$$

retaining only enough terms to obtain the codimension 1 normal forms.

The expressions for Υ_1 and Υ_2 in (7.10) are substituted into the reduced functional (7.9). After some algebra we find

$$\begin{aligned} \widehat{\mathrm{W}}_4(\chi, \Lambda) = \tfrac{1}{2}\lambda u &+ (w_2 + w_3\lambda)\, u^2 + (w_4 + w_5\lambda)\,\Delta + w_6\, u\Delta + w_7\, u^3 \\ &+ w_8\, u^2\Delta + w_9\, \Delta^2 + w_{10}\, u^4 + \cdots \end{aligned} \tag{7.11}$$

where

$$w_2 = \tfrac{1}{16}(b^2 - c)\,, \quad w_3 = \tfrac{1}{32}b^2\,, \quad w_4 = -\tfrac{1}{16}(b^2 + c)\,, \quad w_5 = \tfrac{1}{32}b^2\,,$$
$$w_6 = \tfrac{1}{128}b^4 + \tfrac{3}{128}b^2 c - \tfrac{1}{32}bd - \tfrac{1}{32}e\,, \quad w_7 = \tfrac{1}{384}b^4 - \tfrac{3}{128}b^2 c + \tfrac{1}{32}bd - \tfrac{1}{96}e\,,$$
$$w_8 = -\tfrac{1}{2048}b^6 - \tfrac{3}{2048}b^4 c + \tfrac{3}{256}b^3 d - \tfrac{9}{512}b^2 c^2 - \tfrac{1}{128}d^2 + \tfrac{5}{256}b^2 e - \tfrac{3}{256}g\,,$$
$$w_9 = \tfrac{1}{4096}b^6 + \tfrac{11}{4096}b^4 c - \tfrac{1}{512}b^3 d - \tfrac{9}{1024}b^2 c^2 - \tfrac{5}{512}b^2 e + \tfrac{1}{256}d^2 + \tfrac{3}{256}bcd - \tfrac{1}{128}bf - \tfrac{1}{512}g$$
$$\text{and} \quad w_{10} = \tfrac{1}{4096}b^6 - \tfrac{13}{4096}b^4 c + \tfrac{3}{512}b^3 d + \tfrac{9}{1024}b^2 c^2 - \tfrac{5}{512}b^2 e + \tfrac{1}{256}d^2 - \tfrac{3}{256}bcd + \tfrac{1}{128}bf - \tfrac{1}{512}g\,.$$

Note that we have hidden the dependence on the unfolding parameters in the w_i's. Let $\hat{p}(u,\Delta,\Lambda) = 2\,\widehat{\mathrm{W}}_u$, $\hat{r}(u,\Delta,\Lambda) = -4\,\widehat{\mathrm{W}}_\Delta$, $u = \chi_1^2 + \chi_2^2$, $\delta = \chi_2^2 - \chi_1^2$ and $\Delta = \delta^2$. Then the $\mathbf{D}_4$-equivariant gradient map associated with the reduced functional $\widehat{\mathrm{W}}_4$ is

$$\nabla_\chi \widehat{\mathrm{W}}_4(\chi,\Lambda) = \hat{p}(u,\Delta,\Lambda)\begin{pmatrix}\chi_1\\ \chi_2\end{pmatrix} + \hat{r}(u,\Delta,\Lambda)\,\delta\begin{pmatrix}\chi_1\\ -\chi_2\end{pmatrix} \tag{7.12}$$

where

$$\left.\begin{aligned} \hat{p} &= \lambda + 4(w_2 + w_3\lambda)\,u + 2w_6\,\Delta + 6w_7\,u^2 + 4w_8\,u\Delta + 8w_{10}\,u^3 + \cdots\,, \\ \hat{r} &= -4(w_4 + w_5\lambda) - 4w_6\,u - 4w_8\,u^2 - 8w_9\,\Delta + \cdots\,, \end{aligned}\right\} \tag{7.13}$$

to which the classification theorem for $\mathbf{D}_4$-equivariant gradient maps can now be applied.

Isotropy Subgroup	Equations	Sign of the Hessian	Action of $\mathbf{Z}_2$ on $\mathbf{R}$	Action of $\mathbf{Z}_2$ on $\mathbf{X}_4^1$
$\mathbf{Z}_2^\kappa$	$u + \delta = 0$ $\lambda + (m - \epsilon_1)\,u = 0$	$\mathrm{sign}\,(\epsilon_1 m - 1) =$ $-\mathrm{sign}\,c\,(b^2 + c)$	$\chi \mapsto (\chi_1, -\chi_2)$	$x \mapsto$ (x_1, x_4, x_3, x_2)
$\mathbf{Z}_2^r$	$\delta = 0$ $\lambda + m\,u = 0$	$-\epsilon_1 \mathrm{sign}\, m =$ $\mathrm{sign}\,(c^2 - b^4)$	$\chi \mapsto (\chi_2, \chi_1)$	$x \mapsto$ (x_2, x_1, x_4, x_3)

Table 7.2: Generic period-4 points in the generalized standard map

The basic result is the generic case (technically it is $\mathcal{K}_\lambda^{\mathbf{D}_4}$-codimension 1 but topological codimension 0); that is, if $c \cdot (b^2 \pm c) \neq 0$ then (7.12) is $\mathcal{K}_\lambda^{\mathbf{D}_4}$-equivalent to

$$g(z,\lambda) = (m\,u + \lambda)\begin{pmatrix}\chi_1\\ \chi_2\end{pmatrix} + \epsilon_1\,\delta\begin{pmatrix}\chi_1\\ -\chi_2\end{pmatrix} \tag{7.14}$$

where $\epsilon_1 = \text{sign}\,\hat{r}^o = -\text{sign}\,w_4 = \text{sign}\,(b^2 + c)$ and

$$m = \frac{\hat{p}_u^o}{|\hat{r}^o|} = \frac{b^2 - c}{|b^2 + c|}\,.$$

The two classes of solutions of (7.14) are given in Table 7.2 with their symmetry and stability properties. In Figure 7.6 the bifurcation diagrams (projected onto $\mathbf{R}^2$) are shown in the (b, c) plane.

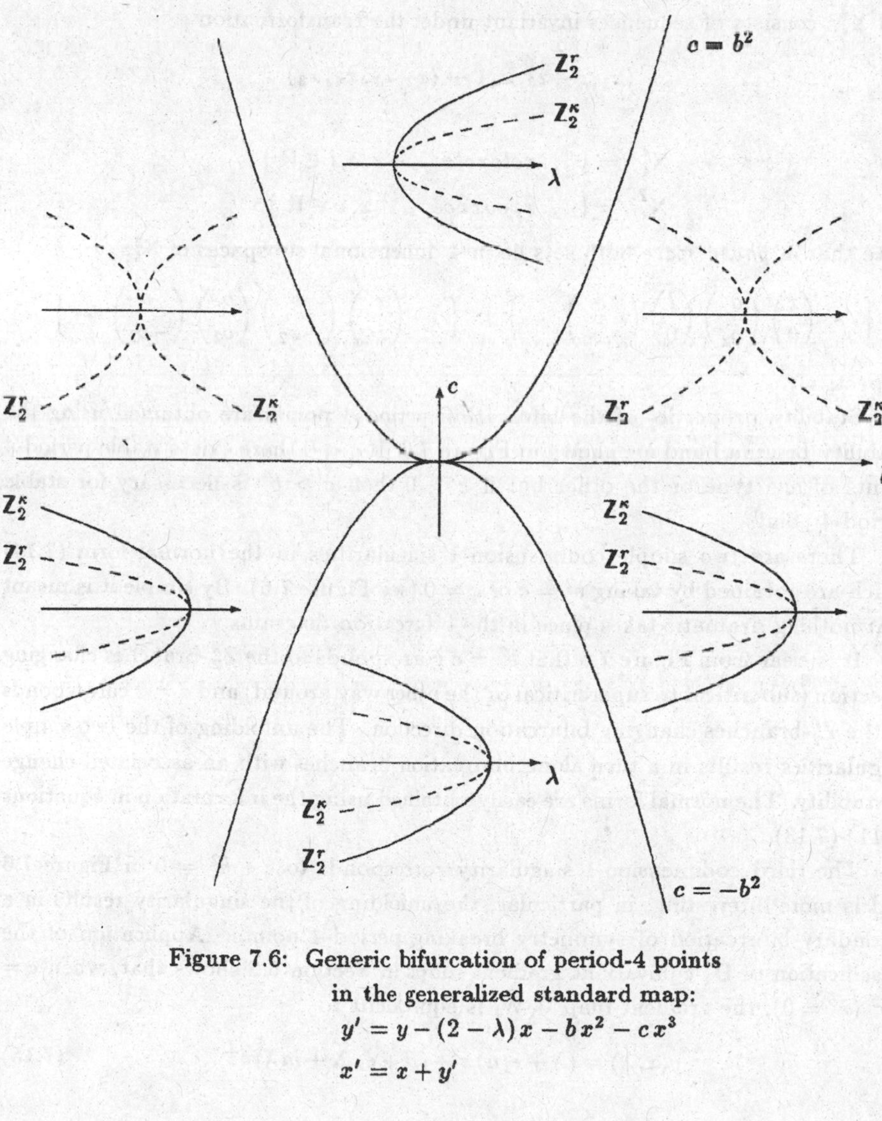

Figure 7.6: Generic bifurcation of period-4 points in the generalized standard map:

$$y' = y - (2 - \lambda)\,x - b\,x^2 - c\,x^3$$
$$x' = x + y'$$

Note that there are two distinct classes of $\mathbf{Z}_2$-symmetric solutions. The solutions with isotropy subgroup $\mathbf{Z}_2^\kappa$ are the usual $\mathbf{Z}_2$-symmetric periodic points that are invariant under $(\chi_1, \chi_2) \mapsto (\chi_1, -\chi_2)$ in the normal form but the $\mathbf{Z}_2^r$-solutions are invariant under $(\chi_1, \chi_2) \mapsto (\chi_2, \chi_1)$.

Therefore the two classes of $\mathbf{Z}_2$-symmetric period-4 points correspond to distinct symmetric sequences; in particular, $\mathsf{X}_4^{\mathbf{Z}_2^\kappa}$ consists of sequences invariant under

$$(x_1, x_2, x_3, x_4) \mapsto (x_1, x_4, x_3, x_2)$$

and $\mathsf{X}_4^{\mathbf{Z}_2^r}$ consists of sequences invariant under the transformation

$$(x_1, x_2, x_3, x_4) \mapsto (x_2, x_1, x_4, x_3)$$

or

$$\mathsf{X}_4^{\mathbf{Z}_2^\kappa} = \{ \ldots xoloxolo \ldots \mid x, o, l \in \mathbf{R} \},$$
$$\mathsf{X}_4^{\mathbf{Z}_2^r} = \{ \ldots xxooxxoo \ldots \mid x, o \in \mathbf{R} \}.$$

Note that in *phase space* both sets lie in 4 dimensional subspaces of X_4^2:

$$\left\{ \cdots \begin{pmatrix} x \\ 0 \end{pmatrix} \begin{pmatrix} o \\ o_2 \end{pmatrix} \begin{pmatrix} l \\ 0 \end{pmatrix} \begin{pmatrix} o \\ -o_2 \end{pmatrix} \cdots \right\}, \quad \left\{ \cdots \begin{pmatrix} x \\ x_2 \end{pmatrix} \begin{pmatrix} x \\ -x_2 \end{pmatrix} \begin{pmatrix} o \\ o_2 \end{pmatrix} \begin{pmatrix} o \\ -o_2 \end{pmatrix} \cdots \right\},$$

respectively.

Stability properties of the bifurcating period-4 points are obtained using the Stability Lemma I and are shown in Figure 7.6. If $c < 0$ there exists stable period-4 points of one type or the other but if $c > 0$ then $c > b^2$ is necessary for stable period-4 points.

There are two simple codimension-1 singularities in the normal form (7.14) which are obtained by taking $b^2 = c$ or $c = 0$ (see Figure 7.6). By simple it is meant that nothing dramatic takes place in the bifurcation diagrams.

It is clear from Figure 7.6 that $b^2 = c$ corresponds to the $\mathbf{Z}_2^\kappa$-branches changing direction (subcritical to supercritical or the other way around) and $c = 0$ corresponds to the $\mathbf{Z}_2^r$-branches changing bifurcation direction. The unfolding of the two simple singularities results in a turn along bifurcation branches with an associated change of stability. The normal forms are easily obtained using the information in equations (7.11)-(7.13).

The third codimension-1 singularity corresponds to $c + b^2 = 0$ in Figure 7.6 and is more interesting; in particular, the unfolding of the singularity results in a secondary bifurcation of symmetry breaking period-4 points. Application of the classification of $\mathbf{D}_4$-equivariant gradient maps in Section 6.2 shows that, when $c = -b^2$ ($\hat{r}^o = 0$), the gradient map $\nabla_\chi \widehat{\mathsf{W}}_4$ is equivalent to

$$g(z, \lambda) = (\lambda + \epsilon_1 u)\, z + (\beta + \epsilon_2 \Delta + m\lambda)\, \delta \bar{z} \qquad (7.15)$$

(with β the unfolding parameter) where

$$\epsilon_1 = \operatorname{sign} \hat{p}_u^o = \operatorname{sign} w_2 = -\operatorname{sign} c = +1$$

$$\epsilon_2 = \epsilon_1 \operatorname{sign}(\hat{p}_u^o \hat{r}_\Delta^o - \hat{p}_\Delta^o \hat{r}_u^o) = \operatorname{sign}(w_6^2 - 4w_2w_9)$$

and

$$m = \frac{\epsilon_1(\hat{p}_u^o \hat{r}_\lambda^o - \hat{r}_u^o)}{|\hat{p}_u^o \hat{r}_\Delta^o - \hat{p}_\Delta^o \hat{r}_u^o|^{1/2}} = -\frac{1}{8\sqrt{8}} \frac{(b^4 + bd + e)}{|w_6^2 - 4w_2w_9|^{1/2}}$$

(using $\hat{p}_\lambda^o = 1$). The non-degeneracy conditions on the normal form are $\hat{p}_u^o \neq 0$, $\hat{p}_u^o \hat{r}_\Delta^o - \hat{p}_\Delta^o \hat{r}_u^o \neq 0$ and $\hat{r}_u^o - \hat{p}_u^o \hat{r}_\lambda^o \neq 0$ or in terms of the coefficients of the map:

$$c \neq 0\,, \quad b^4 + bd + e \neq 0 \quad \text{and} \quad w_6^2 - 4w_2w_9 \neq 0\,.$$

The last expression $w_6^2 - 4w_2w_9$ is a complicated polynomial in $b, c \ldots g$ which we can assume is generically nonzero.

In the unfolding of (7.15) ($\beta \neq 0$) there are three classes of period-4 points; the two symmetric classes ($\mathbf{Z}_2^\kappa$ and $\mathbf{Z}_2^r$) as well as a nonsymmetric class of connecting secondary branches.

The bifurcation diagram has features similar to the symmetry breaking bifurcation found in normal form (7.6), except that the non-symmetric branch of period-4 points connects the two *distinct* ($\mathbf{Z}_2^\kappa$ and $\mathbf{Z}_2^r$) symmetric branches.

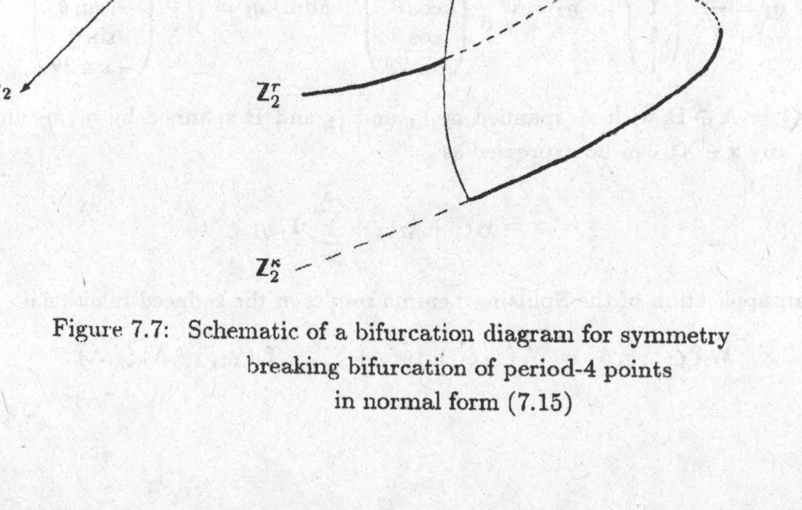

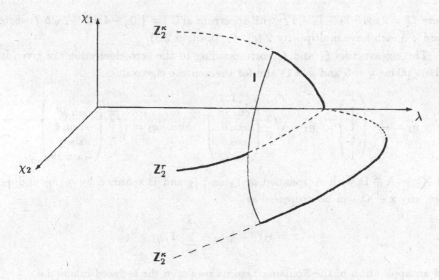

Figure 7.7: Schematic of a bifurcation diagram for symmetry breaking bifurcation of period-4 points in normal form (7.15)

Figure 7.7 shows a schematic in $\mathbf{R}^3$ (in the positive quadrant of (β_1, β_2)) of the symmetry breaking bifurcation. The figure corresponds to parameter values in the normal form (7.15) where the non symmetric branch is stable. The phase portrait will go through a sequence of bifurcations much like that shown in Figure 7.3.

There is enough information in the reduced functional $\widehat{\mathrm{W}}_4$ in (7.11) (and enough parameters) to analyze most of the codimension 2 singularities in the classification of D_4-equivariant gradient maps. Although these bifurcations are 3-parameter bifurcations (including the distinguished parameter λ) they do provide interesting organizing centers for the dynamics observed in the phase portraits.

7.3 Generic period-5 points in the generalized standard map

Suppose the linearization of (7.1) has a period-5 point. Then $2\cos\theta = 2 - a$ and $\theta = \frac{2p}{5}\pi$, $1 \le p \le 4$. Without loss of generality take $p = 1$ and define the bifurcation parameter λ by

$$V_{xx}(0,\lambda) = 2(1-\cos\theta) - \lambda \quad \text{with} \quad \cos\theta = \tfrac{1}{4}(\sqrt{5}-1).$$

On $\mathbf{X}_5^1$, the space of period-5 sequences, the action takes the form

$$\mathrm{W}_5(\mathbf{x},\Lambda) = \sum_{j=1}^{5} h(x_j, x_{j+1}, \Lambda) = \tfrac{1}{2}\langle \mathbf{x}, \mathbf{L}^o\mathbf{x}\rangle + \tfrac{1}{2}\lambda\|\mathbf{x}\|^2 - \sum_{j=1}^{5} \widehat{V}(x_j,\Lambda)$$

where $\mathbf{L}^o = 2\cos\theta\,\mathbf{I}_5 - \Gamma_5 - \Gamma_5^T$ with spectrum $\sigma(\mathbf{L}^o) = \{\,0, -4\sin^2\frac{\theta}{2}, \sqrt{5}\,\}$ where 0 and $\sqrt{5}$ each have multiplicity 2 (cf. Proposition 2.3).

The eigenvectors ξ_1 and ξ_2 corresponding to the zero eigenvalue are given in (2.18) (taking $q = 5$ and $p = 1$) and for the nonzero eigenvalues,

$$\eta_1 = \frac{1}{\sqrt{5}}\begin{pmatrix}1\\1\\1\\1\\1\end{pmatrix}, \quad \eta_2 = \sqrt{\frac{2}{5}}\begin{pmatrix}1\\ \cos 2\theta\\ \cos\theta\\ \cos\theta\\ \cos 2\theta\end{pmatrix} \quad \text{and} \quad \eta_3 = \sqrt{\frac{2}{5}}\begin{pmatrix}0\\ \sin 2\theta\\ -\sin\theta\\ \sin\theta\\ -\sin 2\theta\end{pmatrix}.$$

Let $\mathbf{X}_5^1 = \mathbf{A} \oplus \mathbf{B}$ with $\mathbf{A}$ spanned by ξ_1 and ξ_2 and $\mathbf{B}$ spanned by η_1, η_2 and η_3. Then any $\mathbf{x} \in \mathbf{X}_5^1$ can be expressed as

$$\mathbf{x} = \chi_1\xi_1 + \chi_2\xi_2 + \sum_{j=1}^{3} \Upsilon_j\eta_j$$

and an application of the Splitting Lemma results in the reduced functional

$$\widehat{\mathrm{W}}_5(\chi_1,\chi_2,\Lambda) = \mathrm{W}_5\big(\chi_1\xi_1 + \chi_2\xi_2 + \textstyle\sum_{j=1}^{3}\Upsilon_j(\chi_1,\chi_2,\Lambda)\,\xi_j, \Lambda\big).$$

The solution of the complementary equation for Υ_1, Υ_2 and Υ_3 is simplified by using the $\mathbf{D}_5$-symmetry. The set $(\Upsilon_1, \Upsilon_2, \Upsilon_3)$ as a map from $\mathbf{A}$ to $\mathbf{B}$ is $\mathbf{D}_5$-equivariant. The action of $\mathbf{D}_5|_A$ is the standard action of $\mathbf{D}_5$ on $\mathbf{R}_2$ but the action of $\mathbf{D}_5$ on $\mathbf{B}$ is generated by

$$\mathbf{D}_5\Big|_B = \left\langle \begin{pmatrix} 1 & 0 & 0 \\ 0 & \cos 2\theta & \sin 2\theta \\ 0 & -\sin 2\theta & \cos 2\theta \end{pmatrix}, \begin{pmatrix} 1 & 0 & 0 \\ 0 & 1 & 0 \\ 0 & 0 & -1 \end{pmatrix} \right\rangle .$$

Therefore $\gamma_B \cdot (\Upsilon_1, \Upsilon_2, \Upsilon_3) = (\Upsilon_1, \Upsilon_2, \Upsilon_3) \cdot (\gamma_A)$ for each $\gamma_A \in \mathbf{D}_5|_A$ and $\gamma_B \in \mathbf{D}_5|_B$ resulting in the following form for $(\Upsilon_1, \Upsilon_2, \Upsilon_3)$,

$$\Upsilon_1 = \hat{\Upsilon}_1(u, v) \quad \text{and}$$

$$\begin{pmatrix} \Upsilon_2 \\ \Upsilon_3 \end{pmatrix} = \hat{\Upsilon}_2(u, v) \begin{pmatrix} \chi_1^2 - \chi_2^2 \\ 2\chi_1\chi_2 \end{pmatrix} + \hat{\Upsilon}_3(u, v) \begin{pmatrix} \chi_1(\chi_1^2 - 3\chi_2^2) \\ \chi_2(\chi_2^2 - 3\chi_1^2) \end{pmatrix}$$

where u and v are the $\mathbf{D}_5$-invariants on $\mathbf{R}_2$. An elementary calculation shows that

$$\hat{\Upsilon}_1 = -\frac{(1+\sqrt{5})}{10} b u + \cdots , \quad \hat{\Upsilon}_2 = \frac{1}{5\sqrt{2}} b + \cdots$$

and

$$\hat{\Upsilon}_3 = \frac{1}{\sqrt{5}} \left(\frac{c}{10} + \frac{b^2}{5\sqrt{5}} \right) + \cdots ,$$

which will be sufficient for the generic theory.

Substitution of $(\Upsilon_1, \Upsilon_2, \Upsilon_3)$ into the reduced functional results in

$$\widehat{W}_5(\chi_1, \chi_2, \Lambda) = \tfrac{1}{2}\lambda u + \tfrac{1}{40}(2b^2 - 3c)\, u^2 - \tfrac{1}{100}\sqrt{\tfrac{2}{5}}\,(d + \sqrt{5}bc + b^3)\, v + \cdots .$$

Using complex notation ($z = \chi_1 + i\chi_2$) and a simple scaling, the $\mathbf{D}_5$-equivariant gradient map associated with $\widehat{W}_5$ is

$$\nabla_\chi \widehat{W}_5(\chi, \Lambda) = (\lambda + \epsilon_1 u)\, z + \epsilon_2\, \bar{z}^4 + \cdots$$

with $\epsilon_1 = \operatorname{sign}(2b^2 - 3c)$ and $\epsilon_2 = -\operatorname{sign}(d + \sqrt{5}bc + b^3)$, to which the singularity theory for $\mathbf{D}_5$-equivariant gradient maps is applicable.

If $2b^2 - 3c \neq 0$ and $d + \sqrt{5}bc + b^3 \neq 0$ then there is a generic bifurcation of period-5 points for $|\lambda|$ and $\|\chi\|$ sufficiently small. The bifurcation is supercritical (subcritical) according to whether $\epsilon_1 = -1$ ($\epsilon_1 = 1$) but the stability is determined by the sign of J ($= \operatorname{sign}|\mathrm{Hess}_\chi \widehat{W}_5(\chi, \lambda)|$). Taking the point on the $\mathbf{D}_5$-orbit of period-5 points with $\chi_2 = 0$ we find that

$$J = -\epsilon_1 \epsilon_2 \operatorname{sign} \chi_1 .$$

Therefore stable period-5 points correspond to $\epsilon_1 \epsilon_2 \chi_1 < 0$.

Note that there are two families of period-5 points that join at the origin with one stable and one unstable (recovering the classical result). Which family is stable is determined by the sign of $\epsilon_1\epsilon_2$. Bifurcation diagrams for generic period-5 points as well as degenerate period-5 points can be found in Furter [1991].

The isotropy subgroup of the bifurcating period-5 points is $\mathbf{Z}_2^\kappa$ which is the reflection element in $\mathbf{D}_5$. In the normal form coordinates (χ_1, χ_2) the branch of period-5 points is invariant under $(\chi_1, \chi_2) \mapsto (\chi_1, -\chi_2)$. In the space of sequences the bifurcating period-5 points will be in the subspace $\mathsf{X}_5^{\mathbf{Z}_2^\kappa}$, which is the subspace of X_5^1 that is invariant under the rearrangement $\kappa \cdot x = (x_1, x_5, x_4, x_3, x_2)$ or sequences of the form $\ldots xolloxolloxolloxo\ldots$ with $x, o, l \in \mathbf{R}$.

Note that in general the generic period-q points of reversible maps for any $q \geq 3$ lie in the subspaces $\mathsf{X}_q^{\mathbf{Z}_2^\kappa}$ (and also in $\mathsf{X}_q^{\mathbf{Z}_2^\tau}$ if q is even) which are subspaces of X_q^1 with considerably lower dimension; in particular,

$$\dim \mathsf{X}_q^{\mathbf{Z}_2^\kappa} = \left[\frac{q}{2}\right] + 1 \quad \text{and} \quad \dim \mathsf{X}_q^{\mathbf{Z}_2^\tau} = \left[\frac{q}{2}\right].$$

But in phase space they all generically lie in subspaces of dimension q of X_q^2.

8. Periodic points of equivariant symplectic maps

The dynamics of equivariant symplectic maps is an interesting and largely unexplored area. Moreover symplectic maps with spatial symmetry are of great interest in applications. A simple example is the time-T map of a parametrically forced (T-periodically) spherical pendulum. If the $\mathbf{O}(2)$-spatial symmetry of the spherical pendulum is not broken by the forcing (say by vertically forcing) then the time-T map of the system is identified with an $\mathbf{O}(2)$-equivariant symplectic map on $\mathbf{R}^4$.

In this chapter we present a general theory for the bifurcation of period-q points (from a fixed point) of equivariant symplectic maps on $\mathbf{R}^{2n}$ which are generated by a Lagrangian generating function. In the presence of symmetry, Floquet multipliers of high multiplicity arise generically resulting in a large dimensional kernel of $\mathbf{L}^o$. The bifurcation equations will therefore be posed on a large-dimensional space and our approach – Splitting Lemma, singularity theory, normal forms – becomes rapidly much more complicated.

In this chapter our approach will therefore be broken into two parts. It turns out that we can obtain quite interesting, but rough, results using purely topological methods. We present this theory first. Then, for the case when the group acts absolutely irreducibly on configuration space we return to the more precise methods – Splitting Lemma, singularity theory and normal forms *on fixed point subspaces* – to study the structure of the bifurcating period-q points.

The analysis is considerably simplified by working in configuration space (with no loss of generality in the bifurcation theory, as in Proposition 2.2) through the use of generating functions. Therefore we consider only symmetries that act on configuration space and lift to a symplectic action on the phase space (see the discussion of more general group actions in Appendix I).

In Section 8.1 we analyze the problem using topological techniques that require only the gradient structure and group-theoretic properties of the functional. With minimal non-degeneracy hypotheses, lower bounds on the number of conjugacy classes of bifurcating period-q points *and* their symmetry group are obtained.

For a Σ-equivariant symplectic map, we prove the existence of a conjugacy class of bifurcating period-q points for each isotropy subgroup $\Pi \subset \Sigma \times \mathbf{Z}_q$ (the spatio-temporal group of symmetries) provided the q-resonant eigenspace of $\mathbf{DT}^o$ satisfies some nondegeneracy assumptions (precisely defined in Section 8.1). Clearly it does not mean that for every isotropy subgroup Π we have solutions with *exact* isotropy Π, it means that for every Π we have solutions with *at least* that symmetry. In particular, in case the action of Σ is absolutely irreducible, we find solutions with the symmetry of every *maximal* isotropy subgroups contrary to what happens for general maps where this is true only for maximal subgroups of fixed-point subspaces of dimension 1.

This result generalizes – to bifurcating period-q points near symmetric period-1 points of symplectic maps – the equivariant "Weinstein-Moser" theory for periodic orbits near symmetric equilibria of continuous-time Hamiltonian systems (cf. Montaldi, Roberts & Stewart [1988]).

When the symplectic map is equivariant with respect to a continuous Lie group two new interesting features arise in the bifurcation of period-q points. First, the continuous symmetry induces – by a variant of Noether's Theorem – a conserved quantity (see Appendix E). In which case, the orbits of the symplectic map will lie on an invariant submanifold (or subset when singular) of the phase space.

Second, the symmetry results in geometrically distinct classes of periodic points. For example there are classes of periodic points which "ride on the group orbit". These orbits are reminiscent of relative equilibria of continuous Hamiltonian systems. There are also periodic points that break the symmetry; that is, the orbit of the period-q point is "transverse" to the group orbit. Both of these types of periodic points occur in $\mathbf{O}(2)$-equivariant symplectic maps for example.

In Section 8.2 we illustrate the results of the general framework of Section 8.1 in the important example of a compact Lie group Σ that acts absolutely irreducibly on the configuration space and so as a direct sum of absolutely irreducible representations on the phase space.

With the spherical pendulum as motivation, the results of Section 8.2 are applied to $\mathbf{O}(2)$-equivariant symplectic maps in Section 8.3. It is shown that generically there are three distinct classes of period-q points for each q, $3 \leq q < \infty$.

There is a family of period-q points that rides on the $\mathbf{O}(2)$-group orbit: these are analogous to the conical pendulum solutions in the unforced spherical pendulum.

There are also two families of period-q points with orbits that are transverse to the $\mathbf{O}(2)$-group orbits: these are analogous to the planar pendulum solutions in the unforced spherical pendulum.

There are other interesting orbits of $\mathbf{O}(2)$-equivariant symplectic maps that we do not consider: for example, the period-q points that lie transverse to the group orbit can drift along the group orbit resulting in a "relative invariant circle"; an invariant circle in a frame of reference that drifts along the $\mathbf{SO}(2)$-group orbit.

In Section 8.4 the results on bifurcations of $\mathbf{O}(2)$-equivariant symplectic maps are discussed with reference to the parametrically forced spherical pendulum.

In Section 8.5 we show how to apply the reduction to $\mathbf{O}(2)$-orbit space and give some immediate consequences of that procedure. In particular the reduced phase space is now equipped with a Poisson structure and the symplectomorphism is reduced to a Poisson map.

Finally in Section 8.6 we sketch some preliminary comments on the linearized stability question for $\mathbf{O}(2)$-equivariant symplectic maps.

8.1 Subharmonic bifurcation in equivariant symplectic maps

Consider a Λ-parametrized family of smooth symplectic maps $\mathbf{T}:(x,y,\Lambda)\mapsto(x',y')$ on $\mathbf{R}^{2n}$ generated by $h(x,x',\Lambda)$. As previously we assume that the parameter Λ splits into the main bifurcation parameter λ (usually one dimensional) and the perturbation (unfolding) parameters (α,β).

Suppose, moreover, that there is a compact Lie group Σ acting orthogonally on $\mathbf{R}^n$ such that h is invariant under the diagonal action of Σ on $\mathbf{R}^{2n}$; that is

$$h(\sigma x,\sigma x',\Lambda)=h(x,x',\Lambda),\quad \forall\sigma\in\Sigma.$$

We refer to the Appendix I for further discussion of symmetry, generating functions and maps. We can use the isotypic decomposition of $\mathbf{R}^n$ to block diagonalize the problem. In Chapter XII of Golubitsky, Stewart & Schaeffer II [1988] one can find the relevant background and proofs. The details necessary for the present analysis are as follows.

We can split $\mathbf{R}^n$ into the direct sum of Σ-invariant subspaces $\{\,V_i\,\}_{i=1}^l$, each the direct sum of isomorphic irreducible subspaces $\{\,W_i\,\}_{i=1}^l$. A characterization of the irreducible representations of Σ is the homomorphism type of $\mathrm{Hom}_\Sigma(W_i,W_i)\approx\mathsf{K}_i$ with $\mathsf{K}_i=\mathsf{R},\mathsf{C}$ or H for each $1\leq i\leq l$. And so, we can write that

$$\mathbf{R}^n=\bigoplus_{i=1}^{l}V_i\approx\bigoplus_{i=1}^{l}\mathsf{K}_i^{n_i}\otimes_{\mathsf{K}_i}W_i,\tag{8.1}$$

where $\{n_i\}_{i=1}^l\subset\mathsf{N}$ and the W_i's are pairwise non-isomorphic. Note that, although for notational convenience we sometimes use the K_i-linear structures, we are only interested in the *real* structure associated with the vector spaces (and linear operators) we are working with. Denote by a subscript i the action of Σ on W_i. Then the action of Σ on $\mathbf{R}^n$ decomposes into

$$\sigma=\bigoplus_{i=1}^{l}\mathbf{I}_{n_i}\otimes_{\mathsf{K}_i}\sigma_i\,,\quad\forall\sigma\in\Sigma.$$

Expand the generating function h in a neighborhood of the origin,

$$h(x,x',\Lambda)=\tfrac{1}{2}(x,\mathbf{A}(\lambda)\,x)-(x,\mathbf{B}(\lambda)x')+\tfrac{1}{2}(x',\mathbf{C}(\lambda)x')+\hat{h}(x,x',\Lambda),$$

where $\hat{h}$ is a smooth function beginning with terms of third order in x,x', $\mathbf{A}$ and $\mathbf{C}$ are Σ-equivariant symmetric $n\times n$ matrices and $\mathbf{B}$ is a general Σ-equivariant $n\times n$ matrix satisfying $|\mathbf{B}|\neq0$ (the *twist* condition).

Using the isotypic decomposition (8.1) we can also block diagonalize $\mathbf{A}$, $\mathbf{B}$ and $\mathbf{C}$; that is,

$$\mathbf{A}=\bigoplus_{i=1}^{l}\mathbf{A}_i\otimes_{\mathsf{K}_i}\mathbf{I}_{W_i}\,,$$

where $\mathbf{A}_i \in \mathrm{Hom}_{\mathbf{K}_i}(\mathbf{K}_i^{n_i}, \mathbf{K}_i^{n_i})$ and similarly for $\mathbf{B}$ and $\mathbf{C}$. Depending on the context, we denote by the same $\mathbf{A}_i$ the $\mathbf{K}_i$-matrix or the associated *real* maps. And so, the linear Σ-equivariant map $\mathbf{DT}(\lambda) = \mathbf{DT}(0,0,\lambda)$ can be block-diagonalized as

$$\mathbf{DT} = \bigoplus_{i=1}^{l} \mathbf{DT}_i \otimes_{\mathbf{K}_i} \mathbf{I}_{W_i},$$

where

$$\mathbf{DT}_i = \begin{pmatrix} \mathbf{B}_i^{-1}\mathbf{A}_i & \mathbf{B}_i^{-1} \\ \mathbf{C}_i\mathbf{B}_i^{-1}\mathbf{A}_i - \mathbf{B}_i^T & \mathbf{C}_i\mathbf{B}_i^{-1} \end{pmatrix}. \tag{8.2}$$

The transpose/inverse matrix is to be understood as the $\mathbf{K}_i$-conjugate/ $\mathbf{K}_i$-inverse equivalent to the same concept for the associated real map.

When no group is acting, all the symplectic operators are equivalent (Darboux Theorem) and we choose $\mathbf{J}_{2n}$ for the canonical operator. This forces the relation (1.1) between a map and its generating function. Note that (1.1) also forces the Σ-action on $\mathbf{R}^{2n}$ to be the direct sum of the actions on $\mathbf{R}^n$. In the equivariant case some care is necessary for the symplectic structure as we would like to use only Σ-equivariant changes of coordinates.

The problem of the symplectic structure can be tackled block by block in the isotypic decomposition. For the real and quaternionic blocks the theory is straightforward, all the operators are equivalent by Σ-equivariant changes of coordinates and so we can take $\mathbf{J}_{2n}$. The problem is more subtle for the complex blocks as there are $n_i + 1$ different symplectic operators (cf. Appendix I), but only $\mathbf{J}_{2n}$ is compatible with (1.1). In this chapter we are going to consider only absolutely irreducible representations of Σ; therefore, for simplicity, we use only the symplectic operators induced by $\mathbf{J}_{2n}$.

The following result tells us how to compute the spectrum of $\mathbf{DT}$ in the complex or quaternionic cases. We denote by $\sigma_{\mathbf{K}_i}(\mathbf{M})$ the spectrum of $\mathbf{M}$ as a $\mathbf{K}_i$-linear map.

Lemma 8.1. *Let* $\mathbf{M}$ *be a complex matrix representing a linear map on* $\mathbf{R}^{2n}$ *via the isomorphism* $\mathbf{C}^n \to \mathbf{R}^{2n}$ *defined by* $x + i\,y \mapsto (x, y)$. *Then,*

(a) $\sigma_{\mathbf{R}}(\mathbf{M}) = \sigma_{\mathbf{C}}(\mathbf{M}) \cup \overline{\sigma_{\mathbf{C}}(\mathbf{M})}$,

(b) *the (generalized) eigenspaces for* $\mu \in \sigma_{\mathbf{C}}(\mathbf{M})$, *respectively* $\bar{\mu} \in \overline{\sigma_{\mathbf{C}}(\mathbf{M})}$, *are generated by the vectors* $\{(v, -iv)\}$, *respectively* $\{(\bar{v}, i\bar{v})\}$, *where* v *is in the (generalized)* $\mathbf{C}$*-eigenspace of* $\mathbf{M}$.

Proof. The proofs of (a) and (b) are simple calculations as $\mathbf{M} = \mathbf{A} + i\,\mathbf{B}$ corresponds to the real matrix

$$\begin{pmatrix} \mathbf{A} & -\mathbf{B} \\ \mathbf{B} & \mathbf{A} \end{pmatrix}.$$

■

Lemma 8.2. *Let* $\mathbf{M}$ *be a quaternionic matrix representing a linear map on* $\mathbb{R}^{4n}$ *via the isomorphism* $\mathbb{H}^n \to \mathbb{R}^{4n}$ *defined by* $x + i\,y + j\,z + k\,w \mapsto (x, y, z, w)$. *Moreover, from the decomposition* $\mathbf{M} = \mathbf{A} + i\,\mathbf{B} + j\,\mathbf{C} + k\,\mathbf{D}$, *with* $\mathbf{A}, \mathbf{B}, \mathbf{C}, \mathbf{D} \in M_n(\mathbb{R})$, *define the following complex matrix*

$$\mathbf{N} = \begin{pmatrix} \mathbf{A} + i\,\mathbf{C} & -\mathbf{B} + i\,\mathbf{D} \\ \mathbf{B} + i\,\mathbf{D} & \mathbf{A} - i\,\mathbf{C} \end{pmatrix}.$$

Then,

(a) $\sigma_{\mathbb{R}}(\mathbf{M}) = \sigma_{\mathbb{C}}(\mathbf{N}) \cup \overline{\sigma_{\mathbb{C}}(\mathbf{N})}$,

(b) *the (generalized) eigenspaces for* $\mu \in \sigma_{\mathbb{C}}(\mathbf{N})$, *respectively* $\bar{\mu} \in \overline{\sigma_{\mathbb{C}}(\mathbf{N})}$, *are generated by the vectors* $\{(v, -iv)\}$, *respectively* $\{(\bar{v}, i\bar{v})\}$, *where* v *is in the (generalized)* $\mathbb{C}$*-eigenspace of* $\mathbf{N}$,

(c) *in terms of quaternionic algebra,* $q \stackrel{\text{def}}{=} (\alpha_1 + i\,\alpha_2 + j\,\beta_1 + k\,\beta_2) \in \mathbb{H}^n$ *corresponds to an eigenvector* $(\alpha, -i\,\alpha, \bar{\beta}, i\,\bar{\beta})$ *for* $\eta \in \sigma_{\mathbb{R}}(\mathbf{M})$ *iff* $\mathbf{M}\,q = q\,\eta$.

Proof. As for Lemma 8.1, those results are simple calculations. Note that for (a) and (b) we can use the previous results for complex matrices as $\mathbf{M}$ corresponds to the complex matrix $\mathbf{N} = \mathbf{A}' + i\,\mathbf{B}'$ where

$$\mathbf{A}' = \begin{pmatrix} \mathbf{A} & -\mathbf{B} \\ \mathbf{B} & \mathbf{A} \end{pmatrix} \quad \text{and} \quad \mathbf{B}' = \begin{pmatrix} \mathbf{C} & \mathbf{D} \\ -\mathbf{D} & \mathbf{C} \end{pmatrix}.$$

■

Clearly $\sigma_{\mathbb{R}}(\mathbf{DT}) = \cup_i\, \sigma_{\mathbb{R}}(\mathbf{DT}_i)$ and each eigenvalue of $\sigma_{\mathbb{R}}(\mathbf{DT}_i)$ has its multiplicity multiplied by $\dim_{\mathbb{K}_i} \mathrm{W}_i$. We assume that

$$1 \notin \sigma(\mathbf{DT}^o). \tag{8.3}$$

That is, the set of fixed points $(0, 0, \Lambda)$ is isolated. The second hypothesis concerns the resonant subspace of period-q.

For any integer $q \geq 2$, we define the *q-resonant subspace* of $\mathbf{T}$ at the origin, denoted by N_q, as the real part of the sum of the eigenspaces of $\mathbf{DT}^o$ corresponding to the eigenvalues of the type $\exp\left(\frac{2p}{q}\pi i\right)$, $0 < p < q$. Note that we do *not* require $\frac{p}{q}$ to be reduced; this allows for points of period *dividing* q to be also considered.

For our analysis, it is useful if T and N_q satisfy some conditions.

A symplectomorphism $\mathbf{T}$ is *q-regular* (at the origin) if

(a) N_q is not empty and, moreover, there exists an eigenvalue of the type $\exp(\frac{2p}{q}\pi i)$ with $\frac{p}{q}$ a reduced fraction,

(b) $1 \notin \sigma(\mathbf{DT}^o)$,

(c) each eigenvalue of each block $\mathbf{DT}_i^o$ is simple, $1 \leq i \leq l$.

Condition (a) means that the linearization $\mathbf{DT}^o$ has at least a periodic orbit of exact period q. Condition (c) is an "equivalent" condition to (H0) in Section 2.1. Condition (H0) has been introduced to justify the use of the one-parameter (λ) normal form of $\mathbf{DT}$ (in the sense of symplectomorphisms, Proposition 2.2). Because of the symmetry, we cannot avoid multiplicity here, but (c) means that we only consider the multiplicities *forced* by symmetry. Note that (c) does not exclude having equal eigenvalues in *different* blocks, but that is considered as mode interactions where we can introduce the interaction-breaking parameters as *unfolding* parameters for the variational problem W_q we later define for the period-q points (see also the discussion of simple mode-interactions in Appendix J).

We refer to the comments in Section 2.1 about relaxing condition (c). In particular we can ask $\mathbf{DT}^o|_{N_q}$ to be only *semi-simple.* But then, a one dimensional λ must be thought of as defining a *path* into the multiparameter (let's say m) normal form (in the sense of symplectomorphisms) of $\mathbf{DT}^o$. In that case, it might therefore be more natural for λ to be m-dimensional and the setting of the problem should be m distinguished parameters bifurcation theory. In Chapters 9 and 10 we are going to consider the main case when (c) is not satisfied: the collision of multipliers.

We are looking for the bifurcation from the origin of period-q points of a given family of symplectomorphisms $\mathbf{T}$. A necessary condition is for N_q to be non empty. Actually, we are asking more, that $\mathbf{T}$ is *q-regular* (at the origin).

As in Section 2.1, a variational problem can be defined for the existence of period-q points. Let

$$\mathbf{X}_q^n = \{ \ \{x^i\} \in (\mathbf{R}^n)^{\mathbf{Z}} \ : \ x^{i+q} = x^i \ , \quad \forall i \in \mathbf{Z} \ \} .$$

Then a period-q point of the symplectic map $\mathbf{T}$ is a critical point in $\mathbf{X}_q^n$ of the following functional:

$$\mathrm{W}_q(\mathbf{x}, \Lambda) = \sum_{i=1}^{q} h(x^i, x^{i+1}, \Lambda) = \tfrac{1}{2} \langle \mathbf{x} \, , \, \mathbf{L}(\lambda) \mathbf{x} \rangle + N(\mathbf{x}, \Lambda) ,$$

where

$$\mathbf{L}(\lambda) = \mathbf{I}_q \otimes_{\mathbf{R}} (\mathbf{A} + \mathbf{C})(\lambda) - \Gamma_q \otimes_{\mathbf{R}} \mathbf{B}(\lambda) - \Gamma_q^T \otimes_{\mathbf{R}} \mathbf{B}^T(\lambda) . \tag{8.4}$$

As previously, it is a straightforward calculation to see that W_q is $\Sigma \times \mathbf{Z}_q$-invariant where the action of $\mathbf{Z}_q$ on $\mathbf{X}_q^n$ is generated by $\Gamma_q \otimes_{\mathbf{R}} \mathbf{I}_n$ and Σ acts on $\mathbf{X}_q^n$ as $\mathbf{I}_q \otimes_{\mathbf{R}} \sigma$, $\sigma \in \Sigma$.

The next step is to apply the Equivariant Splitting Lemma to W_q to get the reduced bifurcation equations. For that, we need some information on the Hessian of W_q.

Define $\omega = \exp(\frac{1}{q} 2\pi i)$ and $t_q = \{ \ p \in \mathbf{N} \mid \omega^p \in \sigma_{\mathbf{R}}(\mathbf{DT}^o) \ \}$. As we do not consider the bifurcation of fixed points, $|t_q| \leq q - 1$.

Let $\mathbf{F}$ be the Fourier matrix defined in (2.16) and denote by $<\overline{\mathbf{F}}|_{p+1}>\approx \mathbf{C}$ the vector space generated by the real and imaginary parts of the $(p+1)$-th column of $\overline{\mathbf{F}}$. For $p \in t_q$, denote by $\mathrm{N}^q_{i,p}$ the real part of $\mathrm{Ker}\,(\mathbf{A}^o_i + \mathbf{C}^o_i - \omega^{p-1}\mathbf{B}^o_i - \overline{\omega}^{p-1}\mathbf{B}^{oT}_i)$ (as real matrices).

Lemma 8.3. (a) *Via Lemmas 8.1 and 8.2, $(x,y) \in \mathbf{K}_i^{2n_i}$ corresponds to an eigenvector of $\mathbf{DT}^o_i$ with eigenvalue η iff*

$$((\mathbf{A}^o_i + \mathbf{C}^o_i)x - \mathbf{B}^o_i x\eta - \mathbf{B}^{oT}_i x\eta^{-1}) = 0 \quad and \quad y = \mathbf{B}^o_i x\eta - \mathbf{A}^o_i x .$$

(b) $\sigma(\mathbf{L}^o) = \bigcup_{j=1}^q \cup_{i=1}^l \sigma_{\mathbf{R}}(\mathbf{A}^o_i + \mathbf{C}^o_i - \omega^{j-1}\mathbf{B}^o_i - \overline{\omega}^{j-1}\mathbf{B}^{oT}_i)$.

(c) $\mathrm{Ker}\,\mathbf{L}^o = \bigoplus_{p\in t_q} <\overline{\mathbf{F}}|_{p+1}> \otimes_{\mathbf{R}} (\oplus_{i=1}^l \mathrm{N}^q_{i,p} \otimes_{\mathbf{K}_i} \mathrm{W}_i)$

$$\approx \mathbf{C}^{|t_q|} \otimes_{\mathbf{R}} \bigoplus_{p\in t_q} (\oplus_{i=1}^l \mathrm{N}^q_{i,p} \otimes_{\mathbf{K}_i} \mathrm{W}_i) .$$

Proof. The proof of part (a) is an exercise using the explicit form of $\mathbf{DT}^o_i$ in (8.2). For part (b), consider the eigenvalue problem

$$\mathbf{L}^o z = \eta z . \tag{8.5}$$

The idea is to block diagonalize $\mathbf{L}^o$ using the Fourier matrix $\mathbf{F}$. As in the proof of Proposition 2.3, we define $z = \overline{\mathbf{F}} \otimes_{\mathbf{R}} \mathbf{I}_n \cdot \hat{z} \otimes_{\mathbf{R}} w$. From the explicit form for $\mathbf{L}^o$, the eigenvalue problem (8.5) can be written as

$$[\mathbf{I}_q \otimes_{\mathbf{R}} (\mathbf{A}^o + \mathbf{C}^o) - \Omega \otimes_{\mathbf{R}} \mathbf{B}^o - \overline{\Omega} \otimes_{\mathbf{R}} \mathbf{B}^{oT} - \eta\, \mathbf{I}_q \otimes_{\mathbf{R}} \mathbf{I}_n]\, \hat{z} \otimes_{\mathbf{R}} w = 0, \tag{8.6}$$

where $\Omega = \mathrm{diag}(1, \omega, \ldots, \omega^{q-1})$. As (8.6) is a block diagonal eigenvalue problem, we can conclude.

The proof of part (c) is obtained by explicitly solving (8.6). ■

Following the ideas used previously in Section 2.3 we can construct an action for $\Sigma \times \mathbf{Z}_q$ on $\mathrm{Ker}\,\mathbf{L}^o$; in particular Γ_q acts as

$$(\omega^{p_1} z_1, \ldots, \omega^{p_{|t_q|}} z_{|t_q|}) \otimes_{\mathbf{R}} \mathbf{I} \tag{8.7}$$

for $\{p_1, \ldots, p_{|t_q|}\} = t_q$ with at least one of the p_i's coprime with q, and σ acts as

$$\mathbf{I}_{|t_q|} \otimes_{\mathbf{R}} \bigoplus_{p\in t_q} (\oplus_{i=1}^l \mathbf{I}_{\mathrm{N}^q_{i,p}} \otimes_{\mathbf{K}_i} \sigma_i) . \tag{8.8}$$

We have now satisfied all the requirements of the Equivariant Splitting Lemma of Appendix A. Therefore the solutions of the operator equation on $\mathbf{X}^n_q$

$$\nabla_{\mathbf{x}} W_q(\mathbf{x}, \Lambda) = 0$$

are in 1-1 correspondence with the solutions of the reduced bifurcation equations on $\mathrm{Ker}\,\mathbf{L}^o$:

$$\nabla_\chi \widehat{\mathrm{W}}_q(\chi, \Lambda) = 0\,. \tag{8.9}$$

Moreover, $\widehat{\mathrm{W}}_q$ is $\Sigma \times \mathbf{Z}_q$-invariant with the group action as described in (8.7) and (8.8).

What we want now is to get a-priori estimates on the number of solutions of (8.9) using the group equivariance and the gradient structure. More precisely, we are going to derive bounds on the number of solutions in each fixed-point subspace of $\Sigma \times \mathbf{Z}_q$.

We refer to GSS II [1988] for background information on bifurcation with symmetry. Let Π be a proper isotropy subgroup of $\Sigma \times \mathbf{Z}_q$. From the principle of symmetric criticality (Palais [1979]), the periodic orbits of isotropy containing *at least* Π are critical points of $\widehat{\mathrm{W}}_\Pi$, the restriction of the functional $\widehat{\mathrm{W}}_q$ to $\mathrm{Fix}\,\Pi$.

Denote by $\mathrm{N}\Pi$ the normalizer of Π in $\Sigma \times \mathbf{Z}_q$; that is, the elements of $\Sigma \times \mathbf{Z}_q$ commuting with Π. It is well-known that Π induces the symmetry $\mathrm{W}\Pi \overset{\mathrm{def}}{=} \mathrm{N}\Pi/\Pi$ on $\mathrm{Fix}\,\Pi$ ($\mathrm{W}\Pi$ is the *Weyl subgroup* associated with Π).

We can use the special structure of $\Sigma \times \mathbf{Z}_q$ to good effect. We use the short notation $\mathbf{Z}_r$ for $I \times \mathbf{Z}_r \subset \Sigma \times \mathbf{Z}_q$. Define the *temporal subgroup* corresponding to Π by $\mathbf{Z}_q^r(\Pi) = \Pi \cap \mathbf{Z}_q$, where r is the integer given by $\mathbf{Z}_q^r(\Pi) = \mathbf{Z}_{q/r}$. $\mathrm{Fix}\,\Pi$ contains only period-r (or less) orbits. As $\mathrm{Fix}\,\mathbf{Z}_q$ is trivial and $\Pi \neq \Sigma \times \mathbf{Z}_q$, note that $1 < r \leq q$. Every element of $\mathbf{Z}_q$ commutes with Π. That is, $\mathrm{W}\Pi$ contains $\mathbf{Z}_r$, $\mathbf{Z}_r(\mathrm{W}\Pi) \overset{\mathrm{def}}{=} \mathbf{Z}_q/\mathbf{Z}_q^r(\Pi)$ with $1 < r \leq q$.

Two situations arise: first, if $\Pi = \Sigma' \times \mathbf{Z}_q^r(\Pi)$ with $\Sigma' \subset \Sigma$ then $\mathrm{Fix}\,\Pi$ contains period-r orbits with only spatial symmetries, and we call Π a *r-spatial subgroup.* Otherwise, the orbits in $\mathrm{Fix}\,\Pi$ are invariant by some combined space and time actions, in that case we call such Π a *r-spatio-temporal subgroup.*

From Appendix N (Proposition N.1) all isotropy subgroups are *twisted* subgroups, that is, there exists a subgroup $\Xi \subset \Sigma$ and a group homomorphism $\theta : \Sigma \to \mathbf{Z}_r(\mathrm{W}\Pi)$ such that

$$\Pi = \Xi^\theta \times \mathbf{Z}_q^r(\Pi) \overset{\mathrm{def}}{=} \{\, (\sigma, \theta(\sigma)) \mid \sigma \in \Xi \,\} \times \mathbf{Z}_q^r(\Pi)\,.$$

We list the principal results on twisted subgroups in Appendix N.

Moreover, we need the following fundamental property of the $\mathrm{W}\Pi$-symmetry.

Lemma 8.4. $\mathrm{Fix}\,\mathrm{W}\Pi = \{\,0\,\}$.

Proof. $\mathrm{Fix}\,\mathrm{W}\Pi$ is contained into $\mathrm{Fix}\,\mathbf{Z}_r(\mathrm{W}\Pi) \cap \mathrm{Fix}\,\Pi$. Now, suppose that

$$\mathrm{Fix}\,\mathbf{Z}_r(\mathrm{W}\Pi) \cap \mathrm{Fix}\,\Pi \neq \{\,0\,\},$$

then there exists a nontrivial $v \in \operatorname{Fix}\Pi$ such that $\gamma v = v$, $\forall \gamma \in \mathbf{Z}_r(W\Pi)$. Any $\delta \in \mathbf{Z}_q$ can be decomposed as $\delta = \delta_1\delta_2$ with $\delta_1 \in \mathbf{Z}_r(W\Pi)$ and $\delta_2 \in \mathbf{Z}_q^r(\Pi)$. And so, $\delta v = \delta_1\delta_2 v = \delta_1 v = v$, that is $v \in \operatorname{Fix}\mathbf{Z}_q$, hence $v = 0$. From the contradiction, $\operatorname{Fix}\mathbf{Z}_r(W\Pi) \cap \operatorname{Fix}\Pi = \{0\}$ and so $\operatorname{Fix}W\Pi = \{0\}$. ■

We need a last regularity condition. As $\mathbf{DT}^o|_{N_q}$ is semi-simple we can associate to each pair of multipliers a signature ϵ_j (cf. Appendix B). We then assume that we can find coordinates on Ker $\mathbf{L}^o$ such that, as in (2.30), the quadratic part of $\widehat{\mathbf{W}}_q$ is given by

$$\sum_{j=1}^{\frac{1}{2}\dim N_q} \epsilon_j \lambda |\chi_j|^2, \tag{8.10}$$

where ϵ_j is the signature of the j-th multiplier counted with multiplicity. To get the λ-dependence in (8.10) we might have, as in Proposition 2.2, to choose a particular parameter λ for $\mathbf{L}(\lambda)$ satisfying the transversality condition (8.10). We say that Π is $\mathbf{T}$-*nondegenerate* if (8.10) is satisfied and the trace of

$$\frac{1}{2}\frac{\mathrm{d}}{\mathrm{d}\lambda}\left(\operatorname{Hess}\widehat{\mathrm{W}_\Pi}\right)$$

is nonzero and in that case we call the absolute value of that trace the *critical dimension* of Π, denoted by $c(\Pi)$. The critical dimension of Π is the absolute value of the algebraic sum of the signatures of the relevant multipliers. The transversality condition (8.10) for $\mathbf{L}(\lambda)$ can be relaxed in various ways (cf. Chapter 2). We can now state the main results. The first theorem is the most general in the present context.

Theorem 8.5. *Let* $\mathbf{T} : (x, y, \lambda) \mapsto (x', y')$ *be a smooth one parameter family of symplectic maps on* $\mathbf{R}^{2n}$ *generated by* $h(x, x', \lambda)$. *Let* Σ *be a compact Lie group acting orthogonally on* $\mathbf{R}^n$. *Suppose:*

(a) h *is* Σ*-invariant under the diagonal action of* Σ *on* $\mathbf{R}^{2n}$,

(b) $\mathbf{T}$ *is* q*-regular for some* $q \geq 2$.

Then, for every nondegenerate isotropy subgroup $\Pi \subset \Sigma \times \mathbf{Z}_q$ *(*Π *can be trivial), there exist right and left open neighborhoods* $I_r, I_l \subset \mathbf{R}$ *of* 0*, integers* $i_r, i_l \geq 0$ *and* $d(\Pi) \geq 1$ *such that* $i_r + i_l \geq d(\Pi)$ *and that for* $\lambda \in I_r$*, resp.* I_l*, there are at least* i_r*, resp.* i_l*, distinct* $W\Pi$*-orbits of trajectories of period* q*-points of isotropy at least* Π *bifurcating from the origin (that is, of norm as small as we want).*

Remarks. It is important for λ to be *unidimensional.* Similar results for multidimensional parameters are much harder to get because of the continuity of the Morse index and the connexity of S^n for $n \geq 1$ (cf. Bartsch [1992a]). In particular it means that the perturbation parameters (α, β) are considered as fixed. This approach does not distinguish between the different levels of degeneracy (codimension) of the bifurcation equations and so the bound we obtain is a lower bound for *any* such problem with the same equivariant structure.

The number $d(\Pi)$ is defined in Bartsch [1992b] from the difference when λ crosses 0 of the "length" of the exit-set of an isolated invariant set around the origin. In our context it corresponds to the difference of dimensions of the unstable manifold of the origin when λ change sign (cf. also Bartsch & Clapp [1990]). We refer to those papers for the theory and to the Remark (a) thereafter for some typical values.

Proof. The discussions prior to the statement of the theorem were in preparation for the application of results on the bifurcation of critical points of invariant functionals. In particular, we use Theorem 7.12 of Bartsch [1992b]. From Lemma 8.4 above, Fix $W\Pi$ is trivial. The other assumptions of Theorem 6.3 of Bartsch are verified from our hypotheses. Rewriting the conclusions of that theorem, they become our conclusions of Theorem 8.5. From classical theory, for instance Kielhöfer [1988], we know that $d(\Pi) \geq 1$. ■

Corollaries, remarks. (a) In applications $W\Pi$ is often isomorphic to $(\mathbf{Z}_q)^r$ or $(\mathbf{S}^1)^r$ for some $r \geq 0$. In those cases $d(\Pi)$ is given, respectively, by $d\big((\mathbf{Z}_q)^r\big) = c(\Pi)$ when q is a prime number or by $d\big((\mathbf{S}^1)^r\big) = \frac{1}{2}\, c(\Pi)$.
When the maximal torus of $W\Pi$ (its largest abelian subgroup) acts without fixed points, we can also use the previous results (cf. Bartsch [1992b] or also Proposition 2.6 of Bartsch & Clapp [1990]).

(b) If Π is a *maximal* (with respect to inclusion) isotropy subgroup, then the isotropy of the orbits is exactly Π. An interesting class of maximal isotropy subgroups are those with 2D-fixed points subspaces. Those are always non-degenerate.

(c) Theorem 8.5 applies also when Σ is trivial. In that case, if there is some isotropy subgroup, it must be some $\mathbf{Z}_q^r$, r dividing q, and so $W\Pi = \mathbf{Z}_r$, $1 < r \leq q$, where r is the period of the elements of Fix Π.

(d) As in Montaldi, Roberts & Stewart [1988], we can use that $\widehat{W}_\Pi$ is $\mathbf{Z}_r(W\Pi)$-invariant. Then, from the $\mathbf{Z}_r$-theory, we get $c(\Pi)$ distinct $\mathbf{Z}_r(W\Pi)$-orbits of trajectories of period-q points, but we cannot rule out that some *spatio-temporal* symmetry in $W\Pi$ cannot bring one trajectory into another.

To illustrate the problem touched in (d) we can consider the following simple example. Consider the canonical $\mathbf{SO}(2)$-action on $\mathbf{R}^2$ by rotation. An $\mathbf{SO}(2)$-invariant function f_o on $\mathbf{R}^2$ is also $\mathbf{Z}_2$-invariant with the action being $(x,y) \mapsto (-x,-y)$ (for instance). From (a) we know that f_o has at least *one* $\mathbf{SO}(2)$-orbit of critical points but at least *two* $\mathbf{Z}_2$-orbits. Those two orbits are not $\mathbf{Z}_2$-conjugate, that is, they are any couple of (non antipodal) points but they are $\mathbf{SO}(2)$-conjugate, hence the existence of only one $\mathbf{SO}(2)$-orbit. This example shows that it is useful to keep in mind that we might not have identified the more powerful symmetry framework and if it is so, to interpret cautiously the results.

For the next result we use the classical approach to transform the problem into finding the critical points of an equivalent functional on topological spheres. We get more information but at the cost of more stringent hypotheses on the quadratic part (8.10) (cf. Appendix L).

Theorem 8.6. *Let* $\mathbf{T} : (x, y, \lambda) \mapsto (x', y')$ *be a smooth one parameter family of symplectic maps on* $\mathbf{R}^{2n}$ *generated by* $h(x, x', \lambda)$. *Let* Σ *be a compact Lie group acting orthogonally on* $\mathbf{R}^n$. *Suppose:*

(a) h *is* Σ*-invariant under the diagonal action of* Σ *on* $\mathbf{R}^{2n}$,

(b) $\mathbf{T}$ *is* q*-regular for some* $q \geq 2$.

Then for every isotropy subgroup $\Pi \subset \Sigma\times\mathbf{Z}_q$ *such that the signatures of the multipliers corresponding to* Fix Π *are of the* same sign, *there are at least* $\text{cat}_{W\Pi}\text{Fix }\Pi$ $W\Pi$*-orbits of trajectories of period-q points with isotropy at least* Π *bifurcating from the origin for* finite *(cf. Appendix L)* $\widehat{W_\Pi}$, *the action restricted to* Fix Π.

The integer $\text{cat}_{W\Pi}\text{Fix }\Pi$ *is the* $W\Pi$ *Lusternik-Schnirelman category of the unit sphere in* Fix Π *(cf. Bartsch [1989]). Lower bounds for it are given in the following Remark (a).*

Proof. In this situation, because of the special form of the linearization, we can use Proposition L.2 which shows that we can use invariant critical point theory to count the distinct orbits on a small enough topological sphere around the origin in Ker $\mathbf{L}^o \times \mathbf{R}$ using the equivariant Lusternik-Schnirelman category. ■

We refer to our references for further discussions of the equivariant Lusternik-Schnirelman category.

Corollaries, remarks. (a) Estimates of $\text{cat}_{W\Pi}\text{Fix }\Pi$ are as follows. Order the subsets of $W\Pi$ by conjugation and consider $\{H_i\}_{i=1}^t$, the set of maximal classes of isotropy subgroups of the action of $W\Pi$ on Fix Π. Then

$$\sum_{i=1}^{t} c(H_i) \leq \text{cat}_{W\Pi}\text{Fix }\Pi\,,$$

where $c(H_i) = 2$ if $W_{W\Pi}H_i = \{1\}$ (the Weyl group for $H_i \subset W\Pi$) or

$$c(H_i) = (\dim \text{Fix } H_i)/(1 + \dim W_{W\Pi}H_i)$$

otherwise (cf. Bartsch [1989] or Krawcewicz & Marzantowicz [1990]).

(b) The finiteness of $\widehat{W_\Pi}$ is a technical condition and we conjecture that it could be lifted (we refer to Appendix L for further discussion on the issue).

In particular, $\widehat{W}_\Pi$ is finite if the parameter λ does not appear in terms of degree 3 or higher or if $\nabla_\chi \widehat{W}_\Pi$ is finitely $\mathcal{K}^\Gamma_\lambda$-determined (that is, as a bifurcation problem).

Note that the only fundamental additional hypothesis for Theorem 8.6 – compared with Theorem 8.5 – is that the signatures of the corresponding multipliers be the same. This last theorem will find an application in the next section where the signatures of the multipliers are forced to be equal by *symmetry*. Note that we can think of the condition on the multipliers as the pendant of the condition of definiteness of the Hamiltonian function in the "Weinstein-Moser" theory for bifurcation in Hamiltonian flows.

8.2 Bifurcation of subharmonics when Σ acts absolutely irreducibly on $\mathbf{R}^n$ – the configuration space

An important example of the theory of last section is provided when Σ acts absolutely irreducibly on $\mathbf{R}^n$, the configuration space. In the previous set-up, $i = 1$ and $\mathbf{K}_1 = \mathbf{R}$. The fact that Σ acts absolutely irreducibly severely restricts the form of the Σ-invariant generating function h: the matrices $\mathbf{A}$, $\mathbf{B}$ and $\mathbf{C}$ are forced to be scalar multiples of the identity and it is enough (and so natural) to control the multipliers of $\mathbf{DT}^o$ with one parameter only, say λ, as the coefficients of the higher order terms of h will still depend on $\Lambda = (\lambda, \beta)$ where β represents the perturbation (multi)parameter. Hence, h takes the form:

$$h(x, x', \Lambda) = \tfrac{1}{2} a(\lambda)\,(x, x) - b(\lambda)\,(x, x') + \tfrac{1}{2} c(\lambda)\,(x', x') + \hat{h}(x, x', \Lambda). \tag{8.11}$$

The requirement that $|h^o_{12}| \neq 0$ reduces to $b^o \neq 0$. The linear part of the symplectic map generated by h in (8.11) has the form

$$\mathbf{DT} = \frac{1}{b} \begin{pmatrix} a & 1 \\ ac - b^2 & c \end{pmatrix} \otimes \mathbf{I}_n$$

with characteristic equation

$$|\mathbf{DT} - \mu \mathbf{I}_n| = \left| \tfrac{1}{b} \begin{pmatrix} a & 1 \\ ac - b^2 & c \end{pmatrix} - \mu \mathbf{I}_n \right|^n = \left(\mu^2 - \tfrac{1}{b}(a + c)\mu + 1\right)^n = 0 .$$

Therefore if

$$a^o + c^o - 2\,b^o \cos\theta = 0 \quad \text{for} \quad \theta = \tfrac{2p}{q}\pi, \tag{8.12}$$

the map $\mathbf{DT}^o$ has a rational multiplier $e^{i\theta}$ of *multiplicity* n. This time, without limiting the generality, we require p and q to be *coprime*. Moreover, (8.12) determines uniquely θ as a function of λ near 0.

The operator (8.4) is then

$$\mathbf{L}(\lambda) = \left[(a(\lambda) + c(\lambda))\, \mathbf{I}_q - b(\lambda)(\Gamma_q + \Gamma_q^T) \right] \otimes \mathbf{I}_n . \tag{8.13}$$

From Proposition B.3, the signature of each of the n multipliers at $e^{i\theta}$ is of the same sign; in particular, it is equal to the sign of b^o.

As a direct consequence of Theorem 8.6 we can now state the main result on spontaneous symmetry breaking of bifurcating period-q points.

Theorem 8.7. *Let $\mathbf{T} : (x, y, \Lambda) \mapsto (x', y')$ be a smooth Λ-parametrized family of symplectic maps on $\mathbf{R}^{2n}$ generated by h. We assume that h is Σ-invariant where Σ is a compact Lie group acting absolutely irreducibly on the configuration space. We fix the perturbation parameter β and suppose that*

(a) *h is of the form (8.11) and satisfies (8.12) with $(p, q) = 1$ and $q \geq 3$,*

(b) $\theta^o_\lambda = a^o_\lambda + c^o_\lambda \neq 0$.

Then, for every isotropy subgroup $\Pi \subset \Sigma \times \mathbf{Z}_q$ there are at least $\mathrm{cat}_{W\Pi}\mathrm{Fix}\,\Pi$ $W\Pi$-orbits of trajectories of period-q points whose symmetry group is at least Σ bifurcating from the origin for finite $\widehat{W_\Pi}$ (cf. Theorem 8.6).

Proof. Theorem 8.7 is a direct consequence of Theorem 8.6 as the signatures of the multipliers are equal, forced by symmetry. Condition (b) simply means that the quadratic part of the action is of the form (8.10). It could be replaced by $\theta(\lambda)\cdot\lambda > 0$ for λ small. ∎

Theorem 8.7 makes the discussion of the lattice of the isotropy subgroups and the dimensions of their fixed-point subspaces particularly useful. In Appendix N we have formulas for the dimensions of the fixed-points subspaces for the twisted subgroups when $q = 1, 2, 3, 4$ and 6.

In the remainder of Chapter 8 we are going back to a more precise analysis of the bifurcation equations, for example of $\mathbf{O}(2)$-equivariant 4D-maps.

In that case we expect 2D-fixed point subspaces. And so, as $\mathbf{Z}_r(W\Pi) \subset W\Pi \subset \mathbf{SO}(2)$, $\widehat{W_\Pi}$ will be $\mathbf{Z}_r$-invariant with $r \geq 2$ or $\mathbf{SO}(2)$-invariant. $\mathbf{SO}(2)$-invariant functions on $\mathbf{R}^2$ can be reduced to $\mathbf{Z}_2$-invariant functions by taking a point on the orbit. Therefore, in cases where Fix Π is two-dimensional, finite-determinacy results for the bifurcating period-q points can be obtained by application of singularity theory for $\mathbf{Z}_r$-equivariant gradient maps for some $r \geq 2$.

The case where $\widehat{W_q}$ reduces to an $\mathbf{SO}(2)$-invariant functional on Fix Π is an interesting case and corresponds to a circle of bifurcating period-q points. Rather than q-distinct points, the set of initial conditions resulting in period-q points is a circle and coincides with the continuous symmetry. This occurs for example in $\mathbf{O}(2)$-equivariant symplectic maps.

8.3 O(2)-equivariant symplectic maps

Let $\Sigma = \mathbf{O}(2)$ with the standard representation on $\mathbf{R}^2$ given by

$$\mathbf{O}(2) = \langle \mathbf{R}_\phi\,,\, \kappa\rangle \quad \text{with} \quad \mathbf{R}_\phi = \cos\phi\,\mathbf{I} + \sin\phi\,\mathbf{J} \quad \text{and} \quad \kappa = \begin{pmatrix} 1 & 0 \\ 0 & -1 \end{pmatrix}.$$

Suppose $\mathbf{T} : (x, y, \Lambda) \mapsto (x', y')$ is a Λ-parametrized family of $\mathbf{O}(2)$-equivariant symplectic maps on $\mathbf{R}^4$ generated by h. The $\mathbf{O}(2)$-invariance simplifies the Taylor expansion of h in x, x'.

Proposition 8.8. *Suppose $h : \mathrm{R}^{2+2} \to \mathrm{R}$ is a smooth function, $\mathbf{O}(2)$-invariant with respect to the standard action on R^2. Then h has the form*

$$h(x, x') \stackrel{\text{def}}{=} \bar{h}(U, V, W) \tag{8.14}$$

where

$$U = (x, x)\,, \quad V = (x', x') \quad \text{and} \quad W = (x, x'). \tag{8.15}$$

Proof. Let $z = x_1 + ix_2$ and $z' = x_1' + ix_2'$, then $\mathbf{SO}(2) \subset \mathbf{O}(2)$ acts as $(e^{i\phi}z, e^{i\phi}z')$ and $\kappa \cdot (z, z') = (\overline{z}, \overline{z}')$. The result is then a straightforward calculation (see GSS II [1988, p.450]). ■

It is natural to control the multipliers of $\mathbf{DT}^o$ with only one parameter, say λ (cf. (8.11)). Moreover, with a scaling of x, x' and choosing the sign of the twist to be negative (it only means that we authorize a change of orientation of the y-coordinate), we can express any smooth $\mathbf{O}(2)$-invariant generating function of twist maps in the form

$$h(x, x', \Lambda) = \tfrac{1}{2}a(\lambda)\,(x, x) - (x, x') + \tfrac{1}{2}c(\lambda)(x', x') + \tilde{h}(U, V, W, \Lambda) \tag{8.16}$$

where $a, c : \mathbf{R} \to \mathbf{R}$ and

$$\begin{aligned}\tilde{h}(U, V, W, \Lambda) = &\tfrac{1}{2}\tilde{h}_{UU}(0, \Lambda)\,U^2 + \tfrac{1}{2}\tilde{h}_{VV}(0, \Lambda)\,V^2 + \tfrac{1}{2}\tilde{h}_{WW}(0, \Lambda)\,W^2 \\ &+ \tilde{h}_{UV}(0, \Lambda)\,UV + \tilde{h}_{UW}(0, \Lambda)\,UW + \tilde{h}_{VW}(0, \Lambda)\,VW + \cdots\end{aligned} \tag{8.17}$$

and in particular there are no terms of third order in x, x'.

Comparison of (8.16) with (8.12) shows that when

$$a^o + c^o = 2\cos\theta - 1 \quad \text{for} \quad \theta = \tfrac{2p}{q}\pi\,, \tag{8.18}$$

the linear map $\mathbf{DT}^o$ has a double multiplier at $e^{i\theta}$ with positive definite signature. The transversality condition $\theta^o_\lambda \neq 0$ reduces to

$$a^o_\lambda + c^o_\lambda \neq 0\,. \tag{8.19}$$

In preparation for a more detailed investigation, we need the action of $\mathbf{O}(2)\times\mathbf{Z}_q$ on $\operatorname{Ker}\mathbf{L}^o$. With h as given in (8.16) and the hypothesis (8.18), $\mathbf{L}^o$ takes the form

$$\mathbf{L}^o = (2\cos\theta\,\mathbf{I}_q - \Gamma_q - \Gamma_q^T)\otimes\mathbf{I}_2 .$$

Using Lemma 8.3 we find that $\operatorname{Ker}\mathbf{L}^o$ is four-dimensional with

$$\operatorname{Ker}\mathbf{L}^o = \operatorname{span}\{\xi_1,\ldots,\xi_4\}$$

and

$$\xi_1 + i\xi_2 = (\eta_1 + i\eta_2)\otimes(e_1 + ie_2) \quad \text{and} \quad \xi_3 + i\xi_4 = (\eta_1 + i\eta_2)\otimes(e_1 - ie_2)$$

where

$$\eta = \eta_1 + i\eta_2 = \frac{1}{\sqrt{q}}\begin{pmatrix} 1 \\ e^{i\theta} \\ \vdots \\ e^{i(q-1)\theta} \end{pmatrix}.$$

The action of $\mathbf{O}(2)\times\mathbf{Z}_q$ on $\operatorname{Ker}\mathbf{L}^o$ is then obtained from

$$[\xi_1|\cdots|\xi_4]^T\,\Gamma_q\otimes\sigma\,[\xi_1|\cdots|\xi_4]\,, \quad \forall\sigma\in\mathbf{O}(2)\,,$$

and the result is summarized in the following.

Proposition 8.9. *The action of* $\mathbf{O}(2)\times\mathbf{Z}_q$ *on* $\operatorname{Ker}\mathbf{L}^o$ *is*

$$\mathbf{O}(2)\times\mathbf{Z}_q\Big|_{\operatorname{Ker}\mathbf{L}^o} = \left\langle \begin{pmatrix} 0 & 1 \\ 1 & 0 \end{pmatrix}\otimes\mathbf{R}_\theta\,,\ \mathbf{R}_{\theta+\phi}\oplus\mathbf{R}_{\theta-\phi} \right\rangle \ \text{for } \theta = \tfrac{2p}{q}\pi\,,\ \forall\phi\in\mathbf{SO}(2)\,.$$

In particular, partition $\mathbf{R}^4 = (\hat{\chi}_1, \hat{\chi}_2)$, then the action of $\mathbf{O}(2)\times\mathbf{Z}_q$ on $\operatorname{Ker}\mathbf{L}^o$ is

$$\begin{aligned} \mathbf{Z}_q: &\quad \theta\cdot(\hat{\chi}_1,\hat{\chi}_2) = (\mathbf{R}_\theta\hat{\chi}_1, \mathbf{R}_\theta\hat{\chi}_2) \quad \text{for} \quad \theta = \tfrac{2p}{q}\pi\,, \\ \mathbf{SO}(2): &\quad \phi\cdot(\hat{\chi}_1,\hat{\chi}_2) = (\mathbf{R}_\phi\hat{\chi}_1, \mathbf{R}_{-\phi}\hat{\chi}_2)\,, \quad \forall\phi\in\mathbf{SO}(2)\,, \\ \kappa: &\quad \kappa\cdot(\hat{\chi}_1,\hat{\chi}_2) = (\hat{\chi}_2,\hat{\chi}_1)\,. \end{aligned} \tag{8.20}$$

Although the $\mathbf{SO}(2)$-symmetry is a *spatial* symmetry and the $\mathbf{Z}_q$-symmetry is a *temporal* symmetry, the group-theoretic properties of the lattice of isotropy subgroups of $\mathbf{O}(2)\times\mathbf{Z}_q$ are reminiscent of the lattice for $\mathbf{D}_n\times\mathbf{S}^1$ (see GSS II [1988, p.368-9]). On the other hand the fact that the continuous symmetry is a spatial symmetry and the discrete $\mathbf{Z}_q$ is associated with time leads to rather different orbit structures.

Let $(\theta,\phi)\cdot(\hat{\chi}_1,\hat{\chi}_2) = (\mathbf{R}_{\theta+\phi}\hat{\chi}_1, \mathbf{R}_{\theta-\phi}\hat{\chi}_2)$ represent the action of $\mathbf{SO}(2)\times\mathbf{Z}_q$ on $\mathbf{R}^4$ and define (for $\theta = \frac{2p}{q}\pi$)

$$\begin{aligned} \widetilde{\mathbf{Z}}_q &= \langle (\theta,-\theta) \rangle, \\ \mathbf{Z}_2(\kappa) &= \langle \kappa \rangle, \\ \mathbf{Z}_2(\kappa\pi) &= \langle \kappa(0,\pi) \rangle, \\ \mathbf{Z}_2(\kappa\theta) &= \langle \kappa(0,\theta) \rangle \quad (q \equiv 0 \ (mod\ 4)), \\ \mathbf{Z}_2^c &= \langle (\pi,\pi) \rangle \quad (q\ even). \end{aligned}$$

Then the isotropy subgroups of $\mathbf{O}(2)\times\mathbf{Z}_q$ with 2D-fixed point spaces are given in Table 8.1. The proof can be obtained by adapting the proof given in GSS II [1988, p.368] for $\mathbf{D}_n\times\mathbf{S}^1$.

Isotropy Subgroup	q-dependence	Fix Π	$W\Pi$
$\widetilde{\mathbf{Z}}_q$	$q \geq 3$	$(\hat{\chi}_1, 0)$	$\mathbf{SO}(2)$
$\mathbf{Z}_2(\kappa)$ $\mathbf{Z}_2(\kappa)\oplus\mathbf{Z}_2^c$	q odd q even	$(\hat{\chi}_1, \hat{\chi}_1)$	$\mathbf{Z}_{2q}$ $\mathbf{Z}_q$
$\mathbf{Z}_2(\kappa\pi)$ $\mathbf{Z}_2(\kappa\pi)\oplus\mathbf{Z}_2^c$ $\mathbf{Z}_2(\kappa\theta)\oplus\mathbf{Z}_2^c$	q odd $q\equiv 2\,(\mathrm{mod}\ 4)$ $q\equiv 0\,(\mathrm{mod}\ 4)$	$(\hat{\chi}_1, -\hat{\chi}_1)$ $(\hat{\chi}_1, \mathbf{R}_\theta\hat{\chi}_1)$	$\mathbf{Z}_{2q}$ $\mathbf{Z}_q$ $\mathbf{Z}_q$

Table 8.1: Isotropy subgroups of $\mathbf{O}(2)\times\mathbf{Z}_q$ with 2D-fixed point spaces

Table 8.1 combined with Theorem 8.7 proves the following.

Theorem 8.10. *Let* $\mathbf{T} : (x,y,\Lambda) \mapsto (x',y')$ *be a* Λ*-parametrized smooth family of* $\mathbf{O}(2)$*-equivariant symplectic maps on* $\mathbf{R}^4$ *generated by* h *and suppose* h *is of the form (8.16) and satisfies (8.18) and (8.19). Then for each* $q \geq 3$ *there are generically three distinct classes, listed in Table 8.1, of bifurcating period-*q *points in the neighborhood of the trivial fixed point.*

The classes of period-q points have very interesting geometric properties. In the next section we will discuss a physical interpretation of the different classes of periodic points but a rough idea of their nature is obtainable from the group-theoretic properties alone.

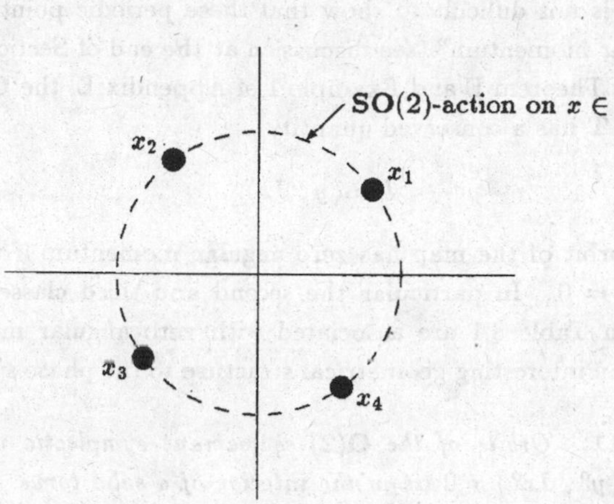

Figure 8.1: Period-4 points in the class $\tilde{Z}_4$

The periodic points with $\tilde{Z}_q$-symmetry ride on the SO(2)-group-orbit. That is, a single discrete time step corresponds to a Z_q ($\subset$ SO(2)) spatial rotation. The fact that the *continuous* group orbit coincides with the *discrete* orbit of the periodic point results in a continuum of periodic points. In essence these periodic points are "integrable" orbits. We will give a more detailed analysis of the normal form shortly, but, the coincidence of the group orbit and the map orbit results in extra symmetry in the normal form.

For example the case $q = 4$ is shown in Figure 8.1. An orbit of a period-4 point is shown but the image of the orbit under $\mathbf{SO(2)} = \langle \mathbf{R}_\theta \rangle$ is also a period-4 point; hence a circle of period-4 points. Similarly for any $q \geq 3$.

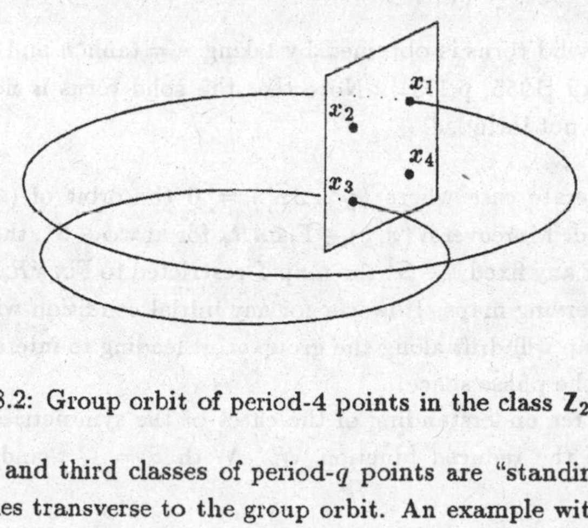

Figure 8.2: Group orbit of period-4 points in the class $Z_2(\kappa) \oplus Z_2^c$

The second and third classes of period-q points are "standing orbits". Their orbits lie in planes transverse to the group orbit. An example with $q = 4$ is shown in Figure 8.2. It is not difficult to show that these periodic points are associated with "zero angular momentum" (see discussion at the end of Section 8.3).

Note that by Theorem E and Example 1 of Appendix E, the **O**(2)-equivariant Λ-family of maps **T** has a conserved quantity

$$I = \langle y , \mathbf{J}x \rangle .$$

We say that an orbit of the map has zero angular momentum if the initial point satisfies $\langle y , \mathbf{J}x \rangle = 0$. In particular the second and third classes of bifurcating period-q points in Table 8.1 are associated with zero angular momentum. The invariant I gives an interesting geometrical structure to the phase space of the map.

Proposition 8.11. *Orbits of the* **O**(2)*-equivariant symplectic map with initial points satisfying* $\langle y^o , \mathbf{J}x^o \rangle \neq 0$ *lie in the interior of a solid torus.*

Proof. By Theorem E the form $I = \langle y , \mathbf{J}x \rangle$ is conserved and when $I \neq 0$ take without loss of generality $I > 0$. Then

$$I = \begin{vmatrix} y_1 & x_1 \\ y_2 & x_2 \end{vmatrix} ,$$

that is, with a suitable scaling of x and y the manifold associated with $I \neq 0$ is reminiscent of $\mathbf{sp}(2,\mathbb{R})$. Now use the classic Gel'fand-Lidskii parametrization of $\mathbf{sp}(2,\mathbb{R})$ scaled by $\sqrt{I}$,

$$\begin{aligned} x_1 &= \sqrt{I}\,(\cosh\tau \sin\psi + \sinh\tau \sin(\theta+\psi)), \\ x_2 &= \sqrt{I}\,(\cosh\tau \cos\psi - \sinh\tau \cos(\theta+\psi)), \\ y_1 &= \sqrt{I}\,(\cosh\tau \cos\psi + \sinh\tau \cos(\theta+\psi)), \\ y_2 &= \sqrt{I}\,(-\cosh\tau \sin\psi + \sinh\tau \sin(\theta+\psi)). \end{aligned}$$

The interior of a solid torus is obtained by taking $r = \tanh^2\tau$ and $(\theta,\psi) \in \mathbf{S}^1 \times \mathbf{S}^1$ (Gel'fand & Lidskii [1955, p.171]). Note that the solid torus is not compact, the boundary $r = 1$ is not included. ∎

In the degenerate case where $\langle y^o , \mathbf{J}x^o \rangle = 0$ the orbit of (x^o, y^o) lies on a singular 3-manifold. Moreover if $(x,y) \in \mathrm{Fix}\,\kappa R_\phi$ for any $\phi \in \mathbf{S}^1$, then $\langle y , \mathbf{J}x \rangle = 0$. In other words, for any fixed $\phi \in \mathbf{S}^1$ the map **T** restricted to $\mathrm{Fix}\,\kappa R_\phi$ is a Λ-family of singular area-preserving maps. However for any initial condition with $\langle x , \mathbf{J}y \rangle \neq 0$ the orbit of the map will drift along the group orbit leading to interesting geometric configurations in the phase space.

To gain a better understanding of the effect of the symmetries on the normal form we compute the reduced function $\widehat{\mathrm{W}}_q$. With $\theta = \frac{2p}{q}\pi$ and $a(\lambda) + c(\lambda) =$

$2\cos\theta - 1 + \lambda$ (say), the linear map $\mathbf{L}$ reduces to

$$\mathbf{L}(\lambda) = (2\cos\theta + \lambda)\mathbf{I}_q \otimes \mathbf{I} - (\Gamma_q + \Gamma_q^T) \otimes \mathbf{I}$$

and the action functional W_q becomes

$$W_q(\mathbf{x}, \Lambda) = \tfrac{1}{2}\langle \mathbf{x}, \mathbf{L}(\lambda)\mathbf{x} \rangle + N(\mathbf{x}, \Lambda)$$

with $N(\mathbf{x}, \Lambda) = \sum_{j=1}^{q} \hat{h}(x^j, x^{j+1}, \Lambda)$ and $\hat{h}$ is given by (8.17).

The kernel of $\mathbf{L}^o$ is spanned by $\xi_1, \ldots, \xi_4$ with

$$\xi_1 + i\xi_2 = \eta \otimes \begin{pmatrix} 1 \\ i \end{pmatrix} \quad \text{and} \quad \xi_3 + i\xi_4 = \eta \otimes \begin{pmatrix} 1 \\ -i \end{pmatrix}.$$

With the decomposition $\mathbf{X}_q^2 = \operatorname{Ker} \mathbf{L}^o \oplus \mathbf{U}$ we can write

$$\mathbf{x} = \sum_{j=1}^{4} \chi_j \xi_j + \Upsilon, \quad \Upsilon \in \mathbf{U}.$$

Application of the Splitting Lemma results in

$$W_q\Big(\sum_{j=1}^{4} \chi_j \xi_j + \phi(\chi, \Upsilon, \Lambda), \Lambda\Big) = \tfrac{1}{2}\langle \Upsilon, \mathbf{L}^o \Upsilon \rangle + \widehat{W}_q(\chi, \Lambda)$$

and a calculation results in

$$\widehat{W}_q(\chi, \Lambda) = \tfrac{1}{2}\lambda I_1 + n_1 I_1^2 + n_2 I_2 + n_3 I_3 + n_4 I_4 + \cdots \tag{8.21}$$

with

$$\begin{aligned} n_1 &= \tfrac{3}{16}(\tilde{h}^o_{UU} + \tilde{h}^o_{VV}) + \tfrac{1}{16}\tilde{h}^o_{WW} + \tfrac{1}{8}\tilde{h}^o_{UV} \\ n_2 &= -\tfrac{1}{16}(\tilde{h}^o_{UU} + \tilde{h}^o_{VV} + \tilde{h}^o_{WW}) + \tfrac{1}{8}\tilde{h}^o_{UV} \\ n_3 &= \tfrac{1}{4}(\tilde{h}^o_{UU} + \tilde{h}^o_{VV} - \tilde{h}^o_{WW}) - \tfrac{1}{2}\tilde{h}^o_{UW} \\ n_4 &= \tfrac{1}{2}(\tilde{h}^o_{UW} - \tilde{h}^o_{VW}). \end{aligned} \tag{8.22}$$

The functions $I_1 \ldots I_4$ are the invariants for $\mathbf{O}(2) \times \mathbf{Z}_q$ which we verify in the following.

Proposition 8.12. *Let $\chi \in \mathbf{R}^4$ and suppose $\mathbf{O}(2) \times \mathbf{Z}_q$ acts on $\mathbf{R}^4$ as in the definition (8.20). Then the ring of invariants is generated by*

$$\begin{aligned} I_1 &= \chi_1^2 + \chi_2^2 + \chi_3^2 + \chi_4^2, \\ I_2 &= (\chi_1^2 + \chi_2^2 - \chi_3^2 - \chi_4^2)^2, \\ I_3 &= \operatorname{Re}[(\chi_1 + i\chi_2)(\chi_3 + i\chi_4)]^{q'}, \\ I_4 &= \operatorname{Im}[(\chi_1 + i\chi_2)(\chi_3 + i\chi_4)]^{q'}. \end{aligned} \tag{8.23}$$

where $q' = q$ if q is odd or $\frac{1}{2}q$ if q is even.

Proof. With complex coordinates $z = (z_1, z_2) \in \mathbf{C}^2$ the group actions reduce to $\theta \cdot z = (e^{i\theta} z_1, e^{i\theta} z_2)$, $\phi \cdot z = (e^{+i\phi} z_1, e^{-i\phi} z_2)$ and $\kappa \cdot z = (z_2, z_1)$. The $\mathbf{SO}(2)$-invariants are $z_1\overline{z}_1$, $z_2\overline{z}_2$, $z_1 z_2$ and $\overline{z}_1\overline{z}_2$. But $\theta \cdot z_1 z_2 = e^{4\pi\frac{p}{q}} z_1 z_2$, therefore the $\mathbf{SO}(2) \times \mathbf{Z}_q$-invariants are $z_1\overline{z}_1$, $z_2\overline{z}_2$, $(z_1 z_2)^{q'}$ and $(\overline{z}_1\overline{z}_2)^{q'}$ (using the relation to eliminate $(z_1 z_2)(\overline{z}_1\overline{z}_2)$). Finally, with $z_1 = \chi_1 + i\chi_2$ and $z_2 = \chi_3 + i\chi_4$ the κ-action reduces the invariants to the four given in (8.23). ∎

Application of singularity theory to the the full $\mathbf{O}(2) \times \mathbf{Z}_q$-invariant functional (8.21) on $\mathbf{R}^4$ is beyond the scope of our work. But we can restrict $\widehat{\mathrm{W}}_q$ to the two-dimensional fixed point spaces of isotropy $\widehat{\Sigma}_i$, $1 \leq i \leq 3$, and study the resulting bifurcation equations (for instance by applying the singularity theory of Chapter 3 or simply by solving explicitly the equations).

Let u, v and w be the usual invariants for $\mathbf{Z}_q$ on $\mathbf{R}^2$ (cf. 4.1). Then Table 8.2 shows the restriction of the $\mathbf{O}(2) \times \mathbf{Z}_q$-invariants $I_1 \ldots I_4$ to Fix $\widehat{\Pi}$. We denote by u, v, w (resp. u, v_{2q}, w_{2q}) the $\mathbf{Z}_q$(resp. $\mathbf{Z}_{2q}$)-invariants (cf. (4.1)).

Isotropy Subgroup	Fix Π	Invariants	WΠ
$\widetilde{\mathbf{Z}}_q$	$(\hat{\chi}_1, 0)$	$I_1 = u,\ I_2 = u^2$ $I_3 = I_4 = 0$	$\mathbf{SO}(2)$
$\mathbf{Z}_2(\kappa)$ $\mathbf{Z}_2(\kappa) \oplus \mathbf{Z}_2^c$	$(\hat{\chi}_1, \hat{\chi}_1)$	$I_1 = 2u,\ I_2 = 0$ $I_3 = v_{2q},\ I_4 = w_{2q}$ $I_3 = v,\ I_4 = w$	$\mathbf{Z}_{2q}$ $\mathbf{Z}_q$
$\mathbf{Z}_2(\kappa\pi)$ $\mathbf{Z}_2(\kappa\pi) \oplus \mathbf{Z}_2^c$ $\mathbf{Z}_2(\kappa\theta) \oplus \mathbf{Z}_2^c$	$(\hat{\chi}_1, -\hat{\chi}_1)$ $(\hat{\chi}_1, \mathbf{R}_\theta\hat{\chi}_1)$	$I_1 = 2u,\ I_2 = 0$ $I_3 = -v_{2q},\ I_4 = -w_{2q}$ $I_3 = -v,\ I_4 = -w$	$\mathbf{Z}_{2q}$ $\mathbf{Z}_q$ $\mathbf{Z}_q$

Table 8.2: Invariants on Fix Π_i $(q \neq 4)$

Denoting by Π_1, Π_2, Π_3, respectively, the three classes of isotropy subgroups, we get the following reduced functionals:

$$\widehat{\mathrm{W}}_q\Big|_{\mathrm{Fix}\,\Pi_1} = \tfrac{1}{2}\lambda u + (n_1 + n_2)u^2 + \cdots,$$

$$\widehat{\mathrm{W}}_q\Big|_{\mathrm{Fix}\,\Pi_2} = \lambda u + 4n_1 u^2 + \begin{cases} n_3 v_{2q} + n_4 w_{2q} + \cdots, & q \text{ odd}, \\ n_3 v + n_4 w + \cdots, & q \text{ even}, \end{cases}$$

$$\widehat{\mathrm{W}}_q\Big|_{\mathrm{Fix}\,\Pi_3} = \lambda u + 4n_1 u^2 - \begin{cases} n_3 v_{2q} - n_4 w_{2q} + \cdots, & q \text{ odd}, \\ n_3 v - n_4 w + \cdots, & q \text{ even}. \end{cases}$$

For $q = 4$, we use the equivalent set of generators u, Δ and w (cf. p.61). It is now straightforward to apply the results of Table 8.2 to the reduced functional (8.21) in the new coordinates, and we find

$$\widehat{\mathrm{W}}_4\Big|_{\mathrm{Fix}\,\Pi_1} = \tfrac{1}{2}\lambda u + (n_1 + n_2)\,u^2 + \cdots$$
$$\widehat{\mathrm{W}}_4\Big|_{\mathrm{Fix}\,\Pi_2} = \lambda u + (4n_1 - n_3)\,u^2 + 2n_3\Delta + n_4 w + \cdots$$
$$\widehat{\mathrm{W}}_4\Big|_{\mathrm{Fix}\,\Pi_3} = \lambda u + (4n_1 + n_3)\,u^2 - 2n_3\Delta - n_4 w + \cdots .$$

Note that for all cases the reduced functional $\widehat{\mathrm{W}}_q$ when restricted to Fix Π_1 is $\mathbf{SO}(2)$-invariant. In particular, if

$$n_1 + n_2 \neq 0$$

that is

$$\tilde{h}^o_{UU} + 2\,\tilde{h}^o_{UV} + \tilde{h}^o_{VV} \neq 0\,,$$

there exists a bifurcating circle of period-q points.

On Fix Π_2 and Fix Π_3 the details of the analysis of the problem depends on the parity of q.

When q is even, in particular when $q = 4$, we get a $\mathbb{Z}_q$-invariant functional on $\mathbb{R}^2$, but when q is odd we get a $\mathbb{Z}_{2q}$-invariant functional, both of which are familiar from the setting of generic bifurcation of period-$q(2q)$ points. Theorem 2.7 is applied to show the existence of bifurcating period-q points with symmetry group Π_2 and Π_3 respectively. From Corollary 4.3, the non-degeneracy conditions for generic bifurcation of period-q points are

$$n_1 \cdot (n_3^2 + n_4^2) \neq 0\,, \quad q \neq 4\,,$$
$$n_3^2 + n_4^2 - 16\,n_1^2 \neq 0\,, \quad q = 4\,.$$

Since the coefficients $n_1 \ldots n_4$ are expressed in terms of the general Taylor expansion of h, explicit non-degeneracy conditions for generic bifurcation can be constructed.

The fact that the reduced functional is $\mathbb{Z}_{2q}$-invariant, when q is odd, is a ramification of Corollary 2.8. Recall that setting: when there exists a $\mathbb{Z}_2$ that acts as minus the identity on $\operatorname{Ker}\mathbf{L}^o$ (or in the present case on $\operatorname{Ker}\mathbf{L}^o\big|_{\mathrm{Fix}\,\Pi_j}$, $j = 2,3$), then the reduced functional is $\mathbb{Z}_{2q}$-invariant (see the proof of Corollary 2.8). The $\mathbb{Z}_2$-action that reduces to $-\mathbf{I}$ for Π_2 and Π_3 in the present case is the restriction of the rotation by π in $\mathbf{SO}(2)$.

And so, there are 3 geometrically distinct classes of period-q points: the family Π_1 has a continuum of period-q points and each of the families Π_2 and Π_3 has two distinct classes of period-q points.

The elements of Fix Π_2 and Fix Π_3 satisfy some spatial symmetries and an explicit calculation shows that those symmetries imply that $I = 0$ on those subspaces.

For the $\Pi_1 (= \widetilde{\mathbf{Z}}_q)$-orbits we can carry out the following calculations. The conserved quantity is $I = \langle\, y^j , \mathbf{J}\, x^j \,\rangle$ but

$$y^j = -h_1(x^j, x^{j+1}) = -2\,\bar{h}_U\, x^j - \bar{h}_W\, x^{j+1}$$

and so

$$I = -\bar{h}_W \langle\, x^{j+1}, \mathbf{J}\, x^j \,\rangle \quad \text{with} \quad \bar{h}_W = -1 + \tilde{h}^o_{WW} W + \tilde{h}^o_{UW} U + \tilde{h}^o_{VW} V + \cdots .$$

To evaluate I on a period-q point orbit use the fact that $\mathbf{x} = \sum_{j=1}^4 \chi_j \xi_j + \Upsilon(\chi, \Lambda)$ to find

$$\begin{aligned} x^1 &= \tfrac{1}{\sqrt{q}} \left[(\chi_1 + \chi_3)\, e_1 + (\chi_2 - \chi_4)\, e_2 \right] + \cdots \\ x^2 &= \tfrac{1}{\sqrt{q}} \{ \cos\theta \left[(\chi_1 + \chi_3)\, e_1 + (\chi_2 - \chi_4)\, e_2 \right] \\ &\qquad + \sin\theta \left[(\chi_2 + \chi_4)\, e_1 + (-\chi_1 + \chi_3)\, e_2 \right] \} + \cdots \end{aligned}$$

and since I is independent of discrete time,

$$I = -\bar{h}_W \langle\, x^2, \mathbf{J}\, x^1 \,\rangle = \tfrac{1}{q} \sin\theta \; (I_2 + \cdots) \left[1 - \tilde{h}^o_{WW} W - \tilde{h}^o_{UW} U - \tilde{h}^o_{VW} V + \cdots \right].$$

For the $\widetilde{\mathbf{Z}}_q$-orbits $\chi_3 = \chi_4 = 0$, therefore

$$I \Big|_{\mathrm{Fix}\, \widetilde{\mathbf{Z}}_q} = \tfrac{1}{q} \sin\theta \; u \left[1 - (\tilde{h}^o_{UW} + \tilde{h}^o_{VW})\, u + \cdots \right] \quad \text{for} \quad u = \chi_1^2 + \chi_2^2 ,$$

which is diffeomorphic to a circle in the configuration space for orbits of sufficiently small amplitude.

8.4 Parametrically forced spherical pendulum

Consider a time-dependent Hamiltonian system of two degrees of freedom with Hamiltonian function:

$$\widehat{H}(q_1, q_2, p_1, p_2, t)$$

and suppose that the system has an $\mathbf{O}(2)$-configuration space symmetry that lifts to the phase space:

$$\widehat{H}(\sigma\, q, \sigma\, p, t) = \widehat{H}(q, p, t)\,, \quad \forall \sigma \in \mathbf{O}(2)\,.$$

Then by application of Proposition 8.6 the Hamiltonian will be a function of

$$Q = q_1^2 + q_2^2\,, \quad P = p_1^2 + p_2^2 \quad \text{and} \quad R = q_1 p_1 + q_2 p_2 \tag{8.24}$$

and t; that is, every (smooth) $\mathbf{O}(2)$-invariant time-dependent Hamiltonian takes the form

$$\widehat{H}(q, p, t) \stackrel{\mathrm{def}}{=} H(q_1^2 + q_2^2, p_1^2 + p_2^2, q_1 p_1 + q_2 p_2, t) \tag{8.25}$$

and moreover there is a conserved quantity, the "angular momentum".

Proposition 8.13. *Let* $\widehat{H}$ *be a time-dependent* $\mathbf{O}(2)$*-invariant Hamiltonian. Then*

$$I = \langle p\,,\, \mathbf{J}q \rangle \tag{8.26}$$

is a first integral.

Proof. As in (8.25), write $\widehat{H}$ as $H(Q,P,R,t)$. Then

$$\nabla_q \widehat{H} = 2\,H_Q\, q + H_R\, p \quad \text{and} \quad \nabla_p \widehat{H} = 2\,H_P\, p + H_R\, q \tag{8.27}$$

and H_Q, H_P and H_R are scalars. Now differentiate I with respect to t,

$$\begin{aligned}
\frac{dI}{dt} &= \langle \dot{p}\,,\, \mathbf{J}q \rangle + \langle p\,,\, \mathbf{J}\dot{q} \rangle \\
&= -\langle \nabla_q \widehat{H}\,,\, \mathbf{J}q \rangle + \langle p\,,\, \mathbf{J}\nabla_p \widehat{H} \rangle \quad (\text{using } \dot{q} = \nabla_p \widehat{H} \text{ and } \dot{p} = -\nabla_q \widehat{H}) \\
&= \langle q\,,\, \mathbf{J}(2\,H_Q\, q + H_R\, p) \rangle + \langle p\,,\, \mathbf{J}(2\,H_P\, p + H_R\, q) \rangle \quad (\text{using (8.27)}) \\
&= 2\,H_Q \langle q\,,\, \mathbf{J}q \rangle + H_R \langle q\,,\, \mathbf{J}p \rangle + 2\,H_P \langle p\,,\, \mathbf{J}p \rangle + H_R \langle p\,,\, \mathbf{J}q \rangle = 0\,.
\end{aligned}$$

■

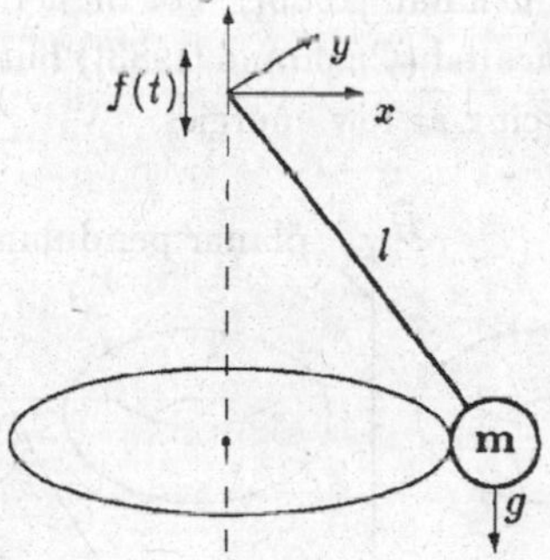

Figure 8.3: The parametrically forced spherical pendulum

The spherical pendulum (see Figure 8.3) is an $\mathbf{O}(2)$-invariant Hamiltonian and when parametrically forced will have the form (8.25) with first integral (8.26). This is demonstrated as follows. Scale the variables and time in the pendulum so that $g = m = l = 1$. Then the Lagrangian for the parametrically forced pendulum (say vertical forcing of the point of attachment) with periodic forcing $f(t)$, is

$$\mathcal{L} = K - V = \tfrac{1}{2}(\dot{x}^2 + \dot{y}^2 + (\dot{f} + \dot{z})^2) - f(t) - z$$

with the constraint $x^2+y^2+z^2=1$. For $z<0$ eliminate z using $z=-\sqrt{1-x^2-y^2}$, then

$$\mathcal{L}=\tfrac{1}{2}(\dot{x}^2+\dot{y}^2)+\tfrac{1}{2}\Big(\frac{x\dot{x}+y\dot{y}}{\sqrt{1-x^2-y^2}}\Big)^2+\sqrt{1-x^2-y^2}+\Big(\frac{x\dot{x}+y\dot{y}}{\sqrt{1-x^2-y^2}}\Big)\dot{f}+\tfrac{1}{2}\dot{f}^2-f.$$

It is evident that the Lagrangian is $\mathbf{O}(2)$-invariant because it depends only on (x^2+y^2), $(\dot{x}^2+\dot{y}^2)$ and $(x\dot{x}+y\dot{y})$.

Now, let $q_1=x$, $q_2=y$ and, with $p_1=\partial\mathcal{L}/\partial\dot{x}$ and $p_2=\partial\mathcal{L}/\partial\dot{y}$ and use the Legendre transform to obtain the Hamiltonian formulation. The Hamiltonian is easily found to be

$$\begin{aligned}\widehat{H}&=\tfrac{1}{2}(p_1^2+p_2^2)-\tfrac{1}{2}(q_1p_1+q_2p_2)^2-\sqrt{1-q_1^2-q_2^2}\\&\quad-\sqrt{1-q_1^2-q_2^2}\,(q_1p_1+q_2p_2)\,\dot{f}(t)-\tfrac{1}{2}(1-q_1^2-q_2^2)\,\dot{f}^2\\&=\tfrac{1}{2}P-\tfrac{1}{2}R^2-\Big[1+\dot{f}(t)\,R+\dot{f}(t)^2\,\sqrt{1-Q}\Big]\sqrt{1-Q}\end{aligned}$$

and by Proposition 8.13 the integral $I=\langle p\,,\,\mathbf{J}q\rangle$ is conserved. For the spherical pendulum, I corresponds to the angular momentum.

When $f(t)=0$ the spherical pendulum is completely integrable and there are (for low amplitudes) three types of orbits. Figure 8.4 shows the energy-momentum space for low energy (see Cushman [1983]). For higher energies the topology of the $(\widehat{H},I)$ space is more complicated (Cushman [1983]) but we will restrict attention to the effect of parametric forcing at low energies.

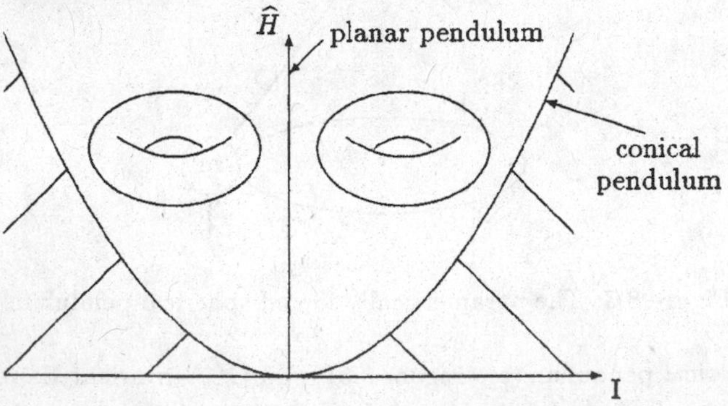

Figure 8.4: Energy-momentum space at low energies for the unforced spherical pendulum (Note that $\widehat{H}$ has been shifted so that $\widehat{H}^o=0$)

For any fixed $\widehat{H}$ sufficiently small, take without loss of generality $I\geq 0$ then the admissible values of I lie in the set $[0,I_{max}]$. The line $I=0$ corresponds to zero angular momentum in which case the pendulum motion is confined to a plane: the system reduces to a one degree of freedom Hamiltonian system. These periodic orbits have $\mathbf{Z}_2^\kappa \oplus \mathbf{Z}_2$ symmetry (see Example 1.3 of Montaldi, Roberts & Stewart [1988]).

When $I = I_{max}$ a family of geometrically distinct periodic orbits occurs: the conical pendulum solutions. For a given value of the energy they correspond to maxima of the angular momentum (or minima for counter-rotating conical pendulum solutions). These solutions have $\widetilde{\mathbf{SO}}(2)$-symmetry; a shift in time corresponds to a shift in the spatial group $\mathbf{SO}(2)$. The conical pendulum solutions ride on the $\mathbf{SO}(2)$-group-orbit. For given $\widehat{H}$, every value of $I \in (0, I_{max})$ corresponds to a flow on a torus (rational or irrational depending on the values of $(\widehat{H}, I)$). The flow is completely integrable so the rotation number varies smoothly between rational and irrational values. For the spherical pendulum, the tori are associated with the drift of the planar pendulum solutions along the $\mathbf{SO}(2)$-group orbit.

When the spherical pendulum is parametrically forced periodically with period-T, the natural Poincare section is the time-T map of the flow. The time-T map will be a symplectic map on $\mathbf{R}^4$ that is equivariant with respect to $\mathbf{O}(2)$ (assuming the forcing is placed so that the $\mathbf{O}(2)$-symmetry is not broken) and depends on three parameters: the amplitude and frequency of the forcing and the angular momentum.

In the case of zero angular momentum the problem reduces to the parametrically forced planar pendulum and the map reduces to an area-preserving map. The second and third classes of periodic points in Table 8.1 correspond to period-q points of the map at zero-angular momentum.

When the angular momentum is perturbed away from zero the whole structure of the area-preserving map, in the fixed point space associated with the planar pendulum, will drift along the group orbit. The orbits of the full symplectic map in this case will behave like orbits of a torus map because the underlying orbit of the autonomous Hamiltonian lies on a torus and, with periodic forcing, the time-T map will behave locally like the map of a torus to itself. In this case we expect the flow of the parametrically forced spherical pendulum to lie on a 3-torus.

For example, the picture on the cover shows a single orbit of the following model $\mathbf{O}(2)$-equivariant map:

$$x' = a\,x + y - d\,(x_1^2 + x_2^2)\,x$$

$$y' = c\,x' - x$$

where $(x, y) \in \mathbf{R}^{2+2}$ and a, c, d are real scalars.

The above map is generated by

$$h(x, x') = \tfrac{1}{2}a\,(x, x) - (x, x') + \tfrac{1}{2}c\,(x', x') - \tfrac{1}{4}d\,(x, x)^2 \,.$$

The picture on the cover is obtained by taking $a = 1$, $c = \frac{9}{10}$ and $d = \frac{1}{10}$. The initial point of the orbit is $((\frac{1}{4}, \frac{8}{10}), (\frac{1}{4}, 0))$. The orbit in this case lies on a pair invariant tori in the phase space; the picture is a projection onto the (x_1, y_1)-plane.

The class of period-q points in Table 8.1 with $\widetilde{\mathbf{Z}}_q$-symmetry are associated with the conical pendulum solutions. They are essentially discrete analogs of the $\widetilde{\mathbf{SO}}(2)$ class of periodic solutions of the unforced problem. When the energy and momentum take the required values and the forcing frequency is in some sense rationally related to the conical pendulum frequency the result is the $\widetilde{\mathbf{Z}}_q$-orbits.

8.5 Reduction to the orbit space

When the map has a continuous symmetry (Lie group of dimension greater than zero) the flow transverse to the group orbit can be studied using orbit space reduction (Bridges & Cushman [1993, Section 5]). In this section orbit space reduction is applied to $\mathbf{O}(2)$-equivariant symplectic maps (the technique applies equally well to equivariant maps with higher dimensional, but compact, symmetry group). $\mathbf{O}(2)$-equivariant maps have an invariant (see Proposition 8.11) and can therefore be reduced to a map on a space of lower dimension. In order to include the singular case we reduce the symplectic map on $\mathbf{R}^4$ to a Poisson map on $\mathbf{R}^3$ – with a Casimir invariant (the fundamental concepts and definitions are available in Bridges & Cushman [1993] and Bridges, Cushman & MacKay [1993]).

First we sketch the reduction using the generating function (8.16) with a potential and then state the general reduction result for $\mathbf{O}(2)$-equivariant maps.

The generating function (8.16) with the higher-order terms in the form of a potential is

$$h(x, x', \Lambda) = \tfrac{1}{2}a(\lambda)(x, x) - (x, x') + \tfrac{1}{2}c(\lambda)(x', x') + \tfrac{1}{2}V((x, x), \Lambda) \tag{8.28}$$

with $a, c : \mathbf{R} \to \mathbf{R}$ and V a smooth potential satisfying $V(0, \Lambda) = V'(0, \Lambda) = 0$ (where the prime indicates differentiation with respect to the first argument). The $\mathbf{O}(2)$-equivariant symplectic map generated by (8.28) is

$$\begin{aligned} x' &= [a(\lambda) + V'(\sigma_1, \Lambda)]\, x + y\,, \\ y' &= [a(\lambda)c(\lambda) + c(\lambda)V'(\sigma_1, \Lambda) - 1]\, x + c(\lambda)\, y\,. \end{aligned} \tag{8.29}$$

To reduce (8.29) to the orbit space introduce coordinates (the Hilbert basis for the $\mathbf{SO}(2)$-invariant functions):

$$\sigma_1 = (x, x)\,, \quad \sigma_2 = (y, y)\,, \quad \sigma_3 = (x, y)\,, \quad \sigma_4 = (y, \mathbf{J}x)\,. \tag{8.30}$$

Note that $\sigma_1 \geq 0$, $\sigma_2 \geq 0$, $\sigma_1\sigma_2 = \sigma_3^2 + \sigma_4^2$ and that $\sigma_4 = I$, the invariant for the map. In terms of the σ-coordinates the map (8.29) is transformed as follows:

$$\begin{aligned}
\sigma_1' &= (x', x') = ((a+V')x + y, (a+V')x + y) \\
&= (a+V')^2\sigma_1 + \sigma_2 + 2(a+V')\sigma_3 \\
\sigma_2' &= (y', y') = ((ac + cV' - 1)x + cy, (ac + cV' - 1)x + cy) \\
&= (ac + cV' - 1)^2\sigma_1 + c^2\sigma_2 + 2c(ac + cV' - 1)\sigma_3 \\
\sigma_3' &= (x', y') = ((a+V')x + y, (ac + cV' - 1)x + cy) \\
&= (a+V')(ac + cV' - 1)\sigma_1 + c\sigma_2 + 2(ac + cV' - \tfrac{1}{2})\sigma_3 \\
\sigma_4' &= (y', \mathbf{J}x') = (y, \mathbf{J}x) = \sigma_4 .
\end{aligned}$$

Let $\sigma = (\sigma_1, \sigma_2, \sigma_3)$ then

$$\sigma' = \mathbf{P}(\sigma, I, \lambda) \stackrel{\text{def}}{=} \mathbf{M}(\sigma_1, I, \lambda)\sigma , \tag{8.31}$$

where

$$\mathbf{M}(\sigma_1, I, \lambda) = \begin{bmatrix} (a+V')^2 & 1 & 2(a+V') \\ (ac+cV'-1)^2 & c^2 & 2c(ac+cV'-1) \\ (a+V')(ac+cV'-1) & c & 2c(a+V') - 1 \end{bmatrix} .$$

Because of the simple form for the potential in (8.28) the linear operator $\mathbf{M}$ does not depend explicitly on σ_2 and σ_3. We verify in the following result that $\mathbf{P}$ is a Poisson map. In fact we prove the more general result that *any* $\mathbf{O}(2)$-equivariant symplectic map on $\mathbf{R}^4$ can be reduced to a Poisson map on $\mathbf{R}^3$ with

$$\psi(\sigma) = \sigma_1\sigma_2 - \sigma_3^2 \tag{8.32}$$

as a Casimir invariant for the Poisson structure.

Theorem 8.14. *Let $\mathbf{T} : (x, y, \Lambda) \mapsto (x', y', \Lambda)$ be a Λ-parametrized smooth family of $\mathbf{O}(2)$-equivariant symplectic maps on $\mathbf{R}^4$ with the standard symplectic action of $\mathbf{SO}(2) \subset \mathbf{O}(2)$. Let $\sigma = (\sigma_1, \sigma_2, \sigma_3)$ be as defined in (8.30). Then $\mathbf{T}$ expressed in terms of the σ-coordinates is a (family of) Poisson maps with respect to the Poisson bracket*

$$\{f, g\} = \langle \nabla\psi, \nabla f \times \nabla g \rangle \tag{8.33}$$

where $f, g \in C^\infty(\mathbf{R}^3, \mathbf{R})$ and $\times$ is the usual cross-product on $\mathbf{R}^3$. Moreover $\psi(\sigma)$ in (8.32) is a Casimir invariant for the Poisson structure.

Proof. The proof of reduction for the general $\mathbf{O}(2)$-equivariant symplectic map (using only the $\mathbf{SO}(2) \subset \mathbf{O}(2)$) follows from Theorem 5.4 in Bridges & Cushman [1993]. Taking the pullback of the standard Poisson bracket on $\mathbf{R}^4$ under the Hilbert map $(x, y) \mapsto (\sigma)$ results in the bracket (8.33). The fact that $\psi(\sigma)$ is a Casimir (that is, $\{\psi, f\} = 0$ for all smooth f) follows from the definition (8.33) and the properties of the vector cross product. ■

Positive values of the Casimir function, $\psi(\sigma) = I^2$, result in a family of smooth 2-manifolds, hyperboloids of a single sheet (cf. Figure 8.5). The foliation is symplectic and each leaf is an invariant set for orbits of the Poisson map $\mathbf{P}$. In other words, when $I \neq 0$, the map can be reduced to a symplectic map on the reduced 2-manifold (which can be flattened and with a suitable change of coordinates it becomes an area-preserving map with respect to the standard area form). When $I = 0$ the level set of the Casimir function degenerates to a half-cone with a singularity at the origin.

Fixed points of the reduced map lift, via reconstruction, to invariant circles, with flow along the group orbit, in the original 4D-map. Note that in general the reduced map is not integrable and will therefore contain all the dynamics expected of a non-integrable area-preserving map, but, in addition the flow along the $\mathbf{SO}(2)$-group orbit will be governed by the complex dynamics on the reduced space.

8.6 Remarks on linear stability for equivariant maps

The linear stability question for periodic points (or more general orbits) of equivariant maps has interesting features forced by the symmetry. In this section we sketch, using an $\mathbf{O}(2)$-equivariant map as an example, the linear stability analysis for orbits confined to a fixed point subspace in phase space. For example we use the $\mathbf{O}(2)$-equivariant map (8.29). Suppressing the parameter dependence the map is

$$\left.\begin{aligned} x' &= ax + y + V'\big((x,x)\big)x \\ y' &= -x + cx' \end{aligned}\right\} \quad (x,y) \in (\mathbf{R}^2 \times \mathbf{R}^2)^{\mathbf{Z}} \tag{8.34}$$

with $a \in \mathbf{R}$ and V a smooth potential satisfying $V(0) = V'(0) = 0$.

Choose an orbit on the fixed point space $(\kappa\, x, \kappa\, y) = (x, y)$ implying $x_2 = y_2 = 0$. Note that such an orbit satisfies zero angular momentum (that is $I = 0$). Restriction of the map (8.34) to Fix κ results in

$$\begin{aligned} \overline{x}_1' &= a\overline{x}_1 + \overline{y}_1 + V'(\overline{x}_1^2)\,\overline{x}_1, \\ \overline{y}_1' &= -\overline{x}_1 + c\overline{x}_1'. \end{aligned} \tag{8.35}$$

Now linearize about the orbit in Fix κ taking $x = \overline{x} + X$ and $y = \overline{y} + Y$ with $\overline{x} = (\overline{x}_1, 0)$ and $\overline{y} = (\overline{y}_1, 0)$,

$$\begin{aligned} X' &= aX + Y + V'(\overline{x}_1^2)\,X + 2\overline{x}_1^2\, V''(\overline{x}_1^2)\,X_1 \begin{pmatrix} 1 \\ 0 \end{pmatrix}, \\ Y' &= -X + cX'. \end{aligned}$$

This equation immediately decouples into

$$\begin{aligned} X_1' &= aX_1 + Y_1 + V'(\overline{x}_1^2)X_1 + 2\overline{x}_1^2 V''(\overline{x}_1^2)X_1, \\ Y_1' &= -X_1 + cX_1', \end{aligned} \tag{8.36}$$

and

$$\begin{aligned} X_2' &= aX_2 + Y_2 + V'(\overline{x}_1^2)\, X_2, \\ Y_2' &= -X_2 + cX_2'. \end{aligned} \tag{8.37}$$

The first set (8.36) is the linear stability problem restricted to Fix κ; that is linearization of the area-preserving map (8.35). When $(\overline{x}_1, \overline{y}_1)$ is a point on a periodic orbit, the linear stability problem is straightforward (for bifurcating period-q points Stability Lemma I applies for example). (8.37) are the *normal variational equations* that determine linear stability normal to the fixed point space. Note that (8.37) has an exact orbit given by taking $X_2 = \overline{x}_1$ (in which case (8.37) reduces to (8.35)). This solution is the tangent vector to the group orbit of solutions of (8.34) (the group orbit is obtained by acting on $(\overline{x}_1, 0, \overline{y}_1, 0)$ with $\mathbf{SO}(2) \subset \mathbf{O}(2)$).

There are two types of "instabilities" here; instability within the fixed point subspace and bifurcation into the group orbit when the angular momentum invariant is perturbed away from zero. It seems that a complete theory for these instabilities and group orbit bifurcations is tractable when the flow in the fixed point subspace is periodic. However it would be interesting to understand the problem when the orbit in Fix κ is ergodic, an invariant circle or other complex orbit and the angular momentum is perturbed away from zero (that is how to characterize the flow along the group orbit in this case).

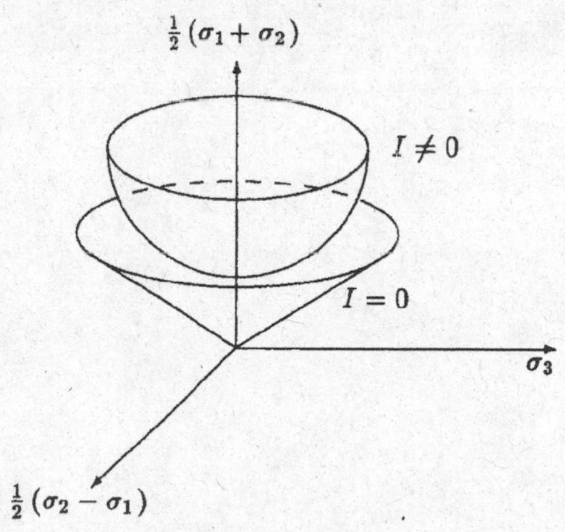

Figure 8.5: Geometry of the Casimir $\psi(\sigma) = I^2$

9. Collision of multipliers at rational points for symplectic maps

Let $\mathbf{T} : (x, y, \lambda, \alpha) \mapsto (x', y')$ be a family of symplectic maps on $\mathbf{R}^4$ parametrized by $(\lambda, \alpha) \in \mathbf{R}^2$. Suppose that the linear map $\mathbf{DT}^o \in \mathbf{Sp}(4, \mathbf{R})$ has eigenvalues $e^{\pm i\theta}$ each of multiplicity two and furthermore that the map $\mathbf{DT}^o$ has a non-trivial nilpotent part. It is shown in Appendix D that there exists a transformation $\widehat{\mathbf{S}} \in \mathbf{Sp}(4, \mathbf{R})$ such that $\widehat{\mathbf{S}}^{-1}\mathbf{DT}^o\widehat{\mathbf{S}} = \mathbf{M}_0$ where $\mathbf{M}_0$ is in Williamson normal form,

$$\mathbf{M}_0 = \begin{pmatrix} \mathbf{R}_\theta & \epsilon \mathbf{R}_\theta \\ 0 & \mathbf{R}_\theta \end{pmatrix} \qquad (9.1)$$

with $\mathbf{R}_\theta = \cos\theta\, \mathbf{I} + \sin\theta\, \mathbf{J}$, $\theta = 2\pi\rho$ with $\rho \in (0, \frac{1}{2})$ and $\epsilon = \pm 1$ is the sign-invariant of the collision (cf. (D.2)).

Consider the nonlinear symplectic map (transform of $\mathbf{T}$ by $\widehat{\mathbf{S}}$)

$$\begin{pmatrix} x' \\ y' \end{pmatrix} = \mathbf{M}_0 \begin{pmatrix} x \\ y \end{pmatrix} + \cdots$$

where the dots refer to terms of higher order in (x, y) and (λ, α).

It is natural in this context to study the dynamics in the neighborhood of the origin by normalizing the nonlinear map. The normalized map is presumably simpler and the dynamics can be completely analyzed (hopefully) and then one studies the persistence of the dynamics in the untransformed map.

For the case when $\theta = 2\pi\rho$ and ρ *is irrational* this approach is used in Bridges & Cushman [1993, Section 4] and Bridges, Cushman & MacKay [1993] to study the dynamics near the irrational collision. The analysis of the normal form is considerably simplified by the fact that the normal-form transformations impose an additional $\mathbf{S}^1$-symmetry resulting in an invariant. Then via reduction the normal form can be completely analyzed (see remarks in Section 9.7).

When ρ is rational however the problem is much more difficult (resonances!). It is still possible to obtain the normal form (in principle) but the construction involves exponentiation of more complicated vectorfields (the resonances producing additional terms). A bigger difficulty however is that the normal form symmetry is discrete ($\mathbf{Z}_q$ when $\rho = p/q$) and therefore there is *no invariant* of the normalized map. On the one hand this implies that resonant collisions will have much more complicated local dynamics and on the other hand it is much harder to analyze.

A complete analysis of periodic points in the unfolding of the rational collision will go a long way towards constructing the backbone of the local dynamics in the phase space. In this chapter we will find that a complete analysis of bifurcating period-q points in the unfolding of a rational collision can be obtained with elegant simplicity using our framework – Splitting Lemma and singularity theory. It turns out that the rational collision shows up in the normal form for bifurcating period-q points as a parameter singularity and so fits neatly into the singularity theory framework.

An important point is that the collision of rational multipliers does not raise the dimension of the kernel of $\mathbf{L}^o$. This can be seen by considering the linear symplectic map

$$\begin{pmatrix} x^{k+1} \\ y^{k+1} \end{pmatrix} = \mathbf{M}_0 \begin{pmatrix} x^k \\ y^k \end{pmatrix} , \quad (x^k, y^k) \in (\mathbf{R}^4)^{\mathbf{Z}}, \tag{9.2}$$

with $\theta = 2\pi\rho$ with $\rho = p/q \in \mathbf{Q}$. Period-$q$ points of the linear map (9.2) lie in a two-dimensional subspace of $\mathbf{R}^4$. To see this note that period-q points of (9.2) satisfy

$$\begin{pmatrix} x^k \\ y^k \end{pmatrix} = \begin{pmatrix} x^{k+q} \\ y^{k+q} \end{pmatrix} = \mathbf{M}_0^q \begin{pmatrix} x^k \\ y^k \end{pmatrix},$$

that is, $(x^0, y^0) \in \mathbf{R}^4$ is an initial condition for a period-q point if and only if $(x^0, y^0) \in \mathrm{Ker}\,(\mathbf{M}_0^q - \mathbf{I}_4)$ but

$$\mathbf{M}_0^q = \begin{pmatrix} 1 & \epsilon \\ 0 & 1 \end{pmatrix}^q \otimes \mathbf{R}_{q\theta} = \begin{pmatrix} 1 & \epsilon q \\ 0 & 1 \end{pmatrix} \otimes \mathbf{I}.$$

It is easily seen that the kernel of $(\mathbf{M}_0^q - \mathbf{I}_4)$ is two dimensional and given by

$$\mathrm{Ker}\,(\mathbf{M}_0^q - \mathbf{I}_4) = \mathrm{span}\left\{ \begin{pmatrix} 1 \\ 0 \\ 0 \\ 0 \end{pmatrix}, \begin{pmatrix} 0 \\ 1 \\ 0 \\ 0 \end{pmatrix} \right\}. \tag{9.3}$$

An additional feature of our analysis is that some surprising linear instability results for the bifurcating period-q points are obtained. We find that – dependent on particular coefficients in the bifurcation equation – the unfolding of the rational collision has a globally connected branch of period-q points. Along the global branches there is *always* a secondary small-angle collision of multipliers.

9.1 Generic theory for the nonlinear rational collision

Before proceeding to an analysis of the nonlinear problem we first construct a two-parameter versal unfolding for $\mathbf{M}_0$ in $\mathbf{Sp}(4, \mathbf{R})$.

Proposition 9.1. *Let* $\mathbf{M} \in \mathbf{Sp}(4,\mathbb{R})$ *and suppose* $\sigma(\mathbf{M}) = \{e^{\pm i2\pi\rho}\}$ *each of algebraic multiplicity two and geometric multiplicity one and* $\rho \in \mathbb{Q} \cap (0, \frac{1}{2})$. *Then the normal form* $\mathbf{M}_0$ *for* $\mathbf{M}$ *is given by (9.1), it is of codimension two in* $\mathbf{Sp}(4,\mathbb{R})$ *and moreover a miniversal two-parameter unfolding of* $\mathbf{M}_0$ *in* $\mathbf{Sp}(4,\mathbb{R})$ *is*

$$\mathbf{M}(\lambda,\alpha) = \begin{bmatrix} (1+\epsilon\alpha)\mathbf{R}_{\theta+\lambda} & \epsilon\mathbf{R}_{\theta+\lambda} \\ \alpha\mathbf{R}_{\theta+\lambda} & \mathbf{R}_{\theta+\lambda} \end{bmatrix} = \begin{pmatrix} 1+\epsilon\alpha & \epsilon \\ \alpha & 1 \end{pmatrix} \otimes \mathbf{R}_{\theta+\lambda}. \tag{9.4}$$

Proof. The fact that $\mathbf{M}_0$ is the normal form under the conditions stated is proved in Appendix D by constructing the necessary symplectic transformation. It is difficult to give a proof of the unfolding working in $\mathbf{Sp}(4,\mathbb{R})$. The idea is to take the logarithm of $\mathbf{M}_0$ and work in the Lie algebra $\mathbf{sp}(4,\mathbb{R})$ where

$$\mathbf{sp}(4,\mathbb{R}) = \{\, \mathbf{A} \in \mathbf{gl}(4,\mathbb{R}) \ :\ \mathbf{A}^T\mathbf{J}_4 + \mathbf{J}_4\mathbf{A} = 0 \,\}.$$

In particular note that

$$\mathbf{M}_0 = \exp \mathbf{A}^o \quad \text{with} \quad \mathbf{A}^o = \begin{pmatrix} \theta\mathbf{J} & \epsilon\mathbf{I} \\ 0 & \theta\mathbf{J} \end{pmatrix} \in \mathbf{sp}(4,\mathbb{R})$$

where $\mathbf{A}^o$ is the Hamiltonian matrix associated with the collision of purely imaginary eigenvalues (Bridges [1990, p.576-7]). A short calculation, using Arnold's theory for versal deformations of matrices (Arnold [1988, p.238]), shows that a spanning set for the normal space of the orbit of $\mathbf{A}^o$ in $\mathbf{sp}(4,\mathbb{R})$ is given by $\{\mathbf{A}_1, \mathbf{A}_2\}$ where

$$\mathbf{A}_1 = \begin{pmatrix} \mathbf{J} & \mathbf{0} \\ \mathbf{0} & \mathbf{J} \end{pmatrix} \quad \text{and} \quad \mathbf{A}_2 = \begin{pmatrix} \mathbf{0} & \mathbf{0} \\ \mathbf{I} & \mathbf{0} \end{pmatrix} \tag{9.5}$$

Given an arbitrary two-parameter unfolding $\mathbf{A}(\lambda,\alpha)$ of $\mathbf{A}^o$, the condition for $\mathbf{A}(\lambda,\alpha)$ to be a versal unfolding is that

$$\operatorname{rank}\begin{pmatrix} [\mathbf{A}^o_\lambda, \mathbf{A}_1] & [\mathbf{A}^o_\lambda, \mathbf{A}_2] \\ [\mathbf{A}^o_\alpha, \mathbf{A}_1] & [\mathbf{A}^o_\alpha, \mathbf{A}_2] \end{pmatrix} = 2 \tag{9.6}$$

where for $\mathbf{U}, \mathbf{V} \in \mathbf{sp}(4,\mathbb{R})$, $[\mathbf{U},\mathbf{V}] = \mathrm{Tr}(\mathbf{U}^T\mathbf{V})$ is an inner-product on $\mathbf{sp}(4,\mathbb{R})$ and

$$\mathbf{A}^o_\lambda = \frac{d}{d\lambda}\bigg|_{\lambda=\alpha=0} \mathbf{A}(\lambda,\mu) \quad \text{and} \quad \mathbf{A}^o_\alpha = \frac{d}{d\alpha}\bigg|_{\lambda=\alpha=0} \mathbf{A}(\lambda,\alpha).$$

Clearly a satisfactory unfolding of $\mathbf{A}^o$ is given by $\hat{\mathbf{A}}(\lambda,\alpha) = \mathbf{A}^o + \lambda\mathbf{A}_1 + \alpha\mathbf{A}_2$. But, although $\hat{\mathbf{A}}(\lambda,\alpha)$ provides a nice unfolding in the Lie algebra, exponentiation of $\hat{\mathbf{A}}(\lambda,\alpha)$ results in a messy unfolding of $\mathbf{M}_0$ in $\mathbf{Sp}(4,\mathbb{R})$. On the other hand the logarithm of $\mathbf{M}(\lambda,\alpha)$ in (9.4) results in a messy function of α in the Lie algebra. But it is $\mathbf{M}(\lambda,\alpha)$ in (9.4) that we use because of its elegant simplicity and therefore all we have to show is that the logarithm satisfies the transversality condition (9.6).

Define $\mathbf{A}(\lambda,\alpha) = \ln \mathbf{M}(\lambda,\alpha)$ with $\mathbf{M}(\lambda,\alpha)$ as given in

$$\mathbf{M}(\lambda,\alpha) = \mathbf{M}_1(\lambda)(\mathbf{I}_4 + \mathbf{M}_2(\alpha))$$

with

$$\mathbf{M}_1(\lambda) = \mathbf{I} \otimes \mathbf{R}_{\theta+\lambda} \quad \text{and} \quad \mathbf{M}_2(\alpha) = \begin{pmatrix} \epsilon\alpha & \epsilon \\ \alpha & 0 \end{pmatrix} \otimes \mathbf{I}$$

and note that $\mathbf{M}_1$ and $\mathbf{M}_2$ commute! Therefore

$$\ln \mathbf{M}(\lambda,\alpha) = \ln \mathbf{M}_1(\lambda) + \ln(\mathbf{I} + \mathbf{M}_2(\alpha))$$

but

$$\ln \mathbf{M}_1(\lambda) = \ln(\mathbf{I} \otimes \mathbf{R}_{\theta+\lambda}) = \mathbf{I} \otimes \ln \mathbf{R}_{\theta+\lambda} = (\theta+\lambda)\mathbf{I} \otimes \mathbf{J}\,.$$

As $\mathbf{M}_2^\circ$ is nilpotent, it is easily verified that the eigenvalues of $\mathbf{M}_2(\alpha)$ go to zero as $\alpha \to 0$. Therefore, the series

$$\ln(\mathbf{I} + \mathbf{M}_2(\alpha)) = \sum_{n=1}^{\infty} \frac{(-1)^{n+1}}{n} (\mathbf{M}_2(\alpha))^n$$

converges absolutely for α sufficiently small. The logarithm of $\mathbf{M}(\lambda,\alpha)$ in (9.4) is then

$$\mathbf{A}(\lambda,\alpha) = (\theta+\lambda)\,\mathbf{I} \otimes \mathbf{J} + \sum_{n=1}^{\infty} \frac{(-1)^{n+1}}{n} (\mathbf{M}_2(\alpha))^n \tag{9.7}$$

from which we compute

$$\left.\frac{d}{d\lambda}\right|_{\lambda=\alpha=0} \mathbf{A}(\lambda,\alpha) = \mathbf{I} \otimes \mathbf{J} \tag{9.8}$$

and

$$\left.\frac{d}{d\alpha}\right|_{\lambda=\alpha=0} \mathbf{A}(\lambda,\alpha) = \begin{pmatrix} \frac{1}{2}\epsilon & -\frac{1}{6} \\ 1 & -\frac{1}{2}\epsilon \end{pmatrix} \otimes \mathbf{I}\,. \tag{9.9}$$

Substitution of (9.8) and (9.9) into (9.6) shows that the 2×2 matrix in (9.6) is diagonal and of full rank. Hence (9.4) is a versal unfolding of $\mathbf{M}_0$. ■

Translation of the linear map $\mathbf{M}(\lambda,\alpha)$ into the **A**,**B**, **C** matrices of the generating function $h(x,x',\Lambda)$ corresponds to

$$\mathbf{A} = (\epsilon+\alpha)\mathbf{I}, \quad \mathbf{B} = \epsilon \mathbf{R}_{\theta+\lambda}^T \quad \text{and} \quad \mathbf{C} = \epsilon\mathbf{I}$$

resulting in the following expression for the quadratic part of the generating function

$$\tfrac{1}{2}(\epsilon+\alpha)(x,x) - \epsilon(x, \mathbf{R}_{\theta+\lambda}^T x') + \tfrac{1}{2}\epsilon(x',x')\,. \tag{9.10}$$

As previously we consider generating functions whose higher order terms (third order or more) might also depend on λ and α. We denote them by $\hat{h}(x,x',\lambda,\alpha)$. One could also consider $\hat{h}$ depending on additional perturbation parameters β to study *degenerate* collisions and their perturbations (cf. Furter [1993] for the reversible case). When the linear map **DT** is given by $\mathbf{M}(\lambda,\alpha)$ in (9.4) or the quadratic part of the generating function is given by (9.10) we say that they are in *general position* for a collision of multipliers.

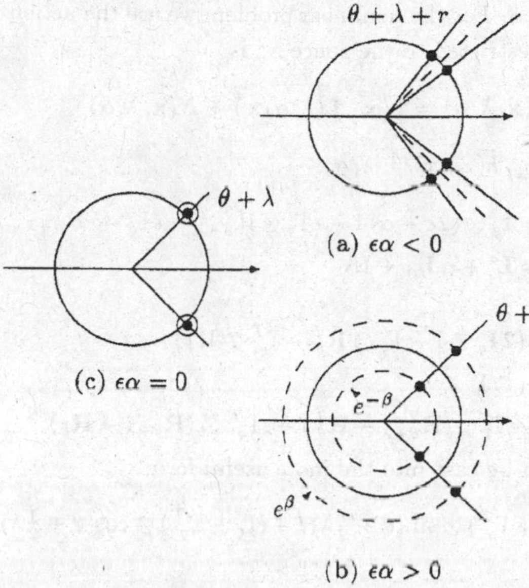

Figure 9.1: Geometry of the Floquet multipliers when the matrix $\mathbf{M}(\lambda,\alpha)\in\mathbf{Sp}(4,\mathbf{R})$ is in general position for a collision of multipliers

The characteristic equation associated with (9.4) is

$$|\mathbf{M}-\mu\mathbf{I}_4|=\mu^4-2(2+\epsilon\alpha)\cos(\theta+\lambda)(\mu+\mu^3)+2(\cos 2(\theta+\lambda)+2+2\epsilon\alpha+\tfrac{1}{2}\alpha^2)\mu^2+1.$$

If $\epsilon\alpha<0$ the multipliers lie at

$$\sigma(\mathbf{M})=\{\,e^{\pm i(\theta+\lambda-r)},e^{\pm i(\theta+\lambda+r)}\,\}\quad\text{with}\quad r=2\sin^{-1}\left(\tfrac{1}{2}\sqrt{-\epsilon\alpha}\right),$$

in particular, the multipliers lie on the unit circle as shown in Figure 9.1(a). When $\alpha=0$ the multipliers coalesce and there is a double (nonsemisimple) multiplier at $e^{\pm i(\theta+\lambda)}$. When $\epsilon\alpha>0$ the multipliers leave the unit circle and

$$\sigma(\mathbf{M})=\{\,e^{-\beta\pm i(\theta+\lambda)},e^{\beta\pm i(\theta+\lambda)}\,\}\quad\text{where}\quad\beta=2\sinh^{-1}\left(\tfrac{1}{2}\sqrt{\epsilon\alpha}\right).$$

In particular they lie on circles of radii $e^{-\beta}$ and e^{β} as shown in Figure 9.1(c). The parameter α is associated with the direction normal to the unit circle while the parameter λ is associated with the tangent space of the unit circle. This completes the analysis of the linear problem.

For analysis of the nonlinear problem near a rational collision we assume henceforth that the multiplier collision takes place at a rational point: $\theta = \frac{2p}{q}\pi$ with p, q in lowest terms and $q \geq 3$. For the nonlinear problem we use the action formulation. The action functional restricted to the space $\mathbf{X}_q^2$ is

$$W_q(\mathbf{x}, \lambda, \alpha) = \tfrac{1}{2}\langle \mathbf{x}, \mathbf{L}(\lambda, \alpha)\mathbf{x} \rangle + N(\mathbf{x}, \lambda, \alpha) \tag{9.11}$$

where $N(\mathbf{x}, \lambda, \alpha) = \sum_{j=1}^{q} \hat{h}(x^j, x^{j+1}, \lambda, \alpha)$,

$$\begin{aligned} \mathbf{L}(\lambda, \alpha) &= \mathbf{I}_q \otimes (2\epsilon + \alpha)\mathbf{I} - \epsilon\Gamma_q \otimes \mathbf{R}_{\theta+\lambda}^T - \epsilon\Gamma_q^T \otimes \mathbf{R}_{\theta+\lambda} \\ &= \mathbf{L}^o + \alpha\,\mathbf{L}_\alpha + \mathbf{L}_\lambda \end{aligned} \tag{9.12}$$

and

$$\begin{aligned} \mathbf{L}^o &= \epsilon\,(2\,\mathbf{I}_q \otimes \mathbf{I} - \Gamma_q \otimes \mathbf{R}_\theta^T - \Gamma_q^T \otimes \mathbf{R}_\theta), \\ \mathbf{L}_\alpha &= \mathbf{I}_q \otimes \mathbf{I} \\ \mathbf{L}_\lambda &= -\epsilon\,\Gamma_q \otimes (\mathbf{R}_{\theta+\lambda}^T - \mathbf{R}_\theta^T) - \epsilon\,\Gamma_q^T \otimes (\mathbf{R}_{\theta+\lambda} - \mathbf{R}_\theta). \end{aligned}$$

Or alternatively $\mathbf{L}_\lambda$ can be cast into the more useful form

$$\mathbf{L}_\lambda = 2\epsilon\,(\sin\tfrac{\lambda}{2})\left[(\Gamma_q + \Gamma_q^T) \otimes \sin(\theta + \tfrac{1}{2}\lambda)\mathbf{I} + (\Gamma_q - \Gamma_q^T) \otimes \cos(\theta + \tfrac{1}{2}\lambda)\,\mathbf{J}\right]. \tag{9.13}$$

Proposition 9.2. *Let $\mathbf{L}^o$ be as defined above with $\theta = \frac{2p}{q}\pi$ and $\epsilon = \pm 1$. Its spectrum (including multiplicities) is given by*

$$\sigma(\mathbf{L}^o) = \left\{ \begin{matrix} 4\epsilon\,(\sin\frac{(p+j-1)}{q}\pi)^2 \\ 4\epsilon\,(\sin\frac{(p-j+1)}{q}\pi)^2 \end{matrix} \right\}, \quad 1 \leq j \leq q. \tag{9.14}$$

In particular, zero is an eigenvalue of multiplicity two ($j = q - p + 1$ and $j = p + 1$) and the nullspace is spanned by $\{\xi_1, \xi_2\}$ with

$$\xi_1 + i\xi_2 = \frac{1}{\sqrt{q}} \begin{pmatrix} 1 \\ e^{i\theta} \\ \vdots \\ e^{i(q-1)\theta} \end{pmatrix} \otimes \begin{pmatrix} 1 \\ i \end{pmatrix}. \tag{9.15}$$

Moreover the spectrum is non-negative and lies in the interval $[0, 4]$ when $\epsilon = +1$ (respectively non-positive and lies in $[-4, 0]$ when $\epsilon = -1$) and the endpoint $4\epsilon \in \sigma(\mathbf{L}^o)$ only when q is even.

Proof. The eigenvalue problem $\mathbf{L}^o z = \eta z$ is block diagonalized by the transformation $z = \overline{\mathbf{F}} \otimes \mathbf{I} \cdot \hat{z} \otimes w$ and with

$$\hat{\mathbf{L}}^o = \mathbf{F} \otimes \mathbf{I} \cdot \mathbf{L}^o \cdot \overline{\mathbf{F}} \otimes \mathbf{I}.$$

The eigenvalue problem $(\hat{\mathbf{L}}^o - \eta\, \mathbf{I}_q \otimes \mathbf{I})\, \hat{z} \otimes w$ reduces to q blocks in $\mathbf{M}_n(2, \mathbf{R})$:

$$(\mathbf{M}_j - \eta_j\, \mathbf{I})\, w_j = 0\,, \quad 1 \le j \le q\,,$$

with

$$\mathbf{M}_j = 2\epsilon \left[1 - (\cos \tfrac{2(j-1)}{q}\pi) \cos\theta\right] \mathbf{I} + 2i\epsilon\,(\sin \tfrac{2(j-1)}{q}\pi) \sin\theta\, \mathbf{J}$$

which has eigenvalues $4\epsilon\,(\sin \frac{(p \pm j \mp 1)}{q}\pi)^2$ verifying (9.14). When $j = p+1$,

$$\mathbf{M}_{p+1} = 2\epsilon\,(\sin\theta)^2 (\mathbf{I} + i\mathbf{J})$$

which has a zero eigenvalue with eigenvector $e_1 + ie_2$. In terms of the original variable z, the eigenfunction is $z = \overline{\mathbf{F}}e_{p+1} \otimes (e_1 + ie_2)$ resulting in (9.15).

The point 4ϵ is in the spectrum when $2(p \pm j \mp 1) = q$ which is satisfied for integer j only when q is even. ■

The basic result on the effect of the rational collision on the bifurcation of period-q points can now be stated.

Theorem 9.3 (Rational Collision Theorem). *Suppose* $\mathbf{T}$ *is a two-parameter smooth family of symplectic maps on* $\mathbf{R}^4$ *generated by* h *with quadratic part in general position for a collision of multipliers at a rational point on the unit circle. Then the bifurcating period-*q *points are in one-to-one correspondence with critical points of a* $\mathbf{Z}_q$*-invariant functional on* $\mathbf{R}^2$ *given by*

$$\widehat{\mathrm{W}}_q(\chi, \lambda, \alpha) = F(u, v, \lambda, \alpha) + w\, G(u, v, \lambda, \alpha) \tag{9.16}$$

with $F(0, 0, \lambda, \alpha) = F^o_u = F^o_{u\lambda} = 0$,

$$F^o_{u\lambda\lambda} = \epsilon \quad \text{and} \quad F^o_{u\alpha} = \tfrac{1}{2}\,. \tag{9.17}$$

Moreover, generically (that is, with the minimal nondegeneracy conditions satisfied) $(\nabla_\chi \widehat{\mathrm{W}}_q \sim g\ ;\ \mathcal{K}^{\mathbf{Z}_q}_\lambda)$ *where*

$$g(z, \lambda, \alpha) = \begin{cases} (\epsilon\lambda^2 + \alpha)\, z + \bar{z}^2, & q = 3\,, \\ (\epsilon\lambda^2 + \alpha + m|z|^2 + \epsilon_2\lambda|z|^2)\, z - \tfrac{1}{2}(z^2 + \bar{z}^2)\,\bar{z}, & q = 4\,, \\ (\epsilon\lambda^2 + \alpha + \epsilon_1|z|^2)\, z + \bar{z}^{q-1}, & q \ge 5\,, \end{cases} \tag{9.18}$$

where $z = \chi_1 + i\chi_2$, $\epsilon_1, \epsilon_2 = \pm 1$ *and* $m \in \mathbf{R}\backslash\{0, 1\}$.

Remark. Note that u, v, and w form a basis for the $\mathbf{Z}_q$ invariants on $\mathbf{R}^2$ (see Proposition 2.6). It is important to note that the normal form for the bifurcating period-q points at a rational collision is identical to that obtained for the generic bifurcation of period-q points in symplectic maps. In particular the period-q points still lie on a two-dimensional subspace due to the presence of a nontrivial nilpotent part in the Jordan normal form. The collision of multipliers manifests itself simply as the singularity $F^o_{u\lambda} = 0$! Therefore the rational collision fits neatly into the singularity theory framework developed for $\mathbf{Z}_q$-equivariant gradient maps on $\mathbf{R}^2$.

Proof. The Splitting Lemma is applied to the action functional W_q in (9.11). First note that in spite of the fact that there are two multipliers (and their conjugates) at the rational point the nullspace of $\mathbf{L}^o$ remains 2-dimensional (Proposition 9.2). Note also that $\nabla_{\mathbf{x}} \mathrm{W}^o_q = 0$ and $\mathrm{Hess}_{\mathbf{x}}\, \mathrm{W}^o_q = \mathbf{L}^o$ with $\mathrm{Ker}\, \mathbf{L}^o = \mathrm{span}\{\xi_1, \xi_2\}$ and that W_q is a $\mathbf{Z}_q$-invariant functional with $\mathbf{Z}_q = \langle \Gamma_q \otimes \mathbf{I} \rangle$. Then with $\mathbf{X}^2_q = \mathbf{A} \oplus \mathbf{B}$, $\mathbf{A} = \mathrm{span}\{\xi_1, \xi_2\}$, all the hypotheses of the Equivariant Splitting Lemma (Appendix A) have been satisfied. Therefore

$$\mathrm{W}_q(\chi_1\xi_1 + \chi_2\xi_2 + \phi(\Upsilon, \chi, \lambda, \alpha), \lambda, \alpha) = \tfrac{1}{2}\langle \Upsilon\, , \mathbf{L}^o\Upsilon \rangle + \widehat{\mathrm{W}}_q(\chi, \lambda, \alpha).$$

Since $\widehat{\mathrm{W}}_q$ is $\mathbf{Z}_q$-invariant with respect to the standard action of $\mathbf{Z}_q$ on $\mathbf{R}^2$ it follows that (see Chapter 2)

$$\widehat{\mathrm{W}}_q(\chi, \lambda, \alpha) = F(u, v, \lambda, \alpha) + w\, G(u, v, \lambda, \alpha),$$

with $F^o = F^o_u = 0$. The fact that $F^o_{u\lambda} = 0$ is a consequence of Proposition 2.11 but it will also be a by-product of the present construction.

The quadratic part of the functional W_q is given by

$$\tfrac{1}{2}\langle \mathbf{x}\, , \mathbf{L}(\lambda, \alpha)\mathbf{x} \rangle = \tfrac{1}{2}\langle \mathbf{x}\, , \mathbf{L}^o\mathbf{x} \rangle + \tfrac{1}{2}\alpha\langle \mathbf{x}\, , \mathbf{L}_\alpha\mathbf{x} \rangle + \tfrac{1}{2}\langle \mathbf{x}\, , \mathbf{L}_\lambda\mathbf{x} \rangle .$$

Therefore, with $\mathbf{x} = \chi_1\xi_1 + \chi_2\xi_2 + \Upsilon$ and $\mathbf{L}_\alpha = \mathbf{I}_q \otimes \mathbf{I}$,

$$\begin{aligned} \tfrac{1}{2}\langle \mathbf{x}\, , \mathbf{L}(\lambda, \alpha)\mathbf{x} \rangle &= \tfrac{1}{2}\langle \Upsilon\, , \mathbf{L}^o\Upsilon \rangle + \tfrac{1}{2}\alpha\langle \chi_1\xi_1 + \chi_2\xi_2\, , \chi_1\xi_1 + \chi_2\xi_2 \rangle \\ &\quad + \tfrac{1}{2}\alpha\langle \Upsilon, \Upsilon \rangle + \tfrac{1}{2}\langle \chi_1\xi_1 + \chi_2\xi_2\, , \mathbf{L}_\lambda(\chi_1\xi_1 + \chi_2\xi_2) \rangle + \cdots \\ &= \tfrac{1}{2}\alpha\, u + \langle \chi_1\xi_1 + \chi_2\xi_2\, , \mathbf{L}_\lambda(\chi_1\xi_1 + \chi_2\xi_2) \rangle + \cdots . \end{aligned} \tag{9.19}$$

The result $F^o_{u\alpha} = \frac{1}{2}$ is now clear. For the result $F^o_{u\lambda\lambda} = \epsilon$, the second term in (9.19) can be simplified,

$$\begin{aligned} \mathbf{L}_\lambda(\xi_1 + i\xi_2) &= 2\epsilon\,(\sin\tfrac{\lambda}{2})\mathbf{I}_q \otimes [2\cos\theta\, \sin(\theta + \tfrac{1}{2}\lambda)\mathbf{I} + 2i\sin\theta\, \cos(\theta + \tfrac{1}{2}\lambda)\mathbf{J}](\xi_1 + i\xi_2) \\ &= 2\epsilon\,(\sin\tfrac{\lambda}{2})\left[2\cos\theta \sin(\theta + \tfrac{1}{2}\lambda) - 2\sin\theta\cos(\theta + \tfrac{1}{2}\lambda)\right] \mathbf{I}_q \otimes \mathbf{I}\,(\xi_1 + i\xi_2) \\ &= 4\epsilon\,(\sin\tfrac{\lambda}{2})^2(\xi_1 + i\xi_2) \end{aligned}$$

from which it follows that

$$\tfrac{1}{2}\langle \chi_1\xi_1 + \chi_2\xi_2\, , \mathbf{L}_\lambda(\chi_1\xi_1 + \chi_2\xi_2) \rangle = 2\epsilon\,(\sin\tfrac{\lambda}{2})^2(\chi_1^2 + \chi_2^2) = \tfrac{1}{2}\epsilon\,\lambda^2 u + \cdots,$$

verifying that $F^o_{u\lambda} = 0$ and $F^o_{u\lambda\lambda} = \epsilon$.

Taking the gradient of $\widehat{W}_q$ using (9.16) and (9.17), and changing to the complex notation $z = \chi_1 + i\chi_2$, results in the following Taylor series expansion

$$g(z,\lambda,\alpha) = (\epsilon\lambda^2 + \alpha + \hat{c}|z|^2 + \cdots)z + (\hat{a} + i\hat{b})\,\bar{z}^{q-1} + \cdots .$$

The normal forms in (9.18) are then obtained by application of the singularity theory for $\mathbf{Z}_q$-equivariant gradient bifurcation maps (cf. Corollary 4.4 in Chapter 4). ■

As an application of Theorem 9.3 we study the rational collision for the class of maps generated by

$$h(x,x',\lambda,\alpha) = \tfrac{1}{2}(\epsilon+\alpha)(x,x) - \epsilon(x,\mathbf{R}^T_{\theta+\lambda}x') + \tfrac{1}{2}\epsilon(x',x') - V(x,\lambda,\alpha) \qquad (9.20a)$$

where we assume that V has the following Taylor expansion to third order (at $\lambda = \alpha = 0$)

$$\begin{aligned} V(x,0,0) &= ax_1^3 + bx_1^2x_2 + cx_1x_2^2 + dx_2^3 \\ &\quad + ex_1^4 + fx_1^3x_2 + gx_1^2x_2^2 + hx_1x_2^3 + ix_2^4 + \cdots \end{aligned} \qquad (9.20b)$$

with $a \ldots i$ arbitrary real numbers. These are the only coefficients in the Taylor expansion of V that contribute essentially to the local bifurcation equations for the rational collision.

9.2 Collision at third root of unity: $\theta = 2\pi/3$

Taking $q = 3$, expanding the reduced functional $\widehat{W}_3$, in equation (9.16), and differentiating, results in the following pre-normal form for a collision of multipliers at a third root of unity,

$$g(z,\lambda,\alpha) = (\epsilon\lambda^2 + \alpha + \cdots)z + (\hat{a} + i\hat{b})\,\bar{z}^2 + \cdots . \qquad (9.21)$$

The singularity theory of Chapter 4 is easily applied to (9.21); in particular, if $\hat{a}^2 + \hat{b}^2 \neq 0$ then

$$\left((\epsilon\lambda^2 + \alpha + \cdots)z + (\hat{a} + i\hat{b})\,\bar{z}^2 + \cdots \sim (\epsilon\lambda^2 + \alpha)z + \bar{z}^2 \;;\; \mathcal{K}^{\mathbf{Z}_3}_\lambda\right).$$

The critical coefficients $(\hat{a},\ \hat{b})$ are contributed by the nonlinear terms in the symplectic map. In this section the coefficients $\hat{a}$ and $\hat{b}$ are computed for the particular class of symplectic maps given by (9.20). The map generated by (9.20) is

$$\begin{aligned} x' &= \mathbf{R}_{\theta+\lambda}\big[(1+\epsilon\alpha)x + \epsilon\, y - \epsilon V_x(x,\lambda,\alpha)\big], \\ y' &= \epsilon\, x' - \epsilon\,\mathbf{R}_{\theta+\lambda}x\,, \end{aligned} \qquad (9.22)$$

and when $q = 3$, $\theta = 2\pi/3$. For the case $q = 3$ the linear operator in (9.12) reduces to $\mathbf{L} = \mathbf{L}^o + \alpha\,\mathbf{L}_\alpha + \mathbf{L}_\lambda$ with

$$\begin{aligned}
\mathbf{L}^o &= \epsilon\left(2\mathbf{I}_3 \otimes \mathbf{I} - \Gamma_3 \otimes \mathbf{R}_\theta^T - \Gamma_3^T \otimes \mathbf{R}_\theta\right) \quad \text{with} \quad \theta = \tfrac{2\pi}{3}\,,\\
\mathbf{L}_\alpha &= \mathbf{I}_3 \otimes \mathbf{I}\,,\\
\mathbf{L}_\lambda &= 2\epsilon\,(\sin\tfrac{\lambda}{2})\left[\sin(\theta + \tfrac{1}{2}\lambda)(\Gamma_3 + \Gamma_3^T) \otimes \mathbf{I} + \cos(\theta + \tfrac{1}{2}\lambda)(\Gamma_3 - \Gamma_3^T) \otimes \mathbf{J}\right].
\end{aligned}$$

A short calculation shows that $\sigma(\mathbf{L}^o) = \{0, 3\epsilon\}$ where 0 has multiplicity 2 and 3ϵ has multiplicity 4. Corresponding to the zero eigenvalues are the eigenvectors

$$\xi_1 = \frac{\sqrt{3}}{6}\begin{pmatrix}2\\-1\\-1\end{pmatrix} \otimes e_1 - \frac{1}{2}\begin{pmatrix}0\\1\\-1\end{pmatrix} \otimes e_2 \quad \text{and} \quad \xi_2 = \frac{\sqrt{3}}{6}\begin{pmatrix}2\\-1\\-1\end{pmatrix} \otimes e_2 + \frac{1}{2}\begin{pmatrix}0\\1\\-1\end{pmatrix} \otimes e_1$$

and corresponding to the eigenvalue 3ϵ are the eigenvectors

$$\xi_3 = \frac{\sqrt{3}}{6}\begin{pmatrix}2\\-1\\-1\end{pmatrix} \otimes e_1 + \frac{1}{2}\begin{pmatrix}0\\1\\-1\end{pmatrix} \otimes e_2\,, \qquad \xi_4 = -\frac{\sqrt{3}}{6}\begin{pmatrix}2\\-1\\-1\end{pmatrix} \otimes e_2 + \frac{1}{2}\begin{pmatrix}0\\1\\-1\end{pmatrix} \otimes e_1,$$

$$\xi_5 = \frac{1}{\sqrt{3}}\begin{pmatrix}1\\1\\1\end{pmatrix} \otimes e_1 \quad \text{and} \quad \xi_6 = \frac{1}{\sqrt{3}}\begin{pmatrix}1\\1\\1\end{pmatrix} \otimes e_2$$

and any $\mathbf{x} \in \mathbf{X}_3^2$ can be expressed as $\mathbf{x} = \sum_{j=1}^{6} \chi_j \xi_j$. Now the action functional on $\mathbf{X}_3^2$ reduces to

$$\mathrm{W}_3(\mathbf{x}, \lambda, \alpha) = \tfrac{1}{2}\langle \mathbf{x}\,,\ \mathbf{L}(\lambda, \alpha)\,\mathbf{x}\rangle - N(\mathbf{x}, \lambda, \alpha)$$

and the leading part of the functional is given by the Rational Collision Theorem to be $(\epsilon\lambda^2 + \alpha)\,u$ and

$$N(\mathbf{x}, \lambda, \alpha) = V(x^1, \lambda, \alpha) + V(x^2, \lambda, \alpha) + V(x^3, \lambda, \alpha)$$

where x^j, $1 \le j \le 3$, are the components of $\mathbf{x}$. Since only χ_1 and χ_2 contribute to the leading order nonlinear term in the normal form we can evaluate $N(\mathbf{x}, 0, 0)$ by taking $\mathbf{x} = \chi_1\xi_1 + \chi_2\xi_2$. A straightforward calculation shows that

$$N(\chi_1\xi_1 + \chi_2\xi_2, 0, 0) = \frac{\sqrt{3}}{12}\left[(a - c)\,v + (d - b)\,w\right]$$

with $v + iw = (\chi_1 + i\chi_2)^3$. Therefore the coefficient $\hat{a} + i\hat{b}$ in the normal form (9.21) is found to be

$$\hat{a} + i\hat{b} = \tfrac{1}{4}\sqrt{3}\left[(c - a) + i\,(d - b)\right].$$

Therefore, supposing that $a \neq c$ *or* $b \neq d$ the normal form for the bifurcating period-3 points in the neighborhood of the collision is $(\epsilon\lambda^2 + \alpha)\,z + \bar{z}^2$.

The bifurcation diagrams are shown below in Figure 9.2 (for $\epsilon = +1$).

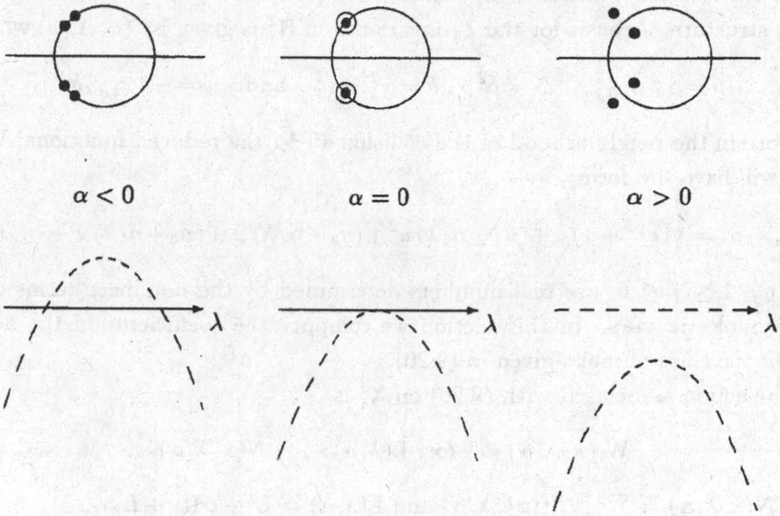

Figure 9.2: Bifurcation of period-3 points in the unfolding of the collision at $\theta = 2\pi/3$

In Figure 9.2, when $\alpha < 0$ the intersection points of the branch with the λ-axis correspond to the two generic bifurcation points of period-3 solution branches which coalesce at $\alpha = 0$ and detach from the origin (as is essential since the origin is linearly unstable). However it is not surprising that the bifurcating period-3 points are all unstable. This is demonstrated as follows.

The normal form $g(z, \lambda, \alpha) = (\epsilon\lambda^2 + \alpha)\, z + \overline{z}^2$ can be written as

$$g(z, \lambda, \alpha) = \begin{bmatrix} \epsilon\lambda^2 + \alpha + \chi_1 & -\chi_2 \\ -\chi_2 & \epsilon\lambda^2 + \alpha - \chi_1 \end{bmatrix} \begin{pmatrix} \chi_1 \\ \chi_2 \end{pmatrix}$$

with non-trivial solutions of $g = 0$ required to satisfy

$$(\epsilon\lambda^2 + \alpha)^2 = (\chi_1^2 + \chi_2^2) = u\,.$$

Now the Jacobian of the normal form is

$$\nabla_z g(z, \lambda, \alpha) = \begin{pmatrix} \epsilon\lambda^2 + \alpha + 2\chi_1 & -2\chi_2 \\ -2\chi_2 & \epsilon\lambda^2 + \alpha - 2\chi_1 \end{pmatrix}$$

and its determinant evaluated at a period-3 point is

$$|\nabla_z g(z, \lambda, \alpha)| = (\epsilon\lambda^2 + \alpha)^2 - 4\,(\chi_1^2 + \chi_2^2) = -3u < 0\,.$$

From the Proposition 9.4 the period-3 points are thus linearly unstable.

9.3 Collision of multipliers at $\pm i$

Of the rational collisions, the collision of multipliers at $\pm i$ has the most interesting structure. A basis for the $\mathbf{Z}_4$-invariants on $\mathbf{R}^2$ is given by (u, Δ, w) where

$$u = \chi_1^2 + \chi_2^2, \quad \Delta = \delta^2, \quad \delta = \chi_2^2 - \chi_1^2 \quad \text{and} \quad w = -4\chi_1\chi_2\delta.$$

Therefore in the neighborhood of the collision at $\pm i$ the reduced functional $\widehat{\mathrm{W}}_4$ in (9.16) will have the form

$$\widehat{\mathrm{W}}_4(z, \lambda, \alpha) = \tfrac{1}{2}(\epsilon\lambda^2 + \alpha)u + (n_1 + n_2\lambda)u^2 + (n_3 + n_4\lambda)\Delta + (n_5 + n_6\lambda)w + \cdots \quad (9.23)$$

where n_j, $1 \le j \le 6$, are real numbers determined by the nonlinear terms of the given symplectic map. In this section we compute the coefficients in the normal form for the class of maps given in (9.20).

The action associated with (9.20) on $\mathbf{X}_4^2$ is

$$\mathrm{W}_4(\mathbf{x}, \lambda, \alpha) = \tfrac{1}{2}\langle \mathbf{x}, \mathbf{L}(\lambda, \alpha)\mathbf{x}\rangle - N(\mathbf{x}, \lambda, \alpha)$$

where $N(\mathbf{x}, \lambda, \alpha) = \sum_{j=1}^4 V(x^j, \lambda, \alpha)$ and $\mathbf{L}(\lambda, \alpha) = \mathbf{L}^o + \alpha\mathbf{L}_\alpha + \mathbf{L}_\lambda$.

When $\theta = \pi/2$ the linear operators forming $\mathbf{L}(\lambda, \alpha)$ reduce to

$$\mathbf{L}^o = \epsilon\,(2\mathbf{I}_4 \otimes \mathbf{I} + (\Gamma_4 - \Gamma_4^T) \otimes \mathbf{J}),$$
$$\mathbf{L}_\alpha = \mathbf{I}_4 \otimes \mathbf{I},$$

and

$$\mathbf{L}_\lambda = \epsilon \sin\lambda\,(\Gamma_4 + \Gamma_4^T) \otimes \mathbf{I} - 2\epsilon\,(\sin\tfrac{\lambda}{2})^2(\Gamma_4 - \Gamma_4^T) \otimes \mathbf{J}.$$

The spectral properties of $\mathbf{L}^o$ are easily obtained using the block diagonalization introduced in Proposition 9.2: $\sigma(\mathbf{L}^o) = \{0, 4\epsilon, 2\epsilon\}$ with $0, 4\epsilon$ each of multiplicity two and 2ϵ of multiplicity four. Corresponding to the zero eigenvalues are the eigenvectors

$$\xi_1 = \frac{1}{2}\begin{pmatrix}1\\0\\-1\\0\end{pmatrix} \otimes e_1 - \frac{1}{2}\begin{pmatrix}0\\1\\0\\-1\end{pmatrix} \otimes e_2 \quad \text{and} \quad \xi_2 = \frac{1}{2}\begin{pmatrix}1\\0\\-1\\0\end{pmatrix} \otimes e_2 + \frac{1}{2}\begin{pmatrix}0\\1\\0\\-1\end{pmatrix} \otimes e_1,$$

corresponding to the eigenvalue 4ϵ are the eigenvectors

$$\xi_3 = \frac{1}{2}\begin{pmatrix}1\\0\\-1\\0\end{pmatrix} \otimes e_1 + \frac{1}{2}\begin{pmatrix}0\\1\\0\\-1\end{pmatrix} \otimes e_2 \quad \text{and} \quad \xi_4 = -\frac{1}{2}\begin{pmatrix}1\\0\\-1\\0\end{pmatrix} \otimes e_2 + \frac{1}{2}\begin{pmatrix}0\\1\\0\\-1\end{pmatrix} \otimes e_1$$

and for the eigenvalues 2ϵ we find

$$\xi_5 = \frac{1}{2}\begin{pmatrix}1\\1\\1\\1\end{pmatrix}\otimes e_1,\ \xi_6 = \frac{1}{2}\begin{pmatrix}1\\1\\1\\1\end{pmatrix}\otimes e_2,\ \xi_7 = \frac{1}{2}\begin{pmatrix}1\\-1\\1\\-1\end{pmatrix}\otimes e_1 \text{ and } \xi_8 = \frac{1}{2}\begin{pmatrix}1\\-1\\1\\-1\end{pmatrix}\otimes e_2.$$

Any $\mathbf{x} \in \mathbf{X}_4^2$ can now be expressed as $\mathbf{x} = \sum_{j=1}^{8} \chi_j \xi_j$. To obtain the coefficients in the reduced functional to fourth order it is necessary to solve the complementary problem (to second order in (χ_1, χ_2)),

$$\mathbf{L}^o\mathbf{w} = \begin{pmatrix} V_x(x^1,\lambda,\alpha)\\ V_x(x^2,\lambda,\alpha)\\ V_x(x^3,\lambda,\alpha)\\ V_x(x^4,\lambda,\alpha)\end{pmatrix} \qquad \text{with} \quad \mathbf{w} \in \{\xi_1,\xi_2\}^{\perp}.$$

The range of $\mathbf{L}^o$ is spanned by $\xi_3 \ldots \xi_8$. A calculation results in

$$\begin{aligned}
&\chi_3 = \chi_4 = 0\\
&\chi_5 = \tfrac{1}{8}\epsilon(1-\lambda)(3a+c)\,u + \cdots, \quad \chi_6 = \tfrac{1}{8}\epsilon(1-\lambda)(3d+b)\,u + \cdots,\\
&\chi_7 = \tfrac{1}{8}\epsilon(1+\lambda)(c-3a)\,\delta + \tfrac{1}{2}\epsilon b(1+\lambda)\,\chi_1\chi_2 + \cdots,\\
&\chi_8 = \tfrac{1}{8}\epsilon(1+\lambda)(3d-b)\,\delta + \tfrac{1}{2}\epsilon c(1+\lambda)\,\chi_1\chi_2 + \cdots,
\end{aligned} \tag{9.24}$$

where a, b, c and d are the Taylor coefficients of $V(x, 0, 0)$. Substitution of (9.24) into W_4 and application of the Rational Collision Theorem results in the following form for the reduced functional

$$\widehat{\mathrm{W}}_4(\chi,\lambda,\alpha) = \tfrac{1}{2}(\epsilon\lambda^2+\alpha)\,u + (n_1+n_2\lambda)\,u^2 + (n_3+n_4\lambda)\,\Delta + (n_5+n_6\lambda)\,w + \cdots \tag{9.25}$$

and the coefficients can be expressed in terms of the potential coefficients $a \ldots i$ as follows

$$\left.\begin{aligned}
n_1 &= -\tfrac{1}{64}\epsilon\left[(3a+c)^2 + (3d+b)^2\right] - \tfrac{1}{16}\,\epsilon\,(b^2+c^2) - \tfrac{1}{16}(e+g+i),\\
n_2 &= \tfrac{1}{64}\epsilon\left[(3a+c)^2 + (3d+b)^2\right] - \tfrac{1}{16}\,\epsilon\,(b^2+c^2),\\
n_3 &= -\tfrac{1}{64}\epsilon\left[(3a-c)^2 + (3d-b)^2\right] + \tfrac{1}{16}\epsilon\,(b^2+c^2) - \tfrac{1}{16}(e-g+i),\\
n_4 &= -\tfrac{1}{64}\epsilon\left[(3a-c)^2 + (3d-b)^2\right] + \tfrac{1}{16}\epsilon\,(b^2+c^2),\\
n_5 &= \tfrac{3}{32}\epsilon\,(cd-ab) - \tfrac{1}{32}\,(f-h),\\
n_6 &= \tfrac{3}{32}\epsilon\,(cd-ab).
\end{aligned}\right\} \tag{9.26}$$

Now we are in a position to resort to the Rational Collision Theorem and Corollary 4.4 of Chapter 4. According to (9.18) with $q = 4$, if $\epsilon_2 m(m-1) \neq 0$, the gradient of $\widehat{\mathrm{W}}_4$ in (9.25) is $\mathcal{K}_\lambda^{\mathbf{Z}_4}$-equivalent to

$$g(z,\lambda,\alpha) = (\epsilon\lambda^2 + \alpha + mu + \epsilon_2\lambda u)\,z + \delta\bar{z}. \tag{9.27}$$

Using the expressions for ϵ_2 and m given in Corollary 4.4 and (9.25)-(9.26) we find

$$m = \frac{1}{2}\left(1 + \frac{F^o_{uu} + F^o_{\Delta}}{\sqrt{F^{o2}_{\Delta} + 4G^{o2}}}\right) = \frac{1}{2}\left(1 + \frac{n_3 + 2n_1}{\sqrt{n_3^2 + 4n_5^2}}\right)$$

and

$$\begin{aligned}\epsilon_2 &= \operatorname{sign}\left(F^o_{\Delta}(F^o_{\Delta}F^o_{uu\lambda} - F^o_{uu}F^o_{\Delta\lambda}) + 4\,G^{o2}(F^o_{\Delta\lambda} + F^o_{uu\lambda}) - 8\,G^oG^o_{\lambda}(F^o_{\Delta} + F^o_{uu})\right)\\ &= \operatorname{sign}\left(2\,n_3(n_2n_3 - n_1n_4) + 4\,n_5^2(n_4 + 2n_2) - 8\,n_5n_6(n_3 + 2n_1)\right).\end{aligned}$$

The non-degeneracy conditions, expressed in terms of $n_1 \ldots n_6$, are

$$\begin{aligned}&0 \neq n_1^2 + n_1n_3 - n_5^2,\\ &0 \neq 2\,n_3(n_2n_3 - n_1n_4)\\ &\qquad\qquad + 4\,n_5^2(n_4 + 2n_2) - 8\,n_5n_6(n_3 + 2n_1).\end{aligned}$$

Substitution of the expressions in (9.26) for $n_1 \ldots n_6$ into the non-degeneracy conditions shows that they are generically satisfied for the map generated by (9.20).

There are *two* $\mathbf{Z}_4$-orbits of bifurcating period-4 points in the unfolding of the collision at $\pm i$ and results are given Table 9.1.

Defining Conditions	Governing Equations	Sign of $\lvert\nabla_z g(z,\lambda,\alpha)\rvert$
$\chi_2 = 0$	$\epsilon\lambda^2 + \alpha + (m - 1 + \epsilon_2\lambda)\,\chi_1^2 = 0$	$(m - 1 + \epsilon_2\lambda)$
$\chi_1^2 = \chi_2^2$	$\epsilon\lambda^2 + \alpha + 2(m + \epsilon_2\lambda)\,\chi_1^2 = 0$	$-(m + \epsilon_2\lambda)$

Table 9.1: Branches of bifurcating period-4 points near the collision at $\pm i$

Note that when the map $\mathbf{T}$ is also reversible, those two orbits correspond to the two isotropy types of bifurcating period-4 points as in Table 7.2. In this situation (that is non-reversible) they do not correspond to any *additional* symmetry but we retain the same notation as in Chapter 7 for convenience in labeling.

Bifurcation diagrams for the bifurcating period-4 points are shown below in Figure 9.3.

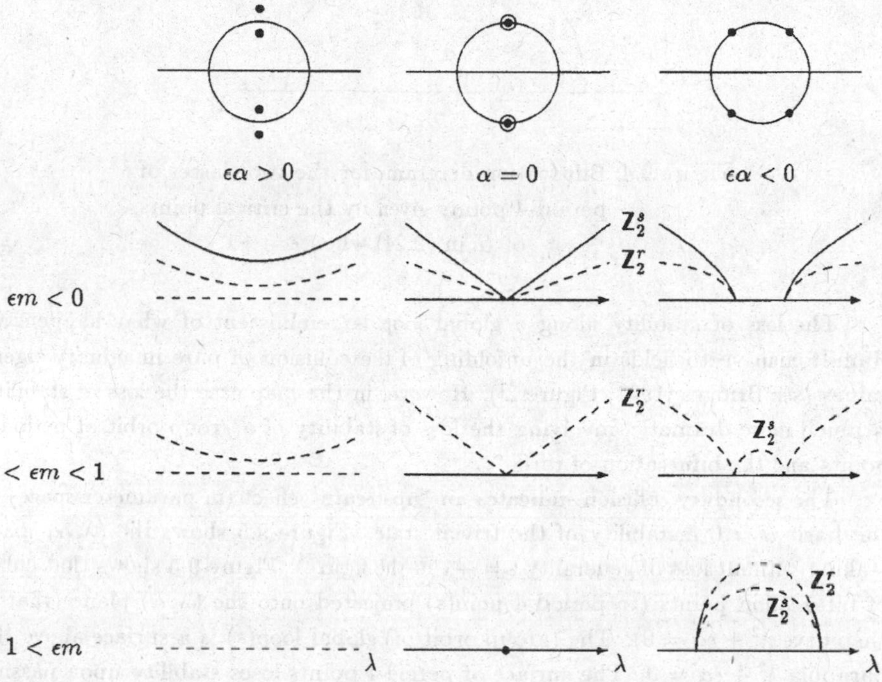

Figure 9.3: Bifurcation of period-4 points in the unfolding of the collision at $\pm i$.

The stability assignments are based on the results in Sections 9.5 and 9.6. When the sign of the determinant of $\nabla_z g$ is negative the period-4 points are linearly unstable (Instability Lemma). When the sign of $\nabla_z g$ is positive additional work is necessary to prove linear stability and this analysis is in Section 9.6. The interesting case is when $\alpha < 0$ and $m \in (1, \infty)$ when both branches of period-4 points form (a group orbit of) globally connected branches (global "loops"). One family of global loops is unstable and the other branch is initially stable but loses stability at finite amplitude through a *secondary (group orbit of) small-angle collision(s)* (a blow-up of this bifurcation diagram is shown in Figure 9.4).

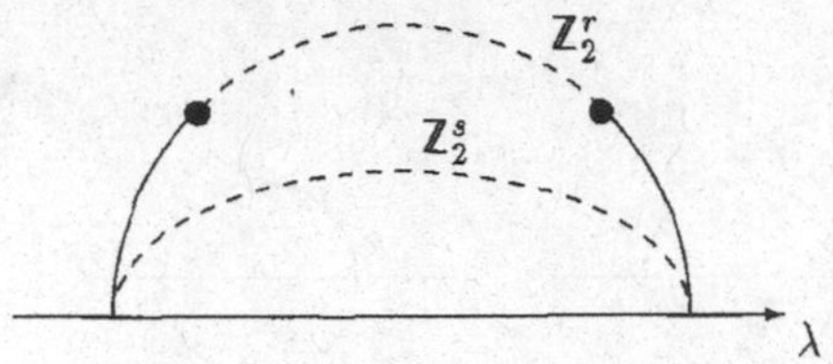

Figure 9.4: Bifurcation diagrams for the two classes of period-4 points given by the critical points of H in (2.24) when $\epsilon = +1$.

The loss of stability along a global loop is reminiscent of what happens in Hamiltonian vectorfields in the unfolding of the collision of pure imaginary eigenvalues (see Bridges [1991, Figure 2]). However in the map case the loss of stability is much more dramatic, involving the loss of stability of a group orbit of periodic points and the bifurcation of tori.

The secondary collision indicates an "upstream" effect (in parameter space) of the basic $\alpha = 0$ instability of the trivial state. Figure 9.5 shows the (λ, α)-space (taking without loss of generality $\epsilon = +1$ in the figure). Figure 9.5 shows that curve of bifurcation points (to period-4 points) projected onto the (λ, α) plane (that is the curve $\lambda^2 + \epsilon\alpha = 0$). The (group orbit of) global loop(s) is a surface above the parabola $\lambda^2 + \epsilon\alpha = 0$. The surface of period-4 points loses stability upon passing into the inner parabola associated with the secondary collision. In other words the period-1 collision effects the bifurcating period-4 points, when $\alpha < 0$, at *finite amplitude*. A study of the small-angle collision in reversible maps is given by Bhowal et al. [1993].

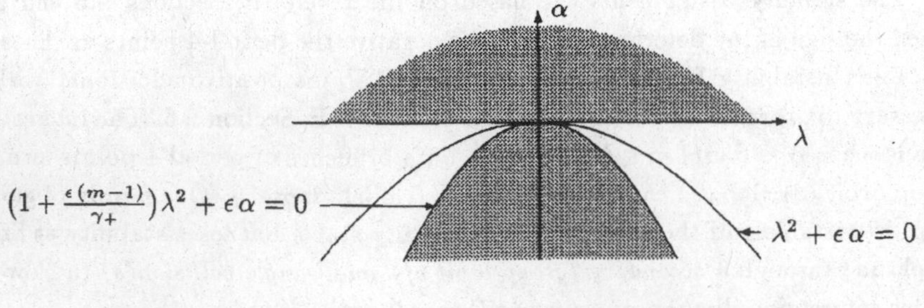

Figure 9.5: Projection of the stable and unstable regions onto the (λ, α)-plane for the Z_2^r-branch when $\epsilon = +1$, $\alpha < 0$ and $\epsilon m > \epsilon$.

9.4 Collision at rational points with $q \geq 5$

The bifurcation of period-q points in the unfolding of the collision at rational points with $\theta = \frac{2p}{q}\pi$ has essentially the same structure for all $q \geq 5$. The normal form is given by

$$g(z, \lambda, \alpha) = (\epsilon\lambda^2 + \alpha + \epsilon_1|z|^2)\,z + \bar{z}^{q-1} \tag{9.28}$$

with $\epsilon = \pm 1$. There are two cases depending on whether ϵ_1 is positive or negative. Note that the only difference between the normal form in (9.28) and the normal form for generic bifurcation of period-q points ($q \geq 0$) is the presence of the *collision singularity*; that is, the bifurcation parameter λ appears quadratically and the unfolding parameter α is necessary for a complete characterization of the local structure.

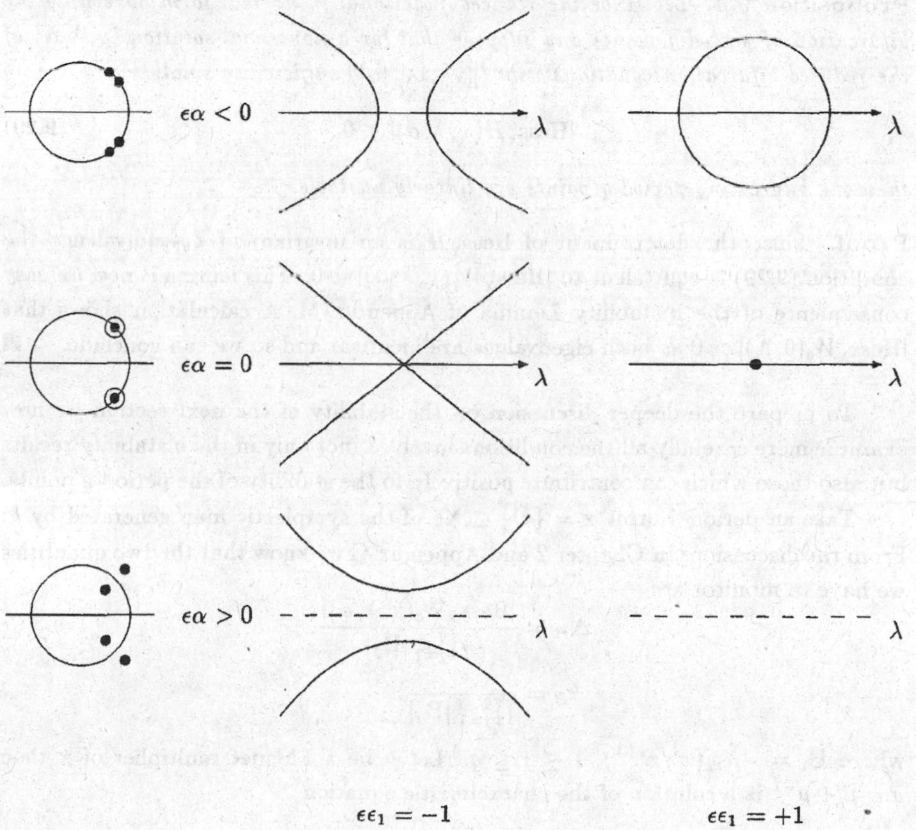

Figure 9.6: Bifurcation of period-q points in the unfolding of the collision at $\theta = \frac{2p}{q}\pi$, $q \geq 5$

Figure 9.6 shows the bifurcation diagrams associated with the normal form (9.28). Note that the role of q is simply to determine the dimension of the group orbit. When $q \geq 5$ we conjecture that the global loops (occurring when $\epsilon\alpha < 0$ and $\epsilon\epsilon_1 = +1$ in Figure 9.6) lose stability at finite amplitude through a secondary collision as in the period-4 case (Figures 9.3 and 9.4).

9.5 Reduced instability for the bifurcating period-q points

The idea of reduced stability for the bifurcating period-q points at a rational collision is to obtain complete information on the linear stability of the period-q points from the reduced functional $\widehat{\mathbf{W}}_q$ in normal form. In this section we will give a complete characterization of the problem and prove a reduced *instability* result.

Proposition 9.4. *Let H be the reduced functional in normal form governing the bifurcation of period-q points and suppose that for a nontrivial solution (χ, λ, α) of the reduced bifurcation equations with $(\|\chi\|, |\lambda|, |\alpha|)$ sufficiently small*

$$|\mathbf{Hess}_\chi H(\chi, \lambda, \alpha)| < 0\,, \tag{9.29}$$

then the bifurcating period-q points are linearly unstable.

Proof. Since the determinant of $\mathbf{Hess}_\chi H$ is an invariant of $\mathbf{Z}_q$-equivalence the condition (9.29) is equivalent to $|\mathbf{Hess}_\chi \widehat{W}_q(\chi, \lambda, \alpha)| < 0$. This lemma is now an easy consequence of the Instability Lemma of Appendix M. A calculation shows that $|\mathbf{Hess}_\chi \widehat{W}_q(0, \Lambda)| > 0$ as both eigenvalues are identical and so we can conclude. ■

To prepare the deeper discussion on the stability of the next section we now examine more carefully all the conditions involved, not only in the instability results but also those which can contribute positively to the stability of the period-q points.

Take an period-q orbit $\mathbf{x} = \{x^i\} \subset \mathbf{X}_q^n$ of the symplectic map generated by h. From the discussions in Chapter 2 and Appendix C we know that the two quantities we have to monitor are

$$\Delta_B = \frac{|\mathbf{Hess_x W}_q(\mathbf{x}, \lambda, \alpha)|}{\prod_{j=1}^q |\mathbf{B}_j|},$$

$$\tau_B = \frac{\tau}{\prod_{j=1}^q |\mathbf{B}_j|},$$

where $\mathbf{B}_i = -h_{12}(x^i, x^{i+1})$, $1 \leq i \leq q$. Let μ be a Floquet multiplier of $\mathbf{x}$ then $\rho = \mu + \mu^{-1}$ is a solution of the characteristic equation

$$\Delta(\rho) = (2-\rho)^2 - \tau_B(2-\rho) + \Delta_B \tag{9.30}$$

or in terms of the residues $R = \frac{1}{4}(2-\rho)$

$$16R^2 - 4\tau_B R + \Delta_B = (4R - R_1)(4R - R_2) = 0\,.$$

The complete stability picture is contained in the coefficients Δ_B and τ_B.

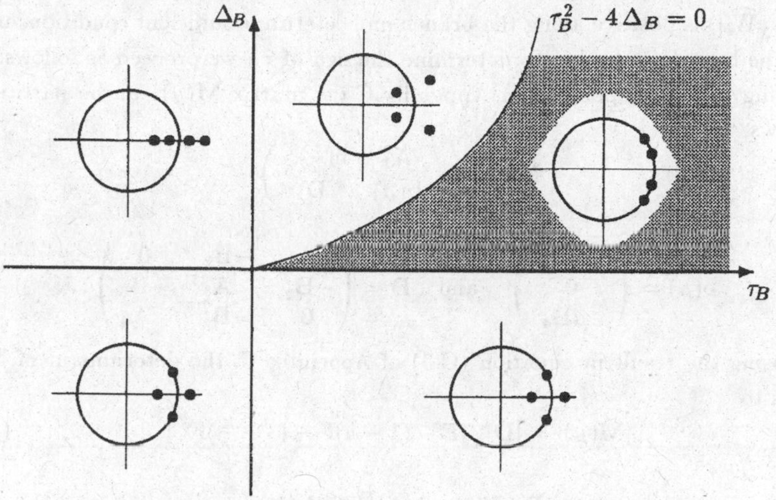

Figure 9.7: Projection of the Floquet multipliers as a function of τ_B and Δ_B along the branches of period-4 points.

Figure 9.7 shows the (τ_B, Δ_B) plane broken into components with different multiplier positions. The shaded region is the only stable region and corresponds to $\Delta_B > 0$, $\tau_B > 0$ and $\tau_B^2 - 4\Delta_B > 0$.

The relation between the residues and the parameters (τ_B, Δ_B) is

$$R_1 R_2 = \Delta_B \qquad \text{and} \qquad R_1 + R_2 = \tau_B \,.$$

When a Floquet multiplier lies at 1, $\rho = 2$ and $R = 0$. At the bifurcation point $(\lambda = \alpha = 0)$ both R_1 and R_2 will be zero; that is, all four Floquet multipliers of the period-q points will be at 1. However, in the unfolding of the codimension-2 point there are 4 possibilities for the configuration of the Floquet multipliers and there is only one stable configuration. Note that a necessary condition for stability is $\Delta_B > 0$ and conversely if $\Delta_B < 0$ the period-q points are unstable (Proposition 9.4).

A complete reduced stability result requires establishment of a relation between the reduced functional $\widehat{W}_q$ and the invariant τ_B in (9.30). For $q = 3$ the sign of τ_B is not needed as the bifurcating period-3 points are unstable. For the case $q = 4$ the sign of τ_B is studied in the next subsection.

9.6 Reduced stability for bifurcating period-4 points

In this section we consider the case when $q = 4$ and Δ_B (equivalently the sign of $|\mathrm{Hess}_\chi \widehat{\mathbf{W}}_4|$) is positive along the branch and determine sufficient conditions under which the branch is stable. To determine the sign of τ_B we proceed as follows.

Using the decomposition in Appendix C the matrix $\mathbf{M}(\mu)$ can be partitioned as follows,

$$\mathbf{M}(\mu) = \begin{pmatrix} \mathbf{A}_1 & \mathbf{b}(\frac{1}{\mu})^T \\ \mathbf{b}(\mu) & \mathbf{D} \end{pmatrix}$$

with

$$\mathbf{b}(\mu) = \begin{pmatrix} -\mathbf{B}_1^T \\ 0 \\ -\mu\mathbf{B}_4 \end{pmatrix} \quad \text{and} \quad \mathbf{D} = \begin{pmatrix} \mathbf{A}_2 & -\mathbf{B}_2 & 0 \\ -\mathbf{B}_2^T & \mathbf{A}_3 & -\mathbf{B}_3 \\ 0 & -\mathbf{B}_3^T & \mathbf{A}_4 \end{pmatrix}.$$

Then using the result in equation (C.5) of Appendix C, the determinant of $\mathbf{M}(\mu)$ reduces to

$$|\mathbf{M}(\mu)| = |\mathbf{D}| \cdot |\mathbf{E} - (1-\mu)\mathbf{F} - (1 - \tfrac{1}{\mu})\mathbf{F}^T| \tag{9.31}$$

where

$$\begin{aligned} \mathbf{E} &= \mathbf{A}_1 - \mathbf{b}(1)^T \mathbf{D}^{-1}\mathbf{b}(1), \\ \mathbf{F} &= [\mathbf{b}(1) + \mathbf{b}']^T \mathbf{D}^{-1}\mathbf{b}', \end{aligned} \tag{9.32}$$

with

$$\mathbf{b}' = \begin{pmatrix} 0 \\ 0 \\ \mathbf{B}_4 \end{pmatrix}.$$

Working out the determinant in (9.31) results in $|\mathbf{M}(\mu)| = |\mathbf{D}| \cdot |\mathbf{F}| \cdot \Delta(\rho)$ with $\Delta(\rho)$ given by (9.30). In terms of $\mathbf{E}$, $\mathbf{F}$ the critical coefficients in the characteristic polynomial (9.30) are

$$\tau_B = \mathrm{Tr}\,(\mathbf{F}^{-1}\mathbf{E}) + \frac{(\mathrm{Tr}\,\mathbf{JF})^2}{|\mathbf{F}|} \quad \text{and} \quad \Delta_B = |\mathbf{F}^{-1}\mathbf{E}|. \tag{9.33}$$

We are now in a position to prove the linear stability results for the bifurcating period-4 points. The idea is to evaluate (9.33) on the existing period-4 points found in Section 9.3. The simplest result is the sufficient condition for linear instability associated with $\Delta_B < 0$ (Proposition 9.4). For τ_B we proceed as follows.

First, a direct calculation verifies

Proposition 9.5. *In terms of the matrices* $\mathbf{A}_j$ *and* $\mathbf{B}_j$, $1 \leq j \leq 4$, *the reduced matrices* $\mathbf{E}$ *and* $\mathbf{F}$ *in (9.32) are given by*

$$\mathbf{F} = -\mathbf{B}_1\mathbf{X}_3\mathbf{B}_4 \tag{9.34}$$

$$\mathbf{E} = \mathbf{A}_1 + \mathbf{F} + \mathbf{F}^T - \mathbf{B}_1\mathbf{X}_1\mathbf{B}_1^T - \mathbf{B}_4^T\mathbf{Z}_3\mathbf{B}_4 \tag{9.35}$$

with

$$\begin{aligned}\mathbf{X}_3^{-1} &= \mathbf{A}_4\mathbf{B}_3^{-1}(\mathbf{A}_3\mathbf{B}_2^{-1}\mathbf{A}_2 - \mathbf{B}_2^T) - \mathbf{B}_3^T\mathbf{B}_2^{-1}\mathbf{A}_2\\ \mathbf{X}_1 &= \mathbf{X}_3[\mathbf{A}_4\mathbf{B}_3^{-1}\mathbf{A}_3\mathbf{B}_2^{-1} - \mathbf{B}_3^T\mathbf{B}_2^{-1}]\\ \mathbf{Z}_3 &= [\mathbf{B}_3^{-1}\mathbf{A}_3\mathbf{B}_2^{-1}\mathbf{A}_2 - \mathbf{B}_3^{-1}\mathbf{B}_2^T]\mathbf{X}_3\,.\end{aligned}$$

The idea is to evaluate the matrices $\mathbf{A}_j$ and $\mathbf{B}_j$, $1 \leq j \leq 4$, at a period-4 point and substitute into (9.34)-(9.35). This is straightforward but lengthy. For simplicity we will give the result for the class of maps studied in Section 9.3. First we note that

$$\begin{aligned}\mathbf{A}_1 &= (2\epsilon + \alpha)\mathbf{I} - \mathbf{U}_1 - \mathbf{U}_2\\ \mathbf{A}_2 &= (2\epsilon + \alpha)\mathbf{I} - \mathbf{V}_1 - \mathbf{V}_2\\ \mathbf{A}_3 &= (2\epsilon + \alpha)\mathbf{I} + \mathbf{U}_1 - \mathbf{U}_2\\ \mathbf{A}_4 &= (2\epsilon + \alpha)\mathbf{I} + \mathbf{V}_1 - \mathbf{V}_2\end{aligned}$$

and

$$\mathbf{B}_j = \epsilon\mathbf{R}_\lambda^T \qquad \text{for } 1 \leq j \leq 4\,.$$

The 2×2 matrices $\mathbf{U}_j$ and $\mathbf{V}_j$ ($j = 1, 2$) are determined from the nonlinearity (the potential),

$$\begin{aligned}\mathbf{U}_1 &= \tfrac{1}{2}[V_{xx}(x^1;\lambda,\alpha) - V_{xx}(x^3;\lambda,\alpha)]\\ \mathbf{U}_2 &= \tfrac{1}{2}[V_{xx}(x^1;\lambda,\alpha) + V_{xx}(x^3;\lambda,\alpha)]\\ \mathbf{V}_1 &= \tfrac{1}{2}[V_{xx}(x^2;\lambda,\alpha) - V_{xx}(x^4;\lambda,\alpha)]\\ \mathbf{V}_2 &= \tfrac{1}{2}[V_{xx}(x^2;\lambda,\alpha) + V_{xx}(x^4;\lambda,\alpha)]\,.\end{aligned}$$

Again a straightforward but lengthy calculation verifies the following.

Proposition 9.6. *The reduced matrices* $\mathbf{E}$ *and* $\mathbf{F}$, *evaluated on a period-4 point of the class of maps generated by* h *as in (9.20a), take the following form*

$$\mathbf{F} = -\tfrac{1}{4}\epsilon\mathbf{I} + \epsilon\lambda\mathbf{J} + \cdots \tag{9.36}$$

$$\mathbf{E} = 4(\epsilon\lambda^2 + \alpha)\mathbf{I} + 2(\mathbf{J}\mathbf{V}_2\mathbf{J} - \mathbf{U}_2) + \epsilon(\mathbf{J}\mathbf{V}_1^2\mathbf{J} - \mathbf{U}_1^2) + \cdots\,. \tag{9.37}$$

Actually it is not difficult to show that to leading order

$$\mathbf{E} = 4\,\mathrm{Hess}_\chi\widehat{\mathbf{W}}_4(u,\Delta,\lambda,\alpha) + \cdots$$

at least up to fourth order terms in the reduced potential. Substituting (9.36) and (9.37) into the expressions for τ_B and Δ_B we find

$$\tau_B = 64\lambda^2 - 4\epsilon \operatorname{Tr} \mathbf{E} + \cdots$$
$$\Delta_B = 16\,|\mathbf{E}| + \cdots .$$

There are two branches of period-4 points in the unfolding of the period-4 collision and they are listed in Table 9.1. Let

$$\mathbf{Z}_2^r = \left\langle \begin{pmatrix} 1 & 0 \\ 0 & -1 \end{pmatrix} \right\rangle \quad \text{and} \quad \mathbf{Z}_2^s = \left\langle \begin{pmatrix} 0 & 1 \\ 1 & 0 \end{pmatrix} \right\rangle .$$

Then the governing equations for the first class (defined by $\chi_2 = 0$ in Table 9.1) are in Fix $\mathbf{Z}_2^r$ and the second class are in Fix $\mathbf{Z}_2^s$. Note that these symmetries are normal form symmetries and only hold locally in phase space (do not persist to all orders in the bifurcation equations). However they are convenient labels for the branches of period-4 points and we will identify the two branches by their normal-form isotropy subgroup $\mathbf{Z}_2^r$ or $\mathbf{Z}_2^s$. We now verify the stability assignments in Figures 9.3 and 9.4. Without loss of generality take $\epsilon = +1$ ($\epsilon = -1$ results in a trivial permutation of the bifurcation diagrams with no qualitative change).

Along the $\mathbf{Z}_2^s$-branch we find that $\Delta_B > 0$ only when $\epsilon m < 0$. We now verify that the $\mathbf{Z}_2^s$-branch is indeed linearly stable when $\epsilon m < 0$. The invariant τ_B evaluated on the $\mathbf{Z}_2^s$-branch is

$$\tau_B = 64(\lambda^2 - 2\epsilon(m-1)\tilde{N} + \cdots),$$

where $\tilde{N}$ is u scaled (multiplied) by $|n_3|$. Since $\epsilon = +1$ and $\epsilon m < 0$ it follows that $\tau_B > 0$ for $(|\lambda|, |\alpha|, u)$ sufficiently small. It remains to check the Krein locus,

$$\begin{aligned} K_B &= \tau_B^2 - 4\Delta_B \\ &= 64^2 \left[-16\epsilon|\epsilon m|\,\tilde{N}^2 + \left(\lambda^2 + 2\left(|\epsilon m| + \epsilon\right)\tilde{N}\right)^2 + \cdots \right] \\ &= 64^2 \left[\lambda^4 + 4\left(|\epsilon m| + \epsilon\right)\lambda^2 \tilde{N} + 4\left(|\epsilon m| - \epsilon\right)^2 \tilde{N}^2 + \cdots \right] \end{aligned}$$

which is clearly positive when $\epsilon m < 0$ and $\epsilon = +1$ and $(u, |\lambda|, |\alpha|)$ are sufficiently small. Therefore the $\mathbf{Z}_2^s$-branch, when $\epsilon m < 0$ and $\epsilon = +1$, is in the stable region of the (τ_B, Δ_B) plane.

For the $\mathbf{Z}_2^r$-branch we noted that when $\epsilon m < \epsilon$ it is unstable. When $\epsilon m > 1$ we find that the stability coefficients are

$$\tau_B = 64(\lambda^2 - 2\epsilon m\tilde{N} + \cdots)$$
$$\Delta_B = 2(64)^2(m-1)\tilde{N}^2 + \cdots$$

and the Krein locus is

$$\begin{aligned} K_B &= \tau_B^2 - 4\Delta_B \\ &= 16^3\left[(\lambda^2 - 2\epsilon m\tilde{N})^2 - 16(m-1)\tilde{N}^2 + \cdots\right]. \end{aligned}$$

Let

$$\begin{aligned} \gamma_- &= 2\epsilon m - 4\sqrt{(m-1)} \\ \gamma_+ &= 2\epsilon m + 4\sqrt{(m-1)} \end{aligned}$$

then the Krein locus becomes

$$K_B = 16^3(\lambda^2 - \gamma_-\tilde{N})(\lambda^2 - \gamma_+\tilde{N}) + \cdots .$$

We have the following scenario along the global $\mathbf{Z}_2^r$-branch when $m > 1$. When $\tilde{N} \approx 0$ the stability parameters satisfy $\Delta_B > 0$, $\tau_B > 0$ and $K_B > 0$ verifying that the branch is stable. However moving along the branch it is clear that K_B changes sign. The first change of sign (smallest value of $\tilde{N}$) is when $\lambda = \pm\sqrt{\gamma_+\tilde{N}}$. At this point $\tau_B > 0$ so a secondary collision takes place. The second change of sign of K_B takes place after the collision so is not important for linear stability. This verifies the picture in Figure 9.4 and completes the linear stability analysis which we summarize in the following.

Lemma 9.7 (Stability Lemma II). *For the class of symplectic maps generated by h as in (9.80a), the linear stability properties of the bifurcating period-4 points for $(u, |\lambda|, |\alpha|)$ sufficiently small, are as follows.*

(a) *When $\epsilon = +1$ the $\mathbf{Z}_2^s$-branch is stable for $m < 0$ and unstable for $m > 0$. The $\mathbf{Z}_2^r$-branch is unstable for $m < 1$ and when $m > 1$ and $\epsilon\alpha < 0$, the global loop of period-4 points is initially stable but loses stability at finite amplitude through a secondary collision of Floquet multipliers.*

(b) *When $\epsilon = -1$ the $\mathbf{Z}_2^r$-branch is stable for $m > 1$ and unstable for $m < 1$. The $\mathbf{Z}_2^s$-branch is unstable for $m > 0$ and when $m < 0$ and $\alpha > 0$, the global loop of period-4 points is initially stable but loses stability at finite amplitude through a secondary collision of Floquet multipliers.*

9.7 Remarks on the collision at irrational points

Some idea of the basic structure of the bifurcation that takes place in the neighborhood of an *irrational* collision of multipliers of opposite signature can be obtained from the structure of a rational collision when q is large. Although only one parameter is essential for a collision of multipliers at an irrational point our claim is that a two-parameter analysis provides a clearer picture of that structure.

Recall that the linear symplectic map on $\mathbb{R}^4$ in general position for a collision of multipliers is

$$\begin{pmatrix} x' \\ y' \end{pmatrix} = \mathbf{M}(\lambda,\alpha)\begin{pmatrix} x \\ y \end{pmatrix} \quad \text{with} \quad \mathbf{M}(\lambda,\alpha) = \begin{pmatrix} 1+\epsilon\alpha & \epsilon \\ \alpha & 1 \end{pmatrix} \otimes \mathbf{R}_{\theta+\lambda}, \tag{9.38}$$

for which we have the following result.

Proposition 9.8. *Let* $\mathbf{T}$ *be the two parameter family of linear symplectic map on* $\mathbb{R}^4$ *given by (9.38). Then for all* $(\lambda,\alpha)\in\mathbb{R}^2$ *there exists two independent integrals*

$$I(x,y) = (y, \mathbf{J}x)$$
$$E(x,y) = (y,y) + \alpha(x,y) - \epsilon\alpha(x,x).$$

When $(\lambda,\alpha) = 0$ *and* $\theta \in (0,\pi)$ *is at an irrational point, the only bounded orbit of the map is an invariant circle lying in the* x*-plane* $(y=0)$.

Proof. The integral I follows from the $\mathbf{SO}(2)$-invariance of the generating function h for $\mathbf{T}$; h is given in (9.10) and for $\mathbf{SO}(2) = \langle \mathbf{R}_\phi \rangle$ it is clear that $h(\mathbf{R}_\phi x, \mathbf{R}_\phi x') = h(x,x')$. The conservation of I then follows from Theorem E in Appendix E. That E is a second integral can be verified by direct calculation.

When $\lambda = \alpha = 0$ the linear map (9.38) reduces to

$$x^{j+1} = \mathbf{R}_\theta x^j + \epsilon \mathbf{R}_\theta y^j$$
$$y^{j+1} = \mathbf{R}_\theta y^j$$

and the orbit of (x^0, y^0) can be explicitly calculated to be

$$x^j = \mathbf{R}_{j\theta}(x^0 + \epsilon j y^0)$$
$$y^j = \mathbf{R}_{j\theta} y^0.$$

If $y=0$ the orbit is an invariant circle in the x-plane of rotation number $\rho = \theta/2\pi$. But the invariant circle is algebraically unstable; that is any initial condition with $\|y\| \neq 0$ lies in the cylinder $\|y^j\| =$ constant but leaves the x-plane algebraically:

$$\|x^j\|^2 = \|x^0\| + 2\epsilon j(x^0,y^0) + j^2\|y^0\|^2 = j^2\|y^0\|^2\left(1 + \frac{2\epsilon}{j}\frac{(x^0,y^0)}{(y^0,y^0)} + \frac{1}{j^2}\frac{\|x^0\|^2}{\|y^0\|^2}\right);$$

that is $\|x^j\| \sim j$ as $j\to\infty$. ■

The primary structure of an irrational collision is a simple – algebraically unstable – invariant circle lying in a 2-dimensional subspace of the four-dimensional phase space. Therefore the primary question is the persistence or non-persistence of the invariant circle in the unfolding – including nonlinear effects – of the singularity associated with the collision. With a two-parameter analysis the rotation number of the invariant circle can be frozen at the collision value, say ρ, and then the idea is to look for curves, in two-dimensional parameter space, along which the invariant circle of rotation number ρ exists.

In such a framework it is roughly clear what is the behavior under perturbation of the invariant circle of rotation number ρ. The bifurcation picture is expected to be similar to that shown in Figure 9.6 for q large. Depending on the sign of a particular coefficient ($\epsilon_1 = \pm 1$) either there exists a global loop ($\epsilon_1 = +1$) along which an invariant circle of rotation number ρ exists for each $\epsilon\alpha < 0$ sufficiently small which vanishes for $\alpha = 0$ or if $\epsilon_1 = -1$ there is no global loop but a curve of invariant circles (in (χ, λ) space) that persists after the collision ($\epsilon\alpha > 0$).

Another approach to studying the irrational collision is to work directly with the symplectic map, apply normal form transformations and analyze the normalized map (Bridges & Cushman [1993, Section 4], Bridges, Cushman & MacKay [1993]). This approach turns out to be particularly successful for the irrational collision because the normal form transformations introduce an $\mathbf{S}^1$-symmetry which simplifies the analysis of the normal form.

Using unipotent normalization the nonlinear normal form for the irrational collision is

$$\begin{pmatrix} x' \\ y' \end{pmatrix} = \begin{pmatrix} (1+\epsilon h_1)\mathbf{R}_{\theta+\phi} & \epsilon \mathbf{R}_{\theta+\phi} \\ h_1 \mathbf{R}_{\theta+\phi} & \mathbf{R}_{\theta+\phi} \end{pmatrix} \begin{pmatrix} x \\ y \end{pmatrix} \tag{9.39}$$

with

$$\psi - h_I - \lambda + b_2\sigma_1 + b_3 I +$$
$$h_1 = h_{\sigma_1} = \alpha + b_1\sigma_1 + b_2 I + \cdots$$

where $\sigma_1 = (x, x)$ $I = (y, \mathbf{J}x)$ and h is a smooth function of σ_1, I and parameters (Bridges & Cushman [1993, equation (4.7)). Note that the normal form is written down to "infinite order" but it is a formal expression and not, in general, convergent (that is, the normal form is obtained by repeated *formal* power series symplectic transformations).

The map (9.39) is analyzed by reducing to the orbit space. Introduce coordinates $\sigma = (\sigma_1, \sigma_2, \sigma_3)$ (as in (8.30)). In terms of the coordinates σ the map (9.39) is

$$\sigma' = \mathbf{M}(\sigma_1, I)\sigma \quad \text{with} \quad \mathbf{M}(\sigma_1, I) = \begin{pmatrix} a^2 & 1 & 2\epsilon a \\ h_1^2 & 1 & -2h_1 \\ -ah_1 & \epsilon & 1 - 2\epsilon h_1 \end{pmatrix} \tag{9.40}$$

where $a = 1 - \epsilon h_1$.

The map (9.40) is a Poisson map (with respect to the Poisson bracket (8.33)) and the Poisson structure has the Casimir $\psi(\sigma)$ defined in (8.32). When $I \neq 0$ the map (9.40) can be reduced to an area-preserving map. Let

$$\sigma_1 = u, \quad \sigma_2 = \tfrac{1}{u}(I^2 + v^2), \quad \sigma_3 = v.$$

Then (9.40) reduces to

$$\begin{aligned} u' &= a^2 u + 2\epsilon a\, u + \tfrac{1}{u}(I^2 + v^2), \\ v' &= -ah_u u + (1 - 2\epsilon h_u)\, v + \epsilon \tfrac{1}{u}(I^2 + v^2), \end{aligned} \tag{9.41}$$

where h depends on u, v, I and parameters.

The symplectic form for (9.41) is $\omega = \frac{1}{2u} du \wedge dv$ but can be flattened by taking $\omega = d(\ln\sqrt{u}) \wedge dv$. In other words the analysis of the dynamics near an irrational collision when $I \neq 0$ can be reduced to the study of an interesting 2-parameter (α, I) family of area-preserving maps (9.41).

Fixed points of the area-preserving map (9.41) lift to invariant circles in the 4D-map with flow along the $\mathbf{S}^1$-group orbit (note that the $\mathbf{S}^1$-symmetry is a normal form symmetry here and is therefore not exact; persistence of the invariant circles in the original map is a separate question). This is a sketch of the analysis using normal form theory. Additional details of the local analysis of the phase space geometry near an irrational collision are given by Bridges, Cushman & MacKay [1993].

10. Equivariant maps and the collision of multipliers

The effect of symmetry on the collision of multipliers is twofold. If there is "too much" symmetry, the collision is prevented; this happens for example in $\mathbf{O}(2)$-equivariant maps when the group acts absolutely irreducibly on the configuration space. On the other hand when a collision of multipliers of opposite signature occurs in an equivariant map there is additional interesting structure in the phase space and bifurcations. In this section, the collision of multipliers in 4D-equivariant symplectic maps is considered with particular attention to symmetries that do not eliminate the collision. A general study of the effect of symmetry on the collision in equivariant maps of arbitrary dimension and general compact Lie group actions would be of great interest but is not considered here (see van der Meer [1990] and Dellnitz et al. [1992] for analogous studies of Hamiltonian vectorfields).

We consider first the case of reversible-symplectic maps and show that the bifurcation is roughly the same as the non-reversible case but there are two new features to account for. The reduced functional for a reversible collision has an additional $\mathbf{Z}_2^{\kappa}$-symmetry but the generic normal form is equivalent to the non-reversible case (modulo some signs). The other feature of interest in the reversible case is the extra symmetry in the sequences; the bifurcating period-q points lie in subsequences of $\mathbf{X}_q^2$ with additional temporal symmetry.

For 4D-maps the largest group of spatial symmetries is $\mathbf{O}(2)$, assuming the group is a compact Lie group that acts independent of discrete time on the configuration space and lifts to a diagonal action on the phase space. It turns out that, for reversible-symplectic maps, any spatial-symmetry group larger than $\mathbf{Z}_2$ will prevent a collision. Therefore, interesting effects of symmetry on a collision, in the reversible-symplectic case, will require a higher dimensional configuration space.

In the absence of reversibility, however, symplectic maps on $\mathbf{R}^4$ admit more interesting configuration space symmetries in the presence of a collision.

The group $\mathbf{O}(2)$ and the dihedral subgroups $\mathbf{D}_m$, $m \geq 3$, act absolutely irreducibly on $\mathbf{R}^2$, and lift to a diagonal symplectic action on $\mathbf{R}^4$, in which case – as shown in Chapter 8 – the multipliers are double and of the same signature (that is there are no collision possible). However we show that every subgroup of $\mathbf{O}(2)$ that does not act absolutely irreducibly, in particular, $\mathbf{SO}(2)$, $\mathbf{Z}_m$, $m \geq 3$, and $\mathbf{Z}_2 \times \mathbf{Z}_2$, are admissible spatial symmetries.

The groups $\mathbf{Z}_2$ and $\mathbf{Z}_2 \times \mathbf{Z}_2$ reduce the problem essentially to a reversible collision. The interesting spatial symmetries are $\mathbf{Z}_m$, $m \geq 3$, and $\mathbf{SO}(2)$. The analysis for $\mathbf{Z}_m$ and $\mathbf{SO}(2)$ is similar but $\mathbf{SO}(2)$ has the most dramatic effect on the collision of multipliers because of the presence of a continuous symmetry. The collision is not eliminated but the bifurcating periodic points near the collision ride on the $\mathbf{SO}(2)$-group orbit.

10.1 Reversible-symplectic maps on $\mathbf{R}^4$

Let $\mathcal{R} = \kappa \otimes \mathbf{I}$ be the standard reversor on $\mathbf{R}^4$. A matrix $\mathbf{M} \in \mathbf{Gl}(4, \mathbf{R})$ is said to be reversible if $\mathcal{R}\mathbf{M}\mathcal{R} = \mathbf{M}^{-1}$. The subgroup of symplectic matrices that is also reversible, denoted by $\mathbf{Sp}_{\mathcal{R}}(4, \mathbf{R})$, is composed of elements of the form

$$\mathbf{M} = \begin{pmatrix} \mathbf{M}_1 & \mathbf{M}_2 \\ \mathbf{M}_3 & \mathbf{M}_1^T \end{pmatrix}.$$

where $\mathbf{M}_2$ and $\mathbf{M}_3$ are 2×2 *symmetric* matrices, $\mathbf{M}_1$ is a general 2×2 matrix and

$$\mathbf{M}_1\mathbf{M}_2 = \mathbf{M}_2\mathbf{M}_1^T, \quad \mathbf{M}_3\mathbf{M}_1 = \mathbf{M}_1^T\mathbf{M}_3 \quad \text{and} \quad \mathbf{M}_1^2 - \mathbf{M}_2\mathbf{M}_3 = \mathbf{I}.$$

To see this note that $\mathbf{M} \in \mathbf{Sp}(4, \mathbf{R})$ requires $\mathbf{J}^T\mathbf{M}^T\mathbf{J} = \mathbf{M}^{-1}$ which when combined with $\mathcal{R}\mathbf{M}\mathcal{R} = \mathbf{M}^{-1}$ requires $\mathbf{M}^T = -\mathbf{J}\mathcal{R}\mathbf{M}\mathcal{R}\mathbf{J}$.

We now consider the collision of multipliers in reversible-symplectic maps. Suppose $\mathbf{T} : (x, y) \mapsto (x', y')$ is a reversible-symplectic map on $\mathbf{R}^4$ and that $\sigma(\mathbf{DT}) = \{e^{\pm i\theta}\}$, $\theta \in (0, \pi)$ with each multiplier of algebraic multiplicity two and geometric multiplicity one. Without loss of generality we can suppose that $\mathbf{DT}^o$ is in normal form for a collision. The linear normal form

$$\mathbf{M}_0 = \begin{pmatrix} 1 & \epsilon \\ 0 & 1 \end{pmatrix} \otimes \mathbf{R}_\theta \quad \text{for} \quad \theta \in (0, \pi),$$

(cf. (9.1) in Chapter 9) is in fact already reversible-symplectic, but with respect to the involution

$$\widetilde{\mathcal{R}} = \kappa \otimes \begin{pmatrix} 1 & -\epsilon \\ 0 & 1 \end{pmatrix}.$$

In other words $\widetilde{\mathcal{R}}^2 = \mathbf{I}_4$ and $\widetilde{\mathcal{R}}\mathbf{M}_0\widetilde{\mathcal{R}} = \mathbf{M}_0^{-1}$. The reversor $\widetilde{\mathcal{R}}$ is also antisymplectic: $\widetilde{\mathcal{R}}^T\mathbf{J}_4\widetilde{\mathcal{R}} = -\mathbf{J}_4$. A change of basis can be introduced so that $\mathbf{M}_0$ is reversible with respect to the standard reversor.

Introduce the following elementary (symplectic) permutation

$$\mathbf{P} = \begin{pmatrix} 0 & 0 & -1 & 0 \\ 0 & 1 & 0 & 0 \\ 1 & 0 & 0 & 0 \\ 0 & 0 & 0 & 1 \end{pmatrix}.$$

Then,

$$\mathbf{P}^T\mathbf{M}_0\mathbf{P} = \widehat{\mathbf{M}}_0 = \begin{bmatrix} \cos\theta & 0 & 0 & \sin\theta \\ -\epsilon\sin\theta & \cos\theta & \sin\theta & \epsilon\cos\theta \\ -\epsilon\cos\theta & -\sin\theta & \cos\theta & -\epsilon\sin\theta \\ -\sin\theta & 0 & 0 & \cos\theta \end{bmatrix} \tag{10.1}$$

and $\widehat{\mathbf{M}}_0$ satisfies $\widehat{\mathbf{M}}_0\mathcal{R}\widehat{\mathbf{M}}_0\mathcal{R} = \mathbf{I}_4$. With $\theta \in (0,\pi)$ and $\epsilon = \pm 1$, $\widehat{\mathbf{M}}_0$ is the linear normal form for a collision of multipliers in reversible-symplectic maps. It is codimension-2 in $\mathbf{Sp}_{\mathcal{R}}(4,\mathbf{R})$ when the multiplier lies at a rational point on the unit circle.

Proposition 10.1. *The versal unfolding of* $\widehat{\mathbf{M}}_0$ *in* $\mathbf{Sp}_{\mathcal{R}}(4,\mathbf{R})$ *is given by*

$$\widehat{\mathbf{M}}(\lambda,\alpha) = \begin{bmatrix} (1+\epsilon\alpha)\cos\hat\theta & 2\alpha\sin\hat\theta & -2\alpha\cos\hat\theta & (1+\epsilon\alpha)\sin\hat\theta \\ -(\epsilon+\frac{1}{2}\alpha)\sin\hat\theta & (1+\epsilon\alpha)\cos\hat\theta & (1+\epsilon\alpha)\sin\hat\theta & (\epsilon+\frac{1}{2}\alpha)\cos\hat\theta \\ -(\epsilon+\frac{1}{2}\alpha)\cos\hat\theta & -(1+\epsilon\alpha)\sin\hat\theta & (1+\epsilon\alpha)\cos\hat\theta & -(\epsilon+\frac{1}{2}\alpha)\sin\hat\theta \\ -(1+\epsilon\alpha)\sin\hat\theta & 2\alpha\cos\hat\theta & 2\alpha\sin\hat\theta & (1+\epsilon\alpha)\cos\hat\theta \end{bmatrix} \tag{10.2}$$

where $\hat\theta = \theta + \lambda$.

Proof. Same as proof of Proposition 9.1 but with $\mathbf{Sp}_{\mathcal{R}}(4,\mathbf{R})$ instead of $\mathbf{Sp}(4,\mathbf{R})$.■

The characteristic polynomial for $\widehat{\mathbf{M}}(\lambda,\alpha)$ is (modulo a simple scaling) exactly as in Section 9.1 (see Figure 9.1).

An argument modeled on Proposition 2.1 shows that any reversible-symplectic map, in the neighborhood of a collision has a generating function

$$h(x,x') = \tfrac{1}{2}(x,\mathbf{A}x) - (x,\mathbf{B}x') + \tfrac{1}{2}(x,\mathbf{C}x') + \hat h(x,x',\lambda,\alpha), \tag{10.4}$$

and the reversibility requires that $\mathbf{A} = \mathbf{C}$ and $\mathbf{B}^T = \mathbf{B}$. Let

$$\gamma = \sin^2(\theta+\lambda) + \alpha^2 + 2\epsilon\alpha,$$

$$\mathbf{B} = \frac{1}{\gamma}\begin{pmatrix} -(\epsilon+\frac{1}{2}\alpha)\cos(\theta+\lambda) & (1+\epsilon\alpha)\sin(\theta+\lambda) \\ (1+\epsilon\alpha)\sin(\theta+\lambda) & 2\alpha\cos(\theta+\lambda)\end{pmatrix} \tag{10.5a}$$

and

$$\mathbf{A} = \mathbf{C} = \frac{1}{\gamma}\begin{pmatrix} -(1+\epsilon\alpha)(\epsilon+\frac{1}{2}\alpha) & \sin(\theta+\lambda)\cos(\theta+\lambda) \\ \sin(\theta+\lambda)\cos(\theta+\lambda) & 2\alpha(1+\epsilon\alpha)\end{pmatrix}. \tag{10.5b}$$

Then when h is of the form (10.4) with $\mathbf{A}$, $\mathbf{B}$ and $\mathbf{C}$ given by (10.5), the generating function is in general position for a collision of multipliers at rational points.

The bifurcation of period-q points in the neighborhood of a rational collision is essentially the same as the non-reversible case. Therefore we will just sketch the results. The differences being that the reduced functional is $\mathbf{D}_q$-invariant rather than $\mathbf{Z}_q$-invariant and the sequence of periodic points have nontrivial isotropy subgroups. The normal form for the bifurcating period-q points will be a $\mathbf{D}_q$-equivariant gradient map on $\mathbf{R}^2$ with the "collision singularity"; that is, λ will appear quadratically in the normal form.

Let $\theta = \frac{2p}{q}\pi$ with p, q in lowest terms and $q \geq 3$. The action functional, with h in general position for a collision, on the space $\mathbf{X}_q^2$, is

$$\mathrm{W}_q(\mathbf{x}, \lambda, \alpha) = \tfrac{1}{2}\langle \mathbf{x}, \mathbf{L}(\lambda, \alpha)\mathbf{x} \rangle + N(\mathbf{x}, \lambda, \alpha) \tag{10.6}$$

where $N(\mathbf{x}, \lambda, \alpha) = \sum_{j=1}^{q} \hat{h}(x^j, x^{j+1}, \lambda, \alpha)$ and

$$\mathbf{L}(\lambda, \alpha) = 2\mathbf{I}_q \otimes \mathbf{A} - (\Gamma_q + \Gamma_q^T) \otimes \mathbf{B} \tag{10.7}$$

with $\mathbf{A}$ and $\mathbf{B}$ as given in (10.5). When $\lambda = \alpha = 0$, $\mathbf{L}$ reduces to

$$\mathbf{L}^o = \frac{1}{\sin^2\theta}\left[\mathbf{I}_q \otimes \begin{pmatrix} -2\epsilon & \sin 2\theta \\ \sin 2\theta & 0 \end{pmatrix} - (\Gamma_q + \Gamma_q^T) \otimes \begin{pmatrix} -\epsilon\cos\theta & \sin\theta \\ \sin\theta & 0 \end{pmatrix}\right]$$

and $\mathrm{Ker}\,\mathbf{L}^o = \mathrm{span}\{\xi_1, \xi_2\}$ with

$$\xi_1 + i\xi_2 = \sqrt{\frac{2}{q}} \begin{pmatrix} 1 \\ e^{i\theta} \\ \vdots \\ e^{i(q-1)\theta} \end{pmatrix} \otimes \begin{pmatrix} 0 \\ 1 \end{pmatrix}. \tag{10.8}$$

Therefore write $\mathbf{X}_q^2 = \mathrm{Ker}\,\mathbf{L}^o \oplus \mathbf{U}$ and express any $\mathbf{x} \in \mathbf{X}_q^2$ as $\mathbf{x} = \chi_1\xi_1 + \chi_2\xi_2 + \Upsilon$. Then application of the Splitting Lemma to W_q results in

$$\mathrm{W}_q(\chi_1\xi_1 + \chi_2\xi_2 + \phi(\Upsilon, \chi, \lambda, \alpha), \lambda, \alpha) = \tfrac{1}{2}\langle \Upsilon, \mathbf{L}^o\Upsilon \rangle + \widehat{\mathrm{W}}_q(\chi, \lambda, \alpha). \tag{10.9}$$

The reduced functional $\widehat{\mathrm{W}}_q$ is $\mathbf{D}_q$-invariant with respect to the standard action of $\mathbf{D}_q$ on $\mathbf{R}^2$. The reversor $\mathcal{K}_q$ introduced in Chapter 2, equation (2.21), reduces to κ; that is, with the eigenvectors ξ_1 and ξ_2 given in (10.8),

$$[\xi_1\,\xi_2]^T \mathcal{K}_q [\xi_1\,\xi_2] = \kappa.$$

Therefore it follows from Proposition 2.6 that the reduced functional simplifies to

$$\widehat{\mathrm{W}}_q(\chi, \lambda, \alpha) = F(u, v, \lambda, \alpha),$$

where u and v are the $\mathbf{D}_q$-invariants on $\mathbf{R}^2$, and from Proposition (2.11) it follows that

$$F(u, v, \lambda, \alpha) = \epsilon\lambda^2 u + F_v^o v + \tfrac{1}{2}F_{uu}^o u^2 + F_{u\alpha}^o \alpha u + \cdots$$

and a straightforward calculation using (10.8) shows that $F_{u\alpha}^o = 1$.

In the 2-parameter family (λ, α) we can assume that F_{uu}^o and F_v^o take generic values. Therefore as a corollary to Theorem 9.3, the normal forms for the rational collisions $q = 3, 4 \ldots$ are as given in Section 6.2. Note that the only difference between the normal forms for the reversible and non-reversible rational collisions is the additional signs in the reversible case. The $\mathbf{D}_q$ equivariant changes of coordinates are not sufficient to eliminate the signs, but the signs result in only trivial changes in the bifurcation diagrams. In particular the bifurcation diagrams are similar to those shown in Figures 9.2 to 9.6. The collision of multipliers at $\pm i$ in reversible-symplectic maps is treated in more detail in Bridges, Furter & Lahiri [1991].

10.2 Symplectic maps on $\mathbf{R}^4$ with spatial symmetry

When a symplectic map on $\mathbf{R}^4$ is not reversible the class of admissible configuration space symmetries that does not eliminate the collision is larger. Any subgroup of $\mathbf{O}(2)$ that does not act absolutely irreducibly is an admissible spatial symmetry. This eliminates $\mathbf{O}(2)$ and $\mathbf{D}_m$, $m \geq 3$, but allows for $\Sigma = \mathbf{SO}(2)$, $\mathbf{Z}_m$, $m \geq 2$, and $\mathbf{Z}_2 \times \mathbf{Z}_2$. The collision with $\mathbf{Z}_2$ or $\mathbf{Z}_2 \times \mathbf{Z}_2$-symmetry will be similar to the reversible collision. We will treat the collision with $\mathbf{SO}(2)$-symmetry in detail as it contains the most interesting features. The analysis is similar for the case $\Sigma = \mathbf{Z}_m$, $m \geq 3$.

Let $\mathbf{T} : (x, y) \mapsto (x', y')$ be a symplectic map on $\mathbf{R}^4$ equivariant with respect to $\mathbf{SO}(2)$. Such maps are quite prevalent in applications. An elementary example that gives rise to an $\mathbf{SO}(2)$-equivariant symplectic map is the time-T map of a parametrically-forced Lagrange top (when the forcing does not break the $\mathbf{SO}(2)$-spatial symmetry). We can assume without loss of generality that, in the neighborhood of a collision, the $\mathbf{SO}(2)$-equivariant symplectic map has a Lagrangian generating function $h(x, x')$ with

$$h(\mathbf{R}_\phi x, \mathbf{R}_\phi x') = h(x, x')$$

where $\mathbf{R}_\phi = \cos\phi \mathbf{I} + \sin\phi \mathbf{J}$ for $\phi \in \mathbf{R}/2\pi\mathbf{Z}$. The ring of $\mathbf{SO}(2)$-invariants on $\mathbf{R}^4 = (x, x')$ is generated by (x, x), (x', x'), (x, x') and $(x, \mathbf{J}x')$ (see Proposition 8.8). Therefore with $\mathbf{B} = b_1\mathbf{I} + b_2\mathbf{J}$, an $\mathbf{SO}(2)$-invariant generating function takes the form

$$h(x, x') = \tfrac{1}{2}a(x, x) - (x, \mathbf{B}x') + \tfrac{1}{2}c(x', x') + \hat{h}((x, x), (x', x'), (x, x'), (x, \mathbf{J}x')). \quad (10.9)$$

where a and c are real numbers. For non-degeneracy we require (twist condition) that $b_1^2 + b_2^2 \neq 0$. Note however that if $b_2 = 0$ the quadratic part of the generating function is $\mathbf{O}(2)$-invariant and a collision is not possible. Hence a necessary condition for a collision is $b_2 \neq 0$.

Proposition 10.2. *Let $\mathbf{T} : (x, y) \mapsto (x', y')$ be an $\mathbf{SO}(2)$-equivariant symplectic map generated by h as in (10.9) and suppose $b_2 \neq 0$. Then a collision of multipliers of opposite signature occurs when*

$$(a + c)^2 - 4(b_1^2 + b_2^2) = 0 \quad \text{and} \quad -1 < \frac{b_1(a + c)}{(b_1^2 + b_2^2)} < 1.$$

Moreover the collision occurs at a rational point when

$$\frac{b_1(a + c)}{(b_1^2 + b_2^2)} = \cos\theta \quad \text{for} \quad \theta = \tfrac{2p}{q}\pi.$$

Proof. The tangent map for the **SO**(2)-invariant generating function in the neighborhood of a trivial fixed point is

$$-\mathbf{B}^T\xi^{j-1} + (a+c)\,\xi^j + \mathbf{B}\,\xi^{j+1} = 0$$

and by Floquet's theorem $\xi^{j+1} = \mu\xi^j$. Therefore the eigenvalue problem for μ is

$$\left((a+c)\mathbf{I} - \mu\mathbf{B} - \tfrac{1}{\mu}\mathbf{B}^T\right)\xi^j = 0$$

with characteristic equation for $\rho = \mu + \mu^{-1}$,

$$\Delta(\rho) = (b_1^2 + b_2^2)\rho^2 - 2b_1(a+c)\rho - 4b_2^2 + (a+c)^2 = 0\,.$$

A collision occurs when $\Delta(\rho) = \Delta'(\rho) = 0$ or

$$(b_1^2 + b_2^2)\,\rho = b_1(a+c) \quad \text{with} \quad \rho = \cos\theta \quad \text{for} \quad \theta \in (0,\pi),$$

and substitution into $\Delta(\rho) = 0$ requires

$$b_2^2\left(\frac{(a+c)^2}{b_1^2 + b_2^2} - 4\right) = 0$$

which completes the proof and shows the necessity of $b_2 \neq 0$. ■

Note that a corollary of this result is that **O**(2)-equivariant maps (which require $b_2 = 0$) are collision-free. Now suppose the conditions of Proposition 10.2 are met and linear map $\mathbf{DT}^o$ has a collision of multipliers at a rational point. There exists a symplectic **SO**(2)-equivariant matrix on $\mathbf{R}^4$ that transforms $\mathbf{DT}^o$ into linear normal form. This transformation turns out to be particularly easy since the quadratic part of the generating function in (10.9) is virtually already in normal form. In fact the normal form for the quadratic part of the generating function, including the unfolding, given in (9.10) is **SO**(2)-invariant. Therefore let

$$U_1 = (x,x), \quad U_2 = (x',x'), \quad U_3 = (x,x') \quad \text{and} \quad U_4 = (x,\mathbf{J}x')$$

then in the neighborhood of a collision any **SO**(2)-invariant generating function h has the form

$$h(x,x') = \tfrac{1}{2}(\epsilon+\alpha)(x,x) - \epsilon(x,\mathbf{R}_{\theta+\lambda}^T x') + \tfrac{1}{2}\epsilon(x',x') + \hat{h}(U_1,U_2,U_3,U_4). \quad (10.10)$$

where $\hat{h}$ begins with terms of degree 4.

The reduction to a gradient map on $\mathbf{R}^2$ follows the usual pattern: construct the action functional on $\mathbf{X}_q^2$ and use the equivariant Splitting Lemma. The action functional on $\mathbf{X}_q^2$ is

$$\mathrm{W}_q(\mathbf{x},\lambda,\alpha) = \tfrac{1}{2}\langle \mathbf{x}, \mathbf{L}(\lambda,\alpha)\mathbf{x}\rangle + N(\mathbf{x},\lambda,\alpha)$$

where $\mathbf{L}(\lambda, \alpha)$ will have exactly the same form as that given in (9.12)-(9.13). But $W_q(\mathbf{x}, \lambda, \alpha)$ has additional symmetry, in particular it is $\mathbf{SO}(2) \times \mathbf{Z}_q$-invariant with $\mathbf{Z}_q = \langle \Gamma_q \otimes \mathbf{I} \rangle$ and $\mathbf{SO}(2) = \langle \mathbf{I}_q \otimes R_\phi \rangle$.

Proposition 10.3. *Let $\chi = (\chi_1, \chi_2)$ be coordinates on* $\operatorname{Ker} \mathbf{L}^o$. *Then the action of* $\mathbf{SO}(2) \times \mathbf{Z}_q$ *on* $\operatorname{Ker} \mathbf{L}^o$ *is generated by* $\theta \cdot \chi \stackrel{\text{def}}{=} \mathbf{R}_\theta \cdot \chi$ *with* $\theta = \frac{2p}{q}\pi$ *and* $\phi \cdot \chi \stackrel{\text{def}}{=} \mathbf{R}_\phi \cdot \chi$ *with* $\phi \in \mathbf{R}/2\pi\mathbf{Z}$. *Let* $(\theta, \phi) \cdot \chi$ *represent the combined action of* $\mathbf{SO}(2) \times \mathbf{Z}_q$.

Then, up to conjugacy, the only isotropy subgroup with a 2D-fixed point space is $\widetilde{\mathbf{Z}}_q \stackrel{\text{def}}{=} (\theta, -\theta)$. *Moreover, let* $\mathrm{N}\widetilde{\mathbf{Z}}_q$ *denote the normalizer of* $\widetilde{\mathbf{Z}}_q$ *in* $\mathbf{SO}(2) \times \mathbf{Z}_q$, *then* $\mathrm{W}\widetilde{\mathbf{Z}}_q = \mathbf{SO}(2)$ *(the Weyl subgroup of* $\widetilde{\mathbf{Z}}_q$*).*

Proof. Let $\operatorname{Ker} \mathbf{L}^o = \operatorname{span}\{\xi_1, \xi_2\}$ with ξ_1 and ξ_2 given in Proposition 9.2. Then

$$[\xi_1\, \xi_2]^T \mathbf{I}_q \otimes \mathbf{R}_\phi\, [\xi_1\, \xi_2] = \mathbf{R}_\phi$$

and a similar construction shows that the action of $\mathbf{Z}_q$ on $\operatorname{Ker} \mathbf{L}^o$ reduces to $\mathbf{R}_\theta$. Now $\chi \in \mathbf{R}^2$ and the only subgroup of $\mathbf{SO}(2) \times \mathbf{Z}_q$, up to conjugacy, that fixes all of $\mathbf{R}^2$ is the subgroup generated by $\mathbf{R}_{\theta+\phi}$ with $\phi = -\theta$. Since $\mathbf{SO}(2) \times \mathbf{Z}_q$ is abelian the result $\mathrm{W}\widetilde{\mathbf{Z}}_q = \mathbf{SO}(2)$ easily follows. ∎

Note that, generically, the period-q points ride on the $\mathbf{SO}(2)$-group orbit; that is an increment of discrete time corresponds to a shift by θ in the group $\mathbf{SO}(2)$. This will have a dramatic effect on the normal form. The normal form for the bifurcating period-q points will have the symmetry of $\mathrm{W}\widetilde{\mathbf{Z}}_q = \mathbf{SO}(2)$. Combining the results of Theorem 9.3 and Proposition 10.3 we have proved the following.

Theorem 10.4 (SO(2)-symmetric rational collision). *Suppose* $\mathbf{T}$ *is an* $\mathbf{SO}(2)$-*equivariant symplectic map on* $\mathbf{R}^4$ *generated by a smooth* h *with quadratic part in general position for a collision of multipliers at a rational point* $\theta = \frac{2p}{q}\pi$. *Then there exists an* $\mathbf{SO}(2)$-*orbit of bifurcating period-q points with symmetry group* $\widetilde{\mathbf{Z}}_q$ *that is in one-to-one correspondence with critical points of an* $\mathbf{SO}(2)$-*invariant function on* $\mathbf{R}^2$ *given by*

$$\widehat{W}_q(\chi, \lambda, \alpha) = F(u, \lambda, \alpha) \quad \text{with} \quad u = \chi_1^2 + \chi_2^2. \tag{10.11}$$

Moreover F has the Taylor expansion

$$F(u, \lambda, \alpha) = (\epsilon\lambda^2 + \tfrac{1}{2}\alpha)\, u + \tfrac{1}{2} F_{uu}^o u^2 + \cdots$$

and if $F_{uu}^o \neq 0$ *then* $\nabla_\chi F$ *is* $\mathcal{K}_\lambda^{\mathbf{Z}_2}$-*equivalent to*

$$(\epsilon\lambda^2 + \alpha + \delta u)\chi = 0 \tag{10.12}$$

where $\delta = \operatorname{sign} F_{uu}^o$ *and* $\epsilon = \pm 1$ *is the sign-invariant of the collision.*

Note that the normal form for the bifurcating period-q points is independent of q. The $\mathbf{SO}(2)$-symmetry essentially reduces the normal form to a $\mathbf{Z}_2$-equivariant gradient map on $\mathbf{R}$. The result in (10.12) then follows from application of the $\mathcal{K}_\lambda^{\mathbf{Z}_2}$-theory. The bifurcation diagrams depend on whether $\epsilon\delta = \pm 1$ and are shown in Figure 10.1. Note that each of the branches in Figure 10.1 is in fact a point on an $\mathbf{SO}(2)$-orbit. For example when $\epsilon\alpha < 0$ and $\epsilon\delta = \pm 1$ there is a bifurcating ellipsoid foliated by branches of period-q points, and this is true for each $q \geq 3$.

The period-q points near the collision are "integrable" in the sense that they ride on the $\mathbf{SO}(2)$-group orbit (although the map is not in general integrable). Recall also that by Example 1 in Appendix E the $\mathbf{SO}(2)$-equivariant symplectic map has the conserved quantity $I = \langle y\,, \mathbf{J}x \rangle$; that is, every orbit of the symplectic iterated $\mathbf{SO}(2)$-equivariant map lies on an invariant submanifold.

An example of the collision of multipliers in an $\mathbf{SO}(2)$-equivariant symplectic map on $\mathbf{R}^4$ can be obtained by parametrically forcing T-periodically the Lagrange top (assuming the forcing does not break the $\mathbf{SO}(2)$-symmetry). The Lagrange top is integrable but has a collision of eigenvalues of opposite sign (van der Meer [1990]).

Further analysis of the dynamics of $\mathbf{SO}(2)$-equivariant symplectic maps on $\mathbf{R}^4$ can be pursued using orbit space reduction. The reduction is identical to that presented for $\mathbf{O}(2)$-equivariant maps in Section 8.5 (indeed it was only the $\mathbf{SO}(2) \subset \mathbf{O}(2)$ that was used in the reduction in Section 8.5). In the non-singular case ($I \neq 0$) the map can be reduced to an area-preserving map (as in (9.50) for the irrational collision normal form). In fact $\mathbf{SO}(2)$-equivariant maps with a collision will have the same local phase space dynamics on the reduced phase space as the normal form for the irrational collision (Section 9.7 and Bridges et al. [1993]). However, since the $\mathbf{SO}(2)$-symmetry is exact in the present case, the map is not near integrable: the reduced phase space will contain all the complex and interesting dynamics typical of area-preserving maps – coupled to the flow along the group orbit.

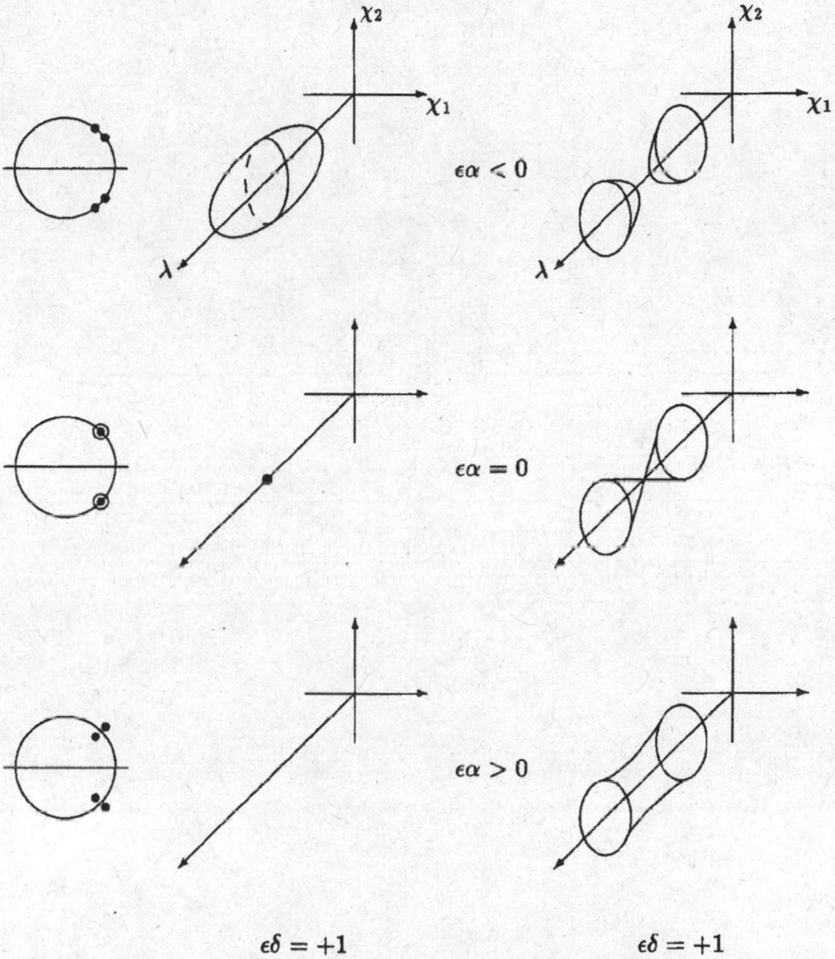

Figure 10.1: Bifurcation of period-q points near a collision of multipliers in **SO**(2)-equivariant maps

Appendices

Appendix A. Equivariant Splitting Lemma

The precise form of the Splitting Lemma needed for the decomposition of the functional $W_q : \mathcal{U}\times\mathcal{V} \to \mathbf{R}$ with $\mathcal{U}\times\mathcal{V}$ a neighborhood of the origin in $\mathbf{X}_q^n\times\mathbf{R}^k$, $k \geq 0$, is presented in the following. Recall that $\mathbf{X}_q^n \approx \mathbf{R}^{nq}$.

Equivariant Splitting Lemma. *Let* $W_q \in C^\infty(\mathcal{U}\times\mathcal{V}, \mathbf{R})$, $\mathcal{U}\times\mathcal{V} \subset \mathbf{X}_q^n \times \mathbf{R}^k$ *a neighborhood of the origin, and suppose that*

$$\nabla_{\mathbf{x}} W_q^o = 0 \quad \text{and} \quad \mathrm{Hess}_{\mathbf{x}}\, W_q^o = \mathbf{L}^o$$

with $\mathrm{Ker}\ \mathbf{L}^o = \mathrm{span}\{\xi_1, \ldots, \xi_l\}$, $l \geq 1$. *Suppose moreover that* W_q *is invariant with respect to the action of a compact Lie group* Γ *acting on* $\mathbf{X}_q^n$ *only:*

$$W_q(\gamma\cdot\mathbf{x}, \Lambda) = W_q(\mathbf{x}, \Lambda)\,, \quad \forall\gamma\in\Gamma \quad \text{and} \quad \forall(\mathbf{x}, \lambda, \alpha) \in \mathcal{U}\times\mathcal{V}.$$

Let $\mathbf{X}_q^n = \mathbf{A}\oplus\mathbf{B}$ *with* $\mathbf{A} = \mathrm{Ker}\,\mathbf{L}^o$ *and for any* $\mathbf{x}\in\mathbf{X}_q^n$ *write* $\mathbf{x} = \chi_1\xi_1 + \cdots + \chi_l\xi_l + \Upsilon$ *with* $\chi \in \mathbf{A} \subset \mathbf{R}^l$ *and* $\Upsilon \in \mathbf{B} \subset \mathbf{R}^{nq-l}$. *Then there exists a neighborhood* $\hat{\mathbf{A}}\times\hat{\mathbf{B}}\times\hat{\mathcal{V}}$ *of the origin in* $\mathbf{A}\times\mathbf{B}\times\mathbf{R}^k$, *a* Γ*-equivariant map* $\phi \in C^\infty(\hat{\mathbf{B}}\times\hat{\mathbf{A}}\times\hat{\mathcal{V}}, \hat{\mathbf{B}})$ *and a* Γ*-invariant map* $\widehat{W}_q \in C^\infty(\hat{\mathbf{A}}\times\hat{\mathcal{V}}, \mathbf{R})$ *such that*

$$W_q(\chi_1\xi_1 + \cdots + \chi_l\xi_l + \phi(\Upsilon, \chi, \Lambda), \Lambda) = \tfrac{1}{2}\langle \Upsilon, \mathbf{L}^o\Upsilon\rangle + \widehat{W}_q(\chi, \Lambda)\,.$$

Moreover ϕ *and* $\widehat{W}_q$ *are* Γ*-invariant, that is, let* $\gamma_A : \Gamma \to \mathbf{O}(l)$, $\gamma_B : \Gamma \to \mathbf{O}(nq-l)$, *be the orthogonal representations for* Γ *induced on* $\mathbf{A}$ *and* $\mathbf{B}$, *respectively, then*

$$\gamma_B\cdot\phi(\Upsilon, \chi, \Lambda) = \phi(\gamma_B\cdot\Upsilon, \gamma_A\cdot\chi, \Lambda)\,,$$
$$\widehat{W}_q(\gamma_A\cdot\chi, \Lambda) = \widehat{W}_q(\chi, \Lambda)\,,$$

for every $\gamma\in\Gamma$ *and* $(\chi, \Upsilon, \Lambda) \in \hat{\mathbf{A}}\times\hat{\mathbf{B}}\times\hat{\mathcal{V}}$.

With the exception of the symmetry properties the above result is the well-known Splitting Lemma or Generalized Morse Lemma. Checking the details of the proof of Magnus [1977] shows that the diffeomorphism ϕ and the reduced functional $\widehat{W}_q$ can be chosen to respect the symmetry of W_q.

Appendix B. Signature on configuration space

Given a symplectic map $\mathbf{T} : (x, y) \mapsto (x', y')$ on $\mathbf{R}^{2n}$, suppose $e^{i\theta}$ with $\theta \in (0, \pi)$ is in the spectrum of $\mathbf{DT}^o$. Suppose moreover that $\mathbf{T}$ is generated by $h(x, x')$. In this appendix, an expression for the signature of the multiplier $e^{i\theta}$ is obtained in terms of the properties of h; that is, the signature in configuration space. In fact we prove a more general result: an expression for the signature of a multiplier of a period-q point.

The idea is roughly as follows. Suppose $\mathbf{x} \in \mathbf{X}_q^n$ is a period-q point of the symplectic map generated by h; that is,

$$h_1(x^i, x^{i+1}) + h_2(x^{i-1}, x^i) = 0\,,\ 1 \le i \le q \quad \text{with} \quad x^{i+q} = x^i.$$

Let $\zeta = (\zeta^1, \ldots, \zeta^q) \in \mathbf{C}^{nq}$ be a sequence in the tangent orbit of $\{x^i\} \in \mathbf{X}_q^n$, with $\zeta^{j+q} = e^{i\theta}\zeta^j$ (for some $\theta \in (0, \pi)$ and any j), and let

$$\mathbf{B}_i = -h_{12}(x^i, x^{i+1}) \quad \text{for} \quad \{x^i\} \in \mathbf{X}_q^n. \tag{B.1}$$

Then the *signature* of the simple multiplier $e^{i\theta}$ is

$$\sigma = \operatorname{sign} Q \tag{B.2}$$

with $Q = \operatorname{Im}(\overline{\zeta}^i, \mathbf{B}_i\zeta^{i+1})$ for any $1 \le i \le q$, noting that $\zeta^{q+1} = e^{i\theta}\zeta^1$.

In other words the signature can be evaluated at *any* point along the period-q point orbit. But this is misleading however as the elements in the sequence $\{\zeta^i\}_{i=1}^q$ are not independent; that is, $\zeta = (\zeta^1, \ldots, \zeta^q) \in \operatorname{Ker} \mathbf{M}(e^{i\theta})$ where

$$\mathbf{M}(\mu) = \begin{bmatrix} \hat{\mathbf{A}}_1 & -\mathbf{B}_1 & \mathbf{0} & \cdots & \mathbf{0} & -\frac{1}{\mu}\mathbf{B}_q^T \\ -\mathbf{B}_1^T & \hat{\mathbf{A}}_2 & -\mathbf{B}_2 & \ddots & & \mathbf{0} \\ \mathbf{0} & -\mathbf{B}_2^T & \hat{\mathbf{A}}_3 & \ddots & \ddots & \vdots \\ \vdots & \ddots & \ddots & \ddots & \ddots & \mathbf{0} \\ \mathbf{0} & & \ddots & \ddots & \hat{\mathbf{A}}_{q-1} & -\mathbf{B}_{q-1} \\ -\mu\mathbf{B}_q & \mathbf{0} & \cdots & \mathbf{0} & -\mathbf{B}_{q-1}^T & \hat{\mathbf{A}}_q \end{bmatrix} \tag{B.3}$$

with $\hat{\mathbf{A}}_j = \mathbf{A}_j + \mathbf{C}_j$,

$$\mathbf{A}_j = h_{11}(x^j, x^{j+1}) \quad \text{and} \quad \mathbf{C}_j = h_{22}(x^{j-1}, x^j) \quad \text{for} \quad \{x^j\} \in \mathbf{X}_q^n. \tag{B.4}$$

In terms of the symplectic map $\mathbf{T}$ there are several definitions for the signature in the literature. For example Howard & MacKay [1987, p.1040] use

$$\sigma = \operatorname{sign}[\operatorname{Re} z, \mathbf{J}_{2n}\mathbf{L}(q)\operatorname{Re} z] \quad \text{for any} \quad z \in \operatorname{Ker}\left(\mathbf{L}(q) - e^{i\theta}\mathbf{I}_{2n}\right), \tag{B.5a}$$

where

$$\mathbf{L}(q) = \mathbf{DT}(x^{i+q-1}, y^{i+q-1}) \circ \cdots \circ \mathbf{DT}(x^i, y^i) \quad \text{with} \quad \{x^i, y^i\} \in \mathbf{X}_q^n \times \mathbf{X}_q^n, \tag{B.5b}$$

$[\cdot,\cdot]$ is a *real* inner product on $\mathbf{R}^{2n}$. Krein's original definition (see Yakubovich & Starzhinskii [1972, p.159]) is

$$\sigma = \operatorname{sign}(i[\overline{z}, \mathbf{J}_{2n}z]) \quad \text{for any} \quad z \in \operatorname{Ker}\left(\mathbf{L}(q) - e^{i\theta}\mathbf{I}_{2n}\right). \tag{B.6}$$

We will use both of these definitions but first we establish that they are equivalent.

Proposition B.1. *Let $e^{i\theta}$ be an eigenvalue of $\mathbf{L}(q)$ with $\theta \in (0,\pi)$ and let $z \in \operatorname{Ker}\left(\mathbf{L}(q) - e^{i\theta}\mathbf{I}_{2n}\right)$. Then*

$$\operatorname{sign}\left[\operatorname{Re} z, \mathbf{J}_{2n}\,\mathbf{L}(q)\operatorname{Re} z\right] = \operatorname{sign}\left(i[\overline{z}, \mathbf{J}_{2n} z]\right).$$

Proof. By the definition of z: $\mathbf{L}(q)\operatorname{Re} z = \operatorname{Re}(e^{i\theta} z)$. Therefore

$$\begin{aligned}
[\operatorname{Re} z, \mathbf{J}_{2n}\,\mathbf{L}(q)\operatorname{Re} z] &= \tfrac{1}{4}\left[z + \overline{z}, \mathbf{J}_{2n}(e^{i\theta} z + e^{-i\theta}\overline{z})\right] \\
&= \tfrac{1}{2}\operatorname{Re}\left(e^{i\theta}[\overline{z}, \mathbf{J}_{2n} z]\right) + \tfrac{1}{2}\operatorname{Re}\left(e^{i\theta}[z, \mathbf{J}_{2n} z]\right) \\
&= \operatorname{Re}\left(ie^{i\theta}[\operatorname{Re} z, \mathbf{J}_{2n}\operatorname{Im} z]\right) \\
&= -\sin\theta\,[\operatorname{Re} z, \mathbf{J}_{2n}\operatorname{Im} z] = \tfrac{1}{2} i\sin\theta\,[\overline{z}, \mathbf{J}_{2n} z].
\end{aligned}$$

Since $\theta \in (0,\pi)$, by hypothesis, $\sin\theta$ is positive and the result follows. ■

Proposition B.2. *Suppose $\{(x^i, y^i)\}_{i=1}^{q}$ is a period-q point of the symplectic map $\mathbf{T}$ on $\mathbb{R}^{2n}$. Let $\mathbf{L}_i = \mathbf{DT}(x^i, y^i)$, define*

$$\Delta(\mu) = \det\left(\mathbf{L}_{q+i-1} \circ \cdots \circ \mathbf{L}_i - \mu \mathbf{I}_{2n}\right),$$

and suppose

$$\Delta(e^{i\theta}) = 0 \quad \text{and} \quad \Delta'(e^{i\theta}) \neq 0 \quad \text{where} \quad \theta \in (0,\pi);$$

that is, the period-q point has a simple Floquet multiplier on the unit circle. Suppose moreover that the symplectic map is generated by h. Then the signature of the multiplier $e^{i\theta}$ is given by (B.2).

Proof. The symplectic map generated by $h(x, x')$ can be written implicitly as

$$y^{i+1} = h_2(x^i, x^{i+1}) \quad \text{and} \quad y^i = -h_1(x^i, x^{i+1}). \tag{B.7}$$

Suppose $\{x^i, y^i\} \in \mathbf{X}_q^n \times \mathbf{X}_q^n$ and perturb about the period-q point,

$$x^i \mapsto x^i + \zeta^i \quad \text{and} \quad y^i \mapsto y^i + \eta^i.$$

Substitution into (B.7) results in

$$\begin{aligned}
\eta^i &= -\mathbf{A}_i \zeta^i + \mathbf{B}_i \zeta^{i+1}, \\
\eta^{i+1} &= -\mathbf{B}_i^T \zeta^i + \mathbf{C}_{i+1}\zeta^{i+1},
\end{aligned} \tag{B.8}$$

with $\mathbf{A}_i$, $\mathbf{B}_i$ and $\mathbf{C}_i$ as defined in (B.1) and (B.4), which can be rewritten as

$$\begin{pmatrix} \zeta^{i+1} \\ \eta^{i+1} \end{pmatrix} = \mathbf{DT}(x^i, y^i)\begin{pmatrix} \zeta^i \\ \eta^i \end{pmatrix}$$

where

$$\mathbf{DT}(x^i, y^i) \stackrel{\text{def}}{=} \mathbf{L}_i = \begin{pmatrix} \mathbf{B}_i^{-1}\mathbf{A}_i & \mathbf{B}_i^{-1} \\ \mathbf{C}_{i+1}\mathbf{B}_i^{-1}\mathbf{A}_i - \mathbf{B}_i^T & \mathbf{C}_{i+1}\mathbf{B}_i^{-1} \end{pmatrix}.$$

Therefore $z^i \in \operatorname{Ker}(\mathbf{L}_{i+q-1} \circ \cdots \circ \mathbf{L}_i - e^{i\theta}\mathbf{I}_{2n})$ has the form

$$z^i = \begin{pmatrix} \zeta^i \\ -\mathbf{A}_i\zeta^i + \mathbf{B}_i\zeta^{i+1} \end{pmatrix}.$$

This follows since $\mathbf{L}_i z^i = z^{i+1}$. Therefore

$$\begin{aligned}[\overline{z}^i, \mathbf{J}_{2n} z^i] &= -(\overline{\zeta}^i, \mathbf{A}_i\zeta^i) + (\overline{\zeta}^i, \mathbf{B}_i\zeta^{i+1}) + (\zeta^i, \mathbf{A}_i\overline{\zeta}^i) - (\zeta^i, \mathbf{B}_i\overline{\zeta}^{i+1}) \\ &= 2i\,\mathrm{Im}\,(\overline{\zeta}^i, \mathbf{B}_i\zeta^{i+1}).\end{aligned}$$

Hence, using Proposition B.1 and (B.6) the signature function Q, for any $1 \le i \le q$, is

$$Q = -\mathrm{Im}\,(\overline{\zeta}^i, \mathbf{B}_i\zeta^{i+1}).$$

To verify that $\zeta = (\zeta_1, \ldots, \zeta_q) \in \operatorname{Ker}\mathbf{M}(e^{i\theta})$ note that by (B.8),

$$-\mathbf{B}_{i-1}^T\zeta^{i-1} + (\mathbf{A}_i + \mathbf{C}_i)\zeta^i - \mathbf{B}_i\zeta^{i+1} = 0$$

which when combined with $\zeta^{i+q} = e^{i\theta}\zeta^i$ results in the matrix equation $\mathbf{M}(\mu)\cdot\zeta = 0$ with $\mu = e^{i\theta}$ and $\mathbf{M}(\mu)$ as defined in equation (B.3). ■

Of special interest is the signature for a multiplier of a period-1 point of a Σ-equivariant map with the compact group Σ acting diagonally on $\mathbf{R}^{2n}$. We consider the *real* blocks of the isotypic decomposition (8.1), that is, those blocks where Σ acts absolutely irreducibly on W_i and the invariant subspace V_i is equal to $\mathbf{R}^{n_i} \otimes_{\mathbf{R}} W_i$.

From (8.2), the block of $\mathbf{DT}$ corresponding to V_i is $\mathbf{DT}_i \otimes_{\mathbf{R}} \mathbf{I}_{W_i}$, say

$$\begin{pmatrix} \mathbf{B}^{-1}\mathbf{A} & \mathbf{B}^{-1} \\ \mathbf{C}\mathbf{B}^{-1}\mathbf{A} - \mathbf{B}^T & \mathbf{C}\mathbf{B}^{-1} \end{pmatrix} \otimes_{\mathbf{R}} \mathbf{I}_{W_i},$$

where $\mathbf{DT}_i$ is generated by $h_i(x, x') = \frac{1}{2}(x, \mathbf{A}x) - (x, \mathbf{B}x') + \frac{1}{2}(x', \mathbf{C}x')$, with $x, x' \in \mathbf{R}^{n_i}$ and $|\mathbf{B}| \neq 0$ (for simplicity we drop the subscript i from $\mathbf{A}$, $\mathbf{B}$ and $\mathbf{C}$).

As we have seen in Chapter 8, the multipliers are then of multiplicity dim W_i, but the fact that the multiplicity is forced by *symmetry* means that they actually behave like multipliers of $\mathbf{DT}_i$ only .

Using the tensor product structure it is clear that we can extend the definitions of the signature to multipliers in the block of $\mathbf{DT}$ corresponding to V_i provided there are *single* multipliers of $\mathbf{DT}_i$. The element z of the kernel are thus $z_1 \otimes w_1$ with z_1 being in the kernel of $(\mathbf{DT}_i - e^{i\theta}\mathbf{I}_{2n_i})$ and w_1 being *any* vector in W_i. And so, the contribution of w_1 drops from the formulas for the signature as it appears as $||w_1||^2 > 0$.

And so, we have the following.

Proposition B.3. *In the previous set-up, suppose that* $e^{i\theta}$, $\theta \in (0, \pi)$, *is a simple multiplier in the spectrum of* $\mathbf{DT}_i$. *Let* $\zeta \in \operatorname{Ker}(\mathbf{A} + \mathbf{C} - e^{i\theta}\mathbf{B} - e^{-i\theta}\mathbf{B}^T)$, *then the signature of the multiplier is*

$$\sigma = \operatorname{sign} Q \quad \text{with} \quad Q = -\operatorname{Im}(e^{i\theta}(\overline{\zeta}, \mathbf{B}\zeta)). \tag{B.9}$$

Moreover, if another multiplier collides with $e^{i\theta}$, *its block in the Jordan normal form of* $\mathbf{DT}_i$ *will have non-trivial nilpotent part iff* $Q = 0$; *that is,*

$$\operatorname{Im}(e^{i\theta}(\overline{\zeta}, \mathbf{B}\zeta)) = 0. \tag{B.10}$$

On the other hand, if $\operatorname{Im}(e^{i\theta}(\overline{\zeta}, \mathbf{B}\zeta)) \neq 0$ *at the collision the Jordan normal form block is diagonal.*

Proof. As we previously mentioned we can discard the contribution of W_i and concentrate upon $\mathbf{DT}_i$ and its simple multiplier $e^{i\theta}$

The first part is a consequence of Proposition B.2 for the special case q=1. But we give a direct proof using (B.4) that will be useful for the second part of the proof.

The signature of $e^{i\theta}$ is defined using (B.5a) with $\mathbf{DT}_i^o = \mathbf{L}(1)$, $q = 1$ and $n = n_i$. An element $z \in \operatorname{Ker}(\mathbf{DT}_i - e^{i\theta}\mathbf{I}_{2n_i})$ is given by

$$z = \begin{pmatrix} \zeta \\ -(\mathbf{A} - e^{i\theta}\mathbf{B})\zeta \end{pmatrix} \quad \text{with} \quad \operatorname{Re} z = \begin{pmatrix} \xi \\ (\cos\theta\,\mathbf{B} - \mathbf{A})\xi - \sin\theta\,\mathbf{B}\eta \end{pmatrix} \tag{B.11}$$

and $\zeta \overset{\text{def}}{=} \xi + i\eta$ satisfying

$$(\mathbf{A} + \mathbf{C} - e^{i\theta}\mathbf{B} - e^{-i\theta}\mathbf{B}^T)\zeta = 0. \tag{B.12}$$

Substitution of $\operatorname{Re} z$ into (B.5a) results in (after some algebra and use of the real and imaginary parts of (B.12))

$$Q = -\sin\theta\left[(\xi, (\mathbf{B} - \mathbf{B}^T)\eta)\cos\theta + ((\xi, \mathbf{B}\xi) + (\eta, \mathbf{B}\eta))\sin\theta\right].$$

Reverting back to complex notation, it is immediate that

$$\operatorname{sign} Q = -\operatorname{sign}\operatorname{Im}(e^{i\theta}(\overline{\zeta}, \mathbf{B}\zeta))$$

verifying (B.9).

If another multiplier collides with $e^{i\theta}$ the spectrum of $\mathbf{DT}_i$ will contain of $e^{i\theta}$ and $e^{-i\theta}$, each with multiplicity two. Their blocks in the Jordan normal form will have nontrivial nilpotent part if and only if there exists a non-trivial solution to $(\mathbf{DT}_i - e^{i\theta}\mathbf{I}_{2n_i})u = z$ where $z \in \operatorname{Ker}(\mathbf{DT}_i - e^{i\theta}\mathbf{I}_{2n_i})$. Let $u = (u_1, u_2)$ and $z = (z_1, z_2)$, then $(\mathbf{DT}_i - e^{i\theta}\mathbf{I}_{2n_i})u = z$ is equivalent to

$$\left.\begin{aligned} (\mathbf{A} + \mathbf{C} - e^{i\theta}\mathbf{B} - e^{-i\theta}\mathbf{B}^T)u_1 &= e^{-i\theta}z_2 - e^{-i\theta}(\mathbf{C} - e^{i\theta}\mathbf{B})z_1, \\ u_2 &= -(\mathbf{A} - e^{i\theta}\mathbf{B})u_1 + \mathbf{B}z_1. \end{aligned}\right\} \tag{B.13}$$

Now $(\mathbf{A}+\mathbf{C}-e^{i\theta}\mathbf{B}-e^{-i\theta}\mathbf{B}^T)$ has non-trivial kernel $\{\zeta\}$ and is Hermitian. Therefore the first of (B.13) has a non-trivial solution if and only if

$$(\overline{\zeta}\,,\, z_2 - (\mathbf{C} - e^{i\theta}\mathbf{B})\, z_1) = 0\,. \tag{B.14}$$

Using (B.11)

$$\begin{aligned} z_2 - (\mathbf{C} - e^{i\theta}\mathbf{B})\, z_1 &= -(\mathbf{A} - e^{i\theta}\mathbf{B})\,\zeta - (\mathbf{C} - e^{i\theta}\mathbf{B})\,\zeta \\ &= -(\mathbf{A} + \mathbf{C} - 2e^{i\theta})\,\zeta \\ &= (e^{i\theta}\mathbf{B} - e^{-i\theta}\mathbf{B}^T)\,\zeta\,. \end{aligned}$$

Therefore the condition (B.14) is

$$(\overline{\zeta}, (e^{i\theta}\mathbf{B} - e^{-i\theta}\mathbf{B}^T)\,\zeta) = 0\,,$$

that is, the block of the Jordan normal form of $\mathbf{DT}_i$ corresponding to $e^{i\theta}$ has non-trivial nilpotent part if and only if (B.10) is satisfied. ■

Appendix C. Linear stability on configuration space

In this appendix a correspondence is established between the $2n$ Floquet multipliers of a period-q orbit of a symplectic map on $\mathbf{R}^{2n}$ and the eigenvalue problem associated with the Hessian of the action on the $\mathbf{R}^q \times \mathbf{R}^n$ dimensional configuration space. In particular we introduce a transformation that reduces the configuration space eigenvalue problem on R^{nq} to a generalized eigenvalue problem on $\mathbf{R}^n$.

Let $\{(x^i, y^i)\}_{i=1}^q$ be a period-q orbit of the symplectic map $\mathbf{T} : (x, y) \mapsto (x', y')$ on $\mathbf{R}^{2n}$ and suppose that $\mathbf{T}$ is generated by h. The Jacobian of $\mathbf{T}$ evaluated on the period-q orbit is denoted $\mathbf{DT}(x^i, y^i)$ and the Floquet multipliers of the period-q orbit are eigenvalues of $\mathbf{L}(q)$ where

$$\mathbf{L}(q) = \mathbf{DT}(x^{i+q-1}, y^{i+q-1}) \circ \cdots \circ \mathbf{DT}(x^i, y^i)\,, \quad \{x^i, y^i\} \in \mathbf{X}_q^n \times \mathbf{X}_q^n\,. \tag{C.1}$$

The matrix $\mathbf{L}(q) \in \mathbf{Sp}(2n, \mathbf{R})$ and the Floquet multipliers satisfy

$$\Delta(\rho) \stackrel{\text{def}}{=} |\mathbf{L}(q) - \mu \mathbf{I}_{2n}| = \rho^n + a_1 \rho^{n-1} + \cdots + a_n = 0 \quad \text{where} \quad \rho = \mu + \tfrac{1}{\mu}\,.$$

For each periodic orbit there are n residues – with residue defined by $R = \frac{1}{4}(2-\rho)$.

When the symplectic map is generated by h, the periodic orbits are obtained as critical points of the action functional on a space of dimension nq. In this context it is useful to express the n residues in terms of the Hessian of the action. The q-dimensional case – period-q points of area-preserving maps – corresponds to a result of MacKay & Meiss [1983]. In this appendix we generalize the MacKay-Meiss result to arbitrary (configuration space) dimension.

Given a period-q orbit in configuration space – $x^{i+q} = x^i, \forall i \in \mathbf{Z}$, – the tangent map is given by

$$-\mathbf{B}_{i-1}\xi^{i-1} + \hat{\mathbf{A}}_i\xi^i - \mathbf{B}_i^T\xi^{i+1} = 0 \tag{C.2}$$

with $\hat{\mathbf{A}}_i = \mathbf{A}_i + \mathbf{C}_i$ and

$$\mathbf{A}_i = h_{11}(x^i, x^{i+1}), \quad \mathbf{B}_i = -h_{12}(x^i, x^{i+1}) \quad \text{and} \quad \mathbf{C}_i = h_{22}(x^{i-1}, x^i).$$

The Floquet multiplier μ satisfies

$$\xi^{i+q} = \mu\xi^i. \tag{C.3}$$

Combining (C.2) and (C.3) results in a matrix system $\mathbf{M}(\mu)\xi = 0$, where $\mathbf{M}(\mu)$ is as defined in (B.3). The polynomial obtained by setting $|\mathbf{M}(\mu)| = 0$ is the characteristic polynomial for the Floquet multipliers. The matrix $\mathbf{M}(\mu)$ is of order nq and the idea is to transform it to a generalized eigenvalue problem on $\mathbf{R}^n$!

Note first of all that $\mathbf{M}(1)$ is the second variation of the action evaluated at a period q point; that is, $\mathbf{M}(1) = \mathrm{Hess}_{\mathbf{x}}\, W_q(\mathbf{x})$. Now, partition $\mathbf{M}(\mu)$ using

$$\mathbf{D} = \begin{pmatrix} \hat{\mathbf{A}}_2 & -\mathbf{B}_2 & \mathbf{0} & \cdots & \mathbf{0} \\ -\mathbf{B}_2^T & \hat{\mathbf{A}}_3 & -\mathbf{B}_3 & \ddots & \vdots \\ \mathbf{0} & -\mathbf{B}_3^T & \hat{\mathbf{A}}_4 & \ddots & \mathbf{0} \\ \vdots & \ddots & \ddots & \ddots & -\mathbf{B}_{q-1} \\ \mathbf{0} & \cdots & \mathbf{0} & -\mathbf{B}_{q-1}^T & \hat{\mathbf{A}}_q \end{pmatrix} \quad \text{and} \quad \mathbf{b}(\mu) = \begin{pmatrix} -\mathbf{B}_1^T \\ \mathbf{0} \\ \vdots \\ \mathbf{0} \\ -\mu\mathbf{B}_q \end{pmatrix} \tag{C.4}$$

and suppose henceforth that $|\mathbf{D}| \neq 0$. Then using (C.4),

$$\begin{aligned} |\mathbf{M}(\mu)| &= \left| \begin{pmatrix} \hat{\mathbf{A}}_1 & \mathbf{b}(\mu^{-1})^T \\ \mathbf{b}(\mu) & \mathbf{D} \end{pmatrix} \right| \\ &= \left| \begin{pmatrix} \mathbf{I}_n & -\mathbf{b}(\mu^{-1})^T\mathbf{D}^{-1} \\ \mathbf{0} & \mathbf{I}_{n(q-1)} \end{pmatrix} \begin{pmatrix} \hat{\mathbf{A}}_1 & \mathbf{b}(\mu^{-1})^T \\ \mathbf{b}(\mu) & \mathbf{D} \end{pmatrix} \right| \\ &= \left| \begin{pmatrix} \hat{\mathbf{A}}_1 - \mathbf{b}(\mu^{-1})^T\mathbf{D}^{-1}\mathbf{b}(\mu) & \mathbf{0} \\ \mathbf{b}(\mu) & \mathbf{D} \end{pmatrix} \right| \\ &= |\mathbf{D}||\hat{\mathbf{A}}_1 - \mathbf{b}(\mu^{-1})^T\mathbf{D}^{-1}\mathbf{b}(\mu)|. \end{aligned}$$

Clearly, if $|\mathbf{D}| \neq 0$ $|\mathbf{M}(\mu)| = 0$ is equivalent to $|\hat{\mathbf{A}}_1 - \mathbf{b}(\mu^{-1})^T\mathbf{D}^{-1}\mathbf{b}(\mu)| = 0$, but $\hat{\mathbf{A}}_1 - \mathbf{b}(\mu^{-1})^T\mathbf{D}^{-1}\mathbf{b}(\mu)$ is a $n\times n$ matrix which can be simplified further as follows. Expand $b(\mu)$ about $\mu = 1$,

$$\mathbf{b}(\mu) = \mathbf{b}(1) - (\mu - 1)\mathbf{b}' \quad \text{with} \quad \mathbf{b}' = \begin{pmatrix} \mathbf{0} \\ \vdots \\ \mathbf{0} \\ \mathbf{B}_q \end{pmatrix}.$$

Then $\mathbf{b}(\mu^{-1})^T = \mathbf{b}(1)^T - (\frac{1}{\mu} - 1)\mathbf{b}'^T$ and

$$\hat{\mathbf{A}}_1 - \mathbf{b}(\mu^{-1})^T\mathbf{D}^{-1}\mathbf{b}(\mu) = \hat{\mathbf{A}}_1 - \mathbf{b}(1)^T\mathbf{D}^{-1}\mathbf{b}(1) + (\mu - 1)(\mathbf{b}(1) + \mathbf{b}')^T\mathbf{D}^{-1}\mathbf{b}' + \left(\frac{1}{\mu} - 1\right)\mathbf{b}'^T\mathbf{D}^{-1}(\mathbf{b}(1) + \mathbf{b}').$$

Therefore, with

$$\mathbf{E} = \hat{\mathbf{A}}_1 - \mathbf{b}(1)^T\mathbf{D}^{-1}\mathbf{b}(1),$$
$$\mathbf{F} = (\mathbf{b}(1) + \mathbf{b}')^T\mathbf{D}^{-1}\mathbf{b}',$$

the eigenvalue problem for μ reduces to

$$|\mathbf{M}(\mu)| = |\mathbf{D}|\,|\mathbf{E} + (\mu - 1)\mathbf{F} + (\tfrac{1}{\mu} - 1)\mathbf{F}^T|. \tag{C.5}$$

In particular, the residues are obtained from the eigenvalues of a generalized eigenvalue problem on $\mathbb{R}^n$ and the reduced matrices **E** and **F** are completely determined from the Hessian of the action. Note that **E** is a symmetric $n \times n$ matrix but **F** is a general $n \times n$ matrix.

Restricting to the case $n = 2$ – 4D symplectic maps – let

$$a = 4 - \operatorname{Tr}(\mathbf{F}^{-1}\mathbf{E}) - |\mathbf{F}|^{-1}(\operatorname{Tr}\mathbf{JF})^2,$$
$$b = 2a - 2 + |\mathbf{F}^{-1}\mathbf{E}|.$$

Then,

$$\begin{aligned}\Delta(\rho) &= \left|\mathbf{E} + (\mu - 1)\mathbf{F} + (\tfrac{1}{\mu} - 1)\mathbf{F}^T\right| \\ &= |\mathbf{F}|(\rho - 2)^2 + \left\{|\mathbf{F}|\operatorname{Tr}(\mathbf{F}^{-1}\mathbf{E}) + (\operatorname{Tr}\mathbf{JF})^2\right\}(\rho - 2) + |\mathbf{E}| \\ &= |\mathbf{F}|(\rho^2 - a\rho + b - 2);\end{aligned}$$

recovering the familiar characteristic equation for the multipliers of an element of $\mathbf{Sp}(4, \mathbb{R})$ (compare with equation (2.43) in Section 2).

Remarks. (a) Complete stability information is obtained from a partition of $\operatorname{Hess}_{\mathbf{x}} W_q(\mathbf{x})$. The only non-trivial numerical calculation is the inversion of **D** and **D** is a block tridiagonal matrix.

(b) If $|\mathbf{D}| = 0$ the above expression will need modification.

(c) Note that $\mathbf{b}'$ and $\mathbf{b}(1) + \mathbf{b}'$ have only one non-zero block. It is then easy to show that $|\mathbf{D}| \cdot |\mathbf{F}| = \prod_{j=1}^{q} |\mathbf{B}_j|$ and $|\mathbf{M}(1)| = |\mathbf{D}| \cdot |\mathbf{E}|$. When $n = 1$, **E** and **F** are scalars (E, F) with $E|\mathbf{D}| = |\mathbf{M}(1)|$ and $F|\mathbf{D}| = \prod_{j=1}^{q} B_j$ (where B_j are scalars) which recovers the result of MacKay-Meiss.

Appendix D. Transformation to linear normal form

Suppose $\mathbf{M} \in \mathbf{Sp}(4, \mathbb{R})$, $\theta \in (0, \pi)$ and $\sigma(\mathbf{M}) = \{e^{\pm i\theta}\}$ each of multiplicity two associated with a collision of multipliers of opposite signature (non-trivial Jordan normal form). A constructive proof of Williamson's normal form theorem is given for the above case by introducing the necessary symplectic basis with particular attention to the sign invariant of the collision (not to be confused with the signature).

Let $z, w \in \mathbb{C}^4$ be the geometric and generalized eigenvectors respectively corresponding to the eigenvalue $e^{i\theta}$, then

$$(\mathbf{M} - e^{i\theta}\mathbf{I}_4)\, z = 0 \qquad \text{and} \qquad (\mathbf{M} - e^{i\theta}\mathbf{I}_4)\, w = e^{i\theta} z\,.$$

Note that the generalized eigenvector w exists iff z is orthogonal to the adjoint geometric eigenvector of $\mathbf{M}$. The existence of w is assured however by the (generic) collision of multipliers of *opposite* signature: with $\mathbf{M} \in \mathbf{Sp}(4, \mathbf{R})$ it follows that

$$(\mathbf{M}^T - e^{-i\theta}\mathbf{I}_4)\,\mathbf{J}_4 z = 0 \qquad \text{and} \qquad (\mathbf{M}^T - e^{-i\theta}\mathbf{I}_4)\,\mathbf{J}_4 w = -e^{-i\theta}\mathbf{J}_4 z\,.$$

That is, the adjoint geometric eigenvector is $\mathbf{J}_4 z$ and $\mathbf{M}$ has non-trivial nilpotent part iff $[\mathbf{J}_4\overline{z}, z] = 0$ (signature function zero (see Appendix B)). The generalized eigenvector w is composed of the unique particular solution plus an arbitrary amount of homogeneous solution z. Therefore w can be adjusted to satisfy $[\overline{w}, \mathbf{J}_4 w] = 0$. The sign invariant of the collision arises because z can only be scaled to within a sign. Without loss of generality we can scale z such that

$$[\overline{z}, \mathbf{J}_4 w] = 2\epsilon \quad \text{for} \quad \epsilon = \pm 1\,. \tag{D.1}$$

Then it is easily verified that the transformation matrix

$$\mathbf{T} = (\operatorname{Re} z \mid \operatorname{Im} z \mid \epsilon \operatorname{Re} w \mid \epsilon \operatorname{Im} w)$$

is symplectic and furthermore

$$\mathbf{T}^{-1}\mathbf{M}\mathbf{T} = \mathbf{M}_0 = \begin{pmatrix} \mathbf{R}_\theta & \epsilon\mathbf{R}_\theta \\ 0 & \mathbf{R}_\theta \end{pmatrix} \quad \text{with} \quad \mathbf{R}_\theta = \cos\theta\,\mathbf{I} + \sin\theta\,\mathbf{J}.$$

In particular there does not exist a symplectic transformation mapping $\left(\begin{smallmatrix}1&1\\0&1\end{smallmatrix}\right) \otimes \mathbf{R}_\theta$ to $\left(\begin{smallmatrix}1&-1\\0&1\end{smallmatrix}\right) \otimes \mathbf{R}_\theta$: ϵ is the *sign invariant* of the collision. The matrix $\mathbf{M}_0$ is the normal form for the collision of multipliers of opposite signature and is equivalent to the normal form originally given by Williamson [1937, p.614].

For use in Chapter 2 we also treat the case where $\mathbf{M} \in \mathbf{Sp}(2n, \mathbf{R})$ and $\sigma(\mathbf{M}) = \{e^{\pm i\theta_1}, \ldots, e^{\pm i\theta_n}\}$ with $\theta_j \in (0, \pi)$ and distinct for $j = 1 \ldots n$.

Corresponding to the multiplier $e^{i\theta_j}$ is the eigenvector $z_j \in \mathbb{C}^{2n}$ with adjoint eigenvector $\mathbf{J}_{2n} z_j$ satisfying

$$(\mathbf{M} - e^{i\theta_j}\mathbf{I}_{2n})\, z_j = 0 \quad \text{and} \quad (\mathbf{M}^T - e^{-i\theta_j}\mathbf{I}_{2n})\,\mathbf{J}_{2n} z_j = 0\,.$$

We can normalize z_j by

$$[\overline{z}_j, \mathbf{J}_{2n} z_j] = 2i\epsilon_j\,, \quad \epsilon_j = \pm 1\,, \quad 1 \le j \le n\,,$$

where ϵ_j is the signature of the multiplier $e^{i\theta_j}$ (see Appendix B).

Now consider the transformation matrix $\widehat{\mathbf{S}}$ defined by

$$\widehat{\mathbf{S}} = (\operatorname{Re} z_1 \,|\, \cdots \,|\, \operatorname{Re} z_n \,|\, \epsilon_1 \operatorname{Im} z_1 \,|\, \cdots \,|\, \epsilon_n \operatorname{Im} z_n).$$

Then it is straightforward to verify that $\widehat{\mathbf{S}}$ is symplectic – with respect to $\mathbf{J}_{2n}$ – and that

$$\widehat{\mathbf{S}}^{-1}\mathbf{M}\widehat{\mathbf{S}} = \widehat{\mathbf{M}}^o = \begin{bmatrix} \cos\theta_1 & & 0 & \epsilon_1 \sin\theta_1 & & 0 \\ & \ddots & & & \ddots & \\ 0 & & \cos\theta_n & 0 & & \epsilon_n \sin\theta_n \\ -\epsilon_1 \sin\theta_1 & & 0 & \cos\theta_1 & & 0 \\ & \ddots & & & \ddots & \\ 0 & & -\epsilon_n \sin\theta_n & 0 & & \cos\theta_n \end{bmatrix} \tag{D.2}$$

or equivalently

$$\widehat{\mathbf{M}}^o = \exp(\mathbf{J} \otimes \mathbf{D}^o) \qquad \text{with} \quad \mathbf{D}^o = \operatorname{diag}(\epsilon_1\theta_1, \ldots, \epsilon_n\theta_n).$$

Note that in this case the normal form $\widehat{\mathbf{M}}^o$ is also reversible with respect to the standard reversor; that is, $\mathcal{R}\widehat{\mathbf{M}}^o\mathcal{R} = \widehat{\mathbf{M}}^{o-1}$ with $\mathcal{R} = \kappa \otimes \mathbf{I}_n$.

Note moreover that $\widehat{\mathbf{M}}^o$ in (D.2) is generated by

$$h(x, x') = \tfrac{1}{2}(x, \mathbf{A}x) - (x, \mathbf{B}x') + \tfrac{1}{2}(x', \mathbf{C}x')$$

with

$$\mathbf{B} = \operatorname{diag}\left(\frac{\epsilon_1}{\sin\theta_1} \cdots \frac{\epsilon_n}{\sin\theta_n}\right)$$
$$\mathbf{A} = \mathbf{C} = \operatorname{diag}(\epsilon_1 \cot\theta_1 \cdots \epsilon_n \cot\theta_n)$$

and that

$$|h_{12}(x, x')| = |-\mathbf{B}| = (-1)^n \prod_{j=1}^{n} \frac{\epsilon_j}{\sin\theta_j} \neq 0\,.$$

Appendix E. Symmetries and conservation laws

The concept of a "momentum map", well known in the setting of continuous Hamiltonian systems, also appears in the discrete setting. That is, associated with every one-parameter group of configuration space symmetries is a conserved quantity. This result is particularly easy to prove when the symplectic map is generated by a Lagrangian generating function. In particular it is a special case of Noether's theorem in a discrete setting.

Theorem E. *Let* $\mathbf{T} : (x, y) \mapsto (x', y')$ *be a symplectic map on* $\mathbf{R}^{2n}$ *generated by* h *and suppose that* h *is invariant with respect to the action of a smooth one-parameter group of spatial symmetries* Σ_θ*:*

$$h(\sigma(\theta)\, x, \sigma(\theta)\, x') = h(x, x'), \quad \forall \sigma(\theta) \in \Sigma_\theta .$$

Then the quadratic form

$$I(x; y) = \langle\, y \,,\, l_o x \,\rangle \quad \text{where} \quad l_o = \frac{d}{d\theta}\bigg|_{\theta=0} \sigma(\theta)$$

is conserved: $I(x', y') = I(x, y)$.

Proof. First note that orbits of the map generated by h are stationary points of the action

$$\mathrm{W}_{(M,N)}(\mathbf{x}) = \sum_{j=M}^{N-1} h(x^j, x^{j+1})$$

with respect to variation segments $\{\xi^M, \ldots, \xi^N\} \subset (\mathbf{R}^n)^{\mathbf{Z}}$ with zero endpoints: $\xi^M = \xi^N = 0$. Setting $\nabla_{\mathbf{x}} \mathrm{W}_{(M,N)}(\mathbf{x}) = 0$ for admissible variation segments requires

$$h_1(x^j, x^{j+1}) + h_2(x^{j-1}, x^j) = 0, \quad \forall j \in \mathbf{Z}. \tag{E.1}$$

The Σ_θ-invariance of h implies that the functional $\mathrm{W}_{(M,N)}$ is Σ_θ-invariant or

$$\sum_{j=M}^{N-1} h(\sigma(\theta)\, x^j, \sigma(\theta)\, x^{j+1}) = \sum_{j=M}^{N-1} h(x^j, x^{j+1}),$$

which when differentiated with respect to θ becomes

$$\sum_{j=M}^{N-1} \langle h_1(\sigma(\theta)\, x^j, \sigma(\theta)\, x^{j+1}), \sigma'(\theta) x^j \rangle + \langle h_2(\sigma(\theta)\, x^j, \sigma(\theta)\, x^{j+1}), \sigma'(\theta) x^{j+1} \rangle = 0 .$$

Setting $\theta = 0$ and using (E.1),

$$\begin{aligned}
0 &= \sum_{j=M}^{N-1} \langle\, h_1(x^j, x^{j+1}),\, l_o x^j \,\rangle + \langle\, h_2(x^j, x^{j+1}),\, l_o x^{j+1} \,\rangle \\
&= \sum_{j=M}^{N-1} -\langle\, h_2(x^{j-1}, x^j),\, l_o x^j \,\rangle + \langle\, h_2(x^j, x^{j+1}),\, l_o x^{j+1} \,\rangle \\
&= \sum_{j=M-1}^{N-2} -\langle\, h_2(x^j, x^{j+1}),\, l_o x^{j+1} \,\rangle + \sum_{j=M}^{N-1} \langle\, h_2(x^j, x^{j+1}),\, l_o x^{j+1} \,\rangle \\
&= -\langle\, h_2(x^j, x^{j+1}),\, l_o x^{j+1} \,\rangle\Big|_{j=M-1} + \langle\, h_2(x^j, x^{j+1}),\, l_o x^{j+1} \,\rangle\Big|_{j=N-1} .
\end{aligned}$$

Since (M, N) are arbitrary integers (with $M < N$) it follows that

$$\langle h_2 \,,\, l_o x' \rangle = \langle y' \,,\, l_o x' \rangle = \text{constant}$$

is independent of discrete time. ■

Example 1. Let $\mathbf{T} : (x, y) \mapsto (x', y')$ on $\mathbb{R}^4$ be generated by the $\mathbf{SO}(2)$-invariant generating function h; that is, $\Sigma_\theta = \mathbf{SO}(2)$ with

$$\sigma(\theta) = \begin{pmatrix} \cos\theta & \sin\theta \\ -\sin\theta & \cos\theta \end{pmatrix} \quad \text{and} \quad \sigma_\theta^o = l_o = \mathbf{J}.$$

and $h(\sigma(\theta)\, x, \sigma(\theta)\, x') = h(x, x')$. It follows from Theorem E that

$$I(x, y) = \langle y \,,\, \mathbf{J}x \rangle = y_1 x_2 - x_1 y_2$$

is independent of discrete time. This conservation law is reminiscent of conservation of angular momentum in classical mechanics. $\mathbf{SO}(2)$-equivariant symplectic maps are considered in Sections 8 and 10.

Example 2. Let $\mathbf{T} : (x, y) \mapsto (x', y')$ on $\mathbb{R}^6$ be a symplectic map generated by a $\mathbf{SO}(3)$-invariant h. The group $\mathbf{SO}(3)$ is a 3-parameter group with Lie algebra

$$\mathbf{so}(3) = \text{span}\left\{ \begin{pmatrix} 0 & 1 & 0 \\ -1 & 0 & 0 \\ 0 & 0 & 0 \end{pmatrix}, \begin{pmatrix} 0 & 0 & 1 \\ 0 & 0 & 0 \\ -1 & 0 & 0 \end{pmatrix}, \begin{pmatrix} 0 & 0 & 0 \\ 0 & 0 & 1 \\ 0 & -1 & 0 \end{pmatrix} \right\} = \text{span}\{l_1, l_2, l_3\}.$$

Theorem E can be applied for each one-parameter subgroup of $\mathbf{SO}(3)$ given by $\exp(l_j \theta)$, $1 \le j \le 3$, to show that the map $\mathbf{T}$ has the three conserved quantities

$$I_j(x, y) = \langle y \,,\, l_j x \rangle, \quad 1 \le j \le 3.$$

Such a map is reminiscent of – a discrete analog of – the rigid body in classical mechanics.

Appendix F. About reversible symplectic maps

When dealing with twist maps it is tempting to work in configuration space only, and then to reconstruct the orbits of the map in the phase space. In particular, it means that one can introduce reversibility in those two contexts, in phase space and also in configuration space. Then, a natural question is how these two notions of reversibility are related. The moral of our story is stated in Proposition F.4 which shows when and how the two notions of reversibility are identical. In this appendix we also aim to start a discussion on the relation between reversibility and symplectic structure. We ignore the parameters.

First, we consider the classical notion of reversibility (in phase space). We suppose that $\widehat{\mathbf{T}} : \mathbb{R}^{2n} \to \mathbb{R}^{2n}$ is a $\widehat{\mathbf{J}}$-symplectic and $\widehat{\mathcal{R}}$-reversible map, that is, $\widehat{\mathbf{T}}^{-1} =$

$\widehat{\mathcal{R}}\widehat{\mathbf{T}}\widehat{\mathcal{R}}$ with $\widehat{\mathcal{R}}^2 = \mathbf{I}_{2n}$ and $\mathbf{D}\widehat{\mathbf{T}}^T \widehat{\mathbf{J}}\, \mathbf{D}\widehat{\mathbf{T}} = \widehat{\mathbf{J}}$ with $\widehat{\mathbf{J}}$ skew-symmetric. Note that $\widehat{\mathcal{R}}$ and $\widehat{\mathbf{J}}$ need not be constant (nor linear).

Around a $\widehat{\mathcal{R}}$-invariant fixed point of $\widehat{\mathbf{T}}$ one can change coordinates and, by Darboux Theorem, flatten $\widehat{\mathbf{J}}$ into the normal form $\mathbf{J}_{2n}$ and so $\widehat{\mathcal{R}}$ is transformed into another involution $\mathcal{R}$ with the transformed $\mathbf{T}$ of $\widehat{\mathbf{T}}$ still being $\mathcal{R}$-reversible.

When $\mathcal{R}$ is *anti-symplectic*, that is, $D\mathcal{R}^T\,\mathbf{J}_{2n}\, D\mathcal{R} = -\mathbf{J}_{2n}$, there exists a $\mathbf{J}_{2n}$-symplectic change of coordinates such that $\mathcal{R} = \kappa \otimes \mathbf{I}_n$ (Meyer [1981]). In the general situation it is not clear whether we can always linearize $\mathcal{R}$.

Nevertheless, at the fixed point, we always have

Proposition f.1. *Let* $\mathbf{T}$ *be a* $\mathcal{R}$*-reversible,* $\mathbf{J}_{2n}$*-symplectic map and suppose that the origin is a* $\mathcal{R}$*-invariant fixed point of* $\mathbf{T}$*. Then* $\mathbf{K} = D\mathcal{R}^o\mathbf{J}_{2n}D\mathcal{R}^{oT}\mathbf{J}_{2n}$ *is a commutator of* $\mathbf{DT}^o$.

Proof. Because the origin is a fixed point of $\mathbf{T}$ and $D\mathcal{R}$, one has $\mathbf{DT}^{o-1} = D\mathcal{R}^o\mathbf{DT}^o D\mathcal{R}^o$ and $\mathbf{DT}^{oT}\mathbf{J}_{2n}\mathbf{DT}^o = \mathbf{J}_{2n}$. Define $\mathbf{L} = \mathbf{J}_{2n}D\mathcal{R}^o$ and $\mathbf{K} = \mathbf{L}^{-1}\mathbf{L}^T$ then it is a simple verification to see that $\mathbf{K}\,\mathbf{DT}^o = \mathbf{DT}^o\mathbf{K}$. ■

Note that $D\mathcal{R}^o$ is symplectic if $\mathbf{K} = -\mathbf{I}_{2n}$ and anti-symplectic if $\mathbf{K} = \mathbf{I}_{2n}$. We conjecture that it is always possible to consider $D\mathcal{R}^o$ as a direct sum of two symplectic and anti-symplectic involutions, respectively.

Lemma F.2. *h generates a* $\mathcal{R}$*-reversible map iff* $h(x,y) = h(y,x)$

Proof. The key to the proof is to remember that the generating function for $\mathbf{T}^{-1}$ is given by $\bar{h}(x,y) = -h(y,x)$ if h is the generating function of $\mathbf{T}$.

Now, suppose that $h(x,y) = h(y,x)$, $\forall\,(x,y) \in \mathcal{U}$ (a neighborhood of the origin). Then, the equations corresponding to (2.3a) are

$$y = h_x(x, \mathbf{T}_1^{-1}(x,y)) \quad \text{and} \quad y = h_x(x, \mathbf{T}_1(x,-y))\,.$$

By uniqueness of the solutions we find that $\mathbf{T}_1^{-1}(x,y) = \mathbf{T}_1(x,-y)$. Writing the components of $\mathbf{T}$, $\mathbf{T}_1$ and $\mathbf{T}_2$, and replacing into (2.3b), we have that

$$\mathbf{T}_2^{-1}(x,y) = -h_x(\mathbf{T}_1^{-1}(x,y), x) = -h_{x'}(x, \mathbf{T}_1^{-1}(x,-y)) = -\mathbf{T}_2(x,-y)\,.$$

Hence, $\mathbf{T}^{-1} = \mathcal{R}\mathbf{T}\mathcal{R}$.

For the reverse implication, note that

$$(x,y) \mapsto (\mathbf{T}_1(x,-y)\,,\, x)$$

is a local diffeomorphism as $\mathbf{T}$ is a twist map. Then, from (2.3), we find that

$$h_x(\mathbf{T}_1(x,-y), x) = h_{x'}(x, \mathbf{T}_1(x,-y))\,.$$

And so, $h_x(x,y) = h_{x'}(y,x)$, $\forall (x,y) \in \mathcal{U}$. As

$$h(x,y) = \int_0^1 \langle h_x(xt,y)\,,\, x\rangle\, dt + h(0,y) = \int_0^1 \langle h_{x'}(y,tx)\,,\, x\rangle\, dt + h(0,y)\,,$$

we find that

$$h(x,y) = h(y,x) - h(y,0) + h(0,y)\,,$$

and so $h(x,0) = h(0,x)$. We have finished, as a similar calculation starting with the other derivative gives

$$h(x,y) = h(y,x) - h(0,x) + h(x,0) = h(y,x)\,.$$

■

There is another concept of reversibility that appears when the action is restricted to period-q sequences. In this case the action is taken to be invariant under the reversal of periodic sequences of points (cf. (2.21)), that is

$$\mathrm{W}_{(x_0,x_q=x_0)}(x_1,\ldots,x_{q-1}) = \mathrm{W}_{(x_0,x_q=x_0)}(x_{q-1},\ldots,x_1)\,, \quad \forall \mathbf{x} \in \mathbf{X}_q^n\,. \tag{F.1}$$

We say that $\mathbf{T}$ is *CS-reversible* (for configuration space) if (F.1) is true for all $q \geq 2$, that is if W_q is $\mathcal{K}_q \otimes \mathbf{I}_n$-invariant (for $\mathcal{K}_q$ as defined in (2.21)). In that case we can prove that h satisfies the following. No greater generality is obtained when $\mathcal{K} \otimes \mathbf{R}$ with $\mathbf{R}^2 = \mathbf{I}_n$.

Lemma F.3. $\mathbf{T}$ *is CS-reversible iff its generating function h satisfies*

$$h(x,y) = h(y,x) - g(x) + g(y) \tag{F.2}$$

for some function g.

Proof. That the condition on h is sufficient is readily verified by direct calculation. To see its necessity, let us take $q = 3$. By twice differentiating the condition on $\mathrm{W}_{(x_0,x_0)}$ with respect to x_0 and x_2, we see that h satisfies

$$h_{xx'}(x,y) = h_{xx'}(y,x), \quad \forall (x,y) \in \mathcal{U}\,.$$

With a calculation similar to the proof of Lemma F.2, we find that this implies

$$h_x(x,y) = h_{x'}(y,x) - h_{x'}(0,x) + h_x(x,0)\,.$$

And so,

$$h(x,y) = h(y,x) + \big(h(0,y) - h(y,0)\big) - \big(h(0,x) - h(x,0)\big)\,,$$

which is our conclusion with $g(x) \stackrel{\text{def}}{=} h(0,x) - h(x,0)$. ■

As a consequence of the previous two lemmas, $\mathcal{R}$-reversibility implies also CS-reversibility. We can now present our conclusion that the two notions of reversibility are equivalent modulo conjugations.

Proposition F.4. *A CS-reversible map* $\mathbf{T}$ *is conjugate to a* $\mathcal{R}$*-reversible map.*

Proof. Let h be a generating function satisfying (F.2) with some g. Define

$$\bar{h}(x,y) = h(x,y) + \tfrac{1}{2}g(x) - \tfrac{1}{2}g(y),$$

then $\bar{h}$ generates a $\mathcal{R}$-reversible map (from Lemma F.2). By construction h and $\bar{h}$ are *dynamically equivalent*, hence their induced maps are conjugate by the symplectic change of variables $(x,y) \mapsto (x, y - \frac{1}{2}\nabla_x g(x))$ (cf. Appendix G). ■

Appendix G. Twist maps and dynamical equivalence

Consider a local diffeomorphism $\mathbf{T}$ on $\mathbf{R}^{2n}$ defined around the origin, a fixed point:

$$x' = \mathbf{T}_1(x,y) \tag{G.1a}$$

$$y' = \mathbf{T}_2(x,y). \tag{G.1b}$$

We suppose that $\mathbf{T}$ is a *twist* map, that is, $|\partial_y \mathbf{T}_1^\circ| \neq 0$. That property means that we can use the Implicit Function Theorem to construct an equivalent map relating the *configuration* space (x-coordinates) to the *phase* space ((x,y)-coordinates): we solve (G.1a) to get $y = \mathbf{K}_1(x,x')$ then $y' = \mathbf{K}_2(x,x') \overset{\text{def}}{=} \mathbf{T}_2(x, \mathbf{K}_1(x,x'))$ by substituting into (G.1b), and so we get another map on $\mathbf{R}^{2n}$

$$y = \mathbf{K}_1(x,x') \tag{G.2a}$$

$$y' = \mathbf{K}_2(x,x'). \tag{G.2b}$$

It is easy to verify that there is a 1-1 correspondence between $\mathbf{T}$ and $\mathbf{K}$. In general there are no relations between $\mathbf{K}_1$ and $\mathbf{K}_2$, but

(a) when $\mathbf{T}$ is symplectic there exist coordinates and a function h such that $\mathbf{K}_1(x,x') = -h_x(x,x')$ and $\mathbf{K}_2(x,x') = h_{x'}(x,x')$.

(b) when $\mathbf{T}$ is reversible, there exists coordinates such that $\mathbf{K}_2(x,x') = -\mathbf{K}_1(x,x')$.

In Dewar & Meiss [1992] the concept of "dynamical equivalence" for area-preserving maps is defined:

Two maps are dynamically equivalent if their orbits restricted to configuration space are identical.

To formalize the idea we proceed as follows.

We can derive from $\mathbf{K}$ a map in configuration space which represents the dynamics of the projection onto configuration space of the dynamics on phase space generated by $\mathbf{T}$.

As $\mathbf{T}$ is twist, the map $\phi : (x, y) \mapsto (x, x' = \mathbf{T}_1(x, y))$ is a local diffeomorphism and so $\mathbf{T_c} = \phi \mathbf{T} \phi^{-1}$ sending $(x, x') \to (x', x'')$ is a local diffeomorphism. Every orbit of $\mathbf{T_c}$ is a projection of an orbit of $\mathbf{T}$ and is defined by any *two* consecutive points.

We say that two twist maps $\mathbf{T}^1, \mathbf{T}^2$ are ***dynamically equivalent*** if the two maps $\mathbf{T}^1_\mathbf{c}, \mathbf{T}^2_\mathbf{c}$ they define on configuration space are *identical.*

The question then arises of the relation between two dynamically equivalent maps or, in other words, what freedom do we have when *reconstructing* $\mathbf{T}$ from $\mathbf{T_c}$.

We believe that for symplectic (or reversible) maps that freedom is constrained to the set of conjugate maps, meaning that the dynamics of $\mathbf{T}$ is wholly captured by $\mathbf{T_c}$.

To justify that conjecture for symplectic maps, observe that orbits $(x_0, \ldots, x_n)$ of $\mathbf{T}$ are stationary points of the actions $\mathrm{W}_{(x_0, x_n)} = \sum_{i=1}^{n-1} h(x_{i-1}, x_i)$ with variables $(x_1, \ldots, x_{n-1})$, that is, they are solutions of the following Euler-Lagrange equations:

$$h_{x'}(x_{i-1}, x_i) + h_x(x_i, x_{i+1}) = 0, \quad 1 \leq i \leq n-1\,. \tag{G.3}$$

Clearly, the multiplication of h by any real number or the addition of a constant has no effect on (G.3) and merely scales (and/or reverses) the y-axis (see(2.3)). Maybe less obvious, once we have fixed the y-axis, is the conjecture that dynamical equivalence is only possible by the addition to the action of a "null-Lagrangian" term $g(x') - g(x)$ with $g : \mathbf{R}^n \to \mathbf{R}$ being any function. It is readily verified that the equations (G.3) are then unchanged and the two maps are related by some simple symplectic transformations in phase space (see Dewar & Meiss [1992]).

In our analysis we have extensively used the twist property of $\mathbf{T}$. How rare is that property? Simple area-preserving examples show that nonzero twist is not an invariant of changes of coordinates. And so the only maps we cannot deal with are those which cannot be put into a twist form. For area-preserving maps there are only two such cases, when $\mathbf{DT}^o = \pm \mathbf{I}$.

Appendix H. Zq-equivariant bifurcation equations and linear stability

The determinant of the Jacobian for $\mathbb{Z}_q$-equivariant maps on $\mathbb{R}^2$ is important for reduced stability (cf. the Stability Lemma I 2.10). In this appendix we give formulae for the determinant. The results are formulated for general (not necessarily gradient) $\mathbb{Z}_q$-equivariant bifurcation equations, the greater generality requiring little additional effort. Restriction of the formulae – for use in Chapter 4 and elsewhere – to the gradient case is then straightforward.

A general $\mathbb{Z}_q$-equivariant bifurcation equation for $q \geq 3$ takes the following form

$$h(z, \Lambda) = f(u, v, \Lambda)\, z + g(u, v, \Lambda)\, \bar{z}^{q-1}, \qquad q \geq 3\,.$$

Note that we use the same coordinates for all $q \geq 3$ (that is also for $q = 4$; illuminating simplifications do not arise when $q = 4$). The functions f and g are *complex-valued* $\mathbf{D}_q$-invariant functions and Λ represents parameter space. The components of h (as a real map) are given by the real and imaginary parts of h.

The only linear stability information worth recording is the sign of the determinant of $D_z h$ as this is the only quantity related to stability which is an invariant of $\mathbf{Z}_q$-equivariant contact equivalence. On the other hand, if h is also $\mathbf{D}_q$-equivariant (forcing f, g to be real-valued), information about the trace of $D_z h$ can be preserved only for those classes of bifurcation diagrams where the leading terms of the trace are invariants of the change of coordinates. The determinant of $D_z h$ is given by the following result:

Lemma H.1. *For any differentiable $\mathbf{Z}_q$-equivariant map h considered as a function of the two independent variables $z, \overline{z}$, the determinant of the derivative map is $|D_z h| = |h_z|^2 - |h_{\overline{z}}|^2$.*

Proof. Using the complex notation, the derivative of h is given by the following linear map

$$D_z h(z)\, w = h_z \cdot w + h_{\overline{z}} \cdot \overline{w}, \quad \forall w \in \mathbf{C}.$$

When represented in real coordinates it becomes

$$D_z h(z) \begin{pmatrix} w_1 \\ w_2 \end{pmatrix} = \begin{pmatrix} \operatorname{Re} h_z + \operatorname{Re} h_{\overline{z}} & -\operatorname{Im} h_z + \operatorname{Im} h_{\overline{z}} \\ \operatorname{Im} h_z + \operatorname{Im} h_{\overline{z}} & \operatorname{Re} h_z - \operatorname{Re} h_{\overline{z}} \end{pmatrix} \begin{pmatrix} w_1 \\ w_2 \end{pmatrix},$$

from which it is immediate that $|D_z h(z)| = |h_z|^2 - |h_{\overline{z}}|^2$. ■

Explicitly, the components of the Jacobian map are

$$h_z = f_u\, u + \tfrac{1}{2} q\, f_v\, (v + iw) + f + g_u\, (v - iw) + \tfrac{1}{2} q\, g_v\, u^{q-1}$$
$$h_{\overline{z}} = f_u\, z^2 + \tfrac{1}{2} q\, f_v\, u \overline{z}^{q-2} + (q-1)\, g\, \overline{z}^{q-2} + g_u\, u \overline{z}^{q-2} + \tfrac{1}{2} q\, g_v\, \overline{z}^{2q-2}.$$

The nonzero solutions of $h = 0$ are given by

$$g(u, v, \Lambda) = -f(u, v, \Lambda) u^{1-q} (v + iw).$$

And so, for these solutions, the general expression for the determinant is

$$\begin{aligned} |D_z h(z)| = &- q(q-2)|f|^2 + qu\,(f\overline{f}_u + \overline{f} f_u) + \tfrac{1}{2} q^2 v\,(f \overline{f}_v + \overline{f} f_v) \\ &+ qv\,(f\overline{g}_u + \overline{f} g_u) + q\left[\tfrac{1}{2} q u^{q-1} - (q-1) w^2/u\right](f\overline{g}_v + \overline{f} g_v) \\ &+ q\, w^2 (f_u \overline{g}_v + \overline{f}_u g_v - \overline{f}_v g_u - f_v \overline{g}_u) \\ &+ i\tfrac{1}{2} q(q-2)\, w\,(f\overline{f}_v - \overline{f} f_v) + iq\, w\,(f\overline{g}_u - \overline{f} g_u) + iq(q-1)\frac{v}{u} w\,(f\overline{g}_v - \overline{f} g_v) \\ &+ iq\, uw\,(\overline{f}_u f_v - f_u \overline{f}_v) + iqu^{q-1} w\,(\overline{g}_u g_v - g_u \overline{g}_v) \\ &+ iq\, vw\,(f_v \overline{g}_u - g_u \overline{f}_v + \overline{f}_u g_v - f_u \overline{g}_v). \end{aligned} \tag{HH.1}$$

Appendix I. About symmetric symplectic operators

Let Σ be a compact Lie group and consider a Σ-equivariant $\widehat{\mathbf{J}}$-symplectic map $\mathbf{T}$, that is:

(a) $\mathbf{DT}(z)^T\,\widehat{\mathbf{J}}(z)\,\mathbf{DT}(z) = \widehat{\mathbf{J}}(z)$, $\forall\, z$ near 0, for a (nonlinearly dependent) skew-symmetric matrix $\widehat{\mathbf{J}}$,

(b) $\mathbf{T}(\sigma z) = \sigma\mathbf{T}(z)$, $\forall\, z$ near 0, $\forall\,\sigma \in \Sigma$.

The general problem, to understand what (a) and (b) together imply for the relationship between $\widehat{\mathbf{J}}$ and the action of Σ, is beyond the scope of this work. Nevertheless we are going to make some remarks about some aspects of the question.

First, as our bifurcations are of a local nature, it should be possible to linearize the symplectic structure and the group-equivariance around fixed-points for both $\mathbf{T}$ and the Σ-action. In that case, note that we could consider – with little extra effort – the initial action of Σ to be *nonlinear* also.

An important, maybe the only important, case where this is possible is when the symplectic form induced by $\widehat{\mathbf{J}}$ is Σ-invariant. Then, the action of Σ is symplectic and the Equivariant Darboux Theorem (Montaldi & Roberts & Stewart [1988], Dellnitz & Melbourne [1992]) tells us that $\widehat{\mathbf{J}}$ can be put – by a Σ-equivariant change of coordinates – into the following normal form $\widetilde{\mathbf{J}}$ (we refer to Chapter 8 for background to the notation).

$\mathbf{R}^{2n}$ decomposes into a direct sum $\oplus_{i=1}^{l} V_i$ of Σ-invariant subspaces (for some integer l). The V_i's are each a direct sum of different isomorphic irreducible subspaces W_i, $1 \le i \le l$, distinguished by the homomorphism type $\mathbf{K}_i$ of the space of Σ-commuting endomorphisms of W_i. There are three types $\mathbf{K}_i = \mathbf{R}$, $\mathbf{C}$ and $\mathbf{H}$. And so the symplectic form decomposes into

$$\widetilde{\mathbf{J}} = \oplus_{i=1}^{l}\,\widetilde{\mathbf{J}}_i$$

with an unique class when $\mathbf{K}_i = \mathbf{R}$ or $\mathbf{H}$ and so we can take $\widetilde{\mathbf{J}}_i = \mathbf{J}_{2n_i}$ in those cases. When $\mathbf{K}_i = \mathbf{C}$ the problem is more subtle, there are $n_i + 1$ (see (8.1) for n_i) nonequivalent classes of symplectic forms given by

$$\widetilde{\mathbf{J}}_i = i\,\mathbf{K}_k^{n_i} \oplus_{\mathbf{C}} \mathbf{I}_{W_i}$$

for some $0 \le k \le n_i$ and

$$\mathbf{K}_k^{n_i} = \begin{pmatrix} \mathbf{I}_k & \mathbf{0} \\ \mathbf{0} & -\mathbf{I}_{n_i-k} \end{pmatrix}.$$

For simplicity, in this work the implications of the above result are not considered; in all cases the canonical form $\widetilde{\mathbf{J}} = \mathbf{J}_{2n}$ is used.

Now suppose that the symplectic form is locally constant, generated by $\widetilde{\mathbf{J}}$, and that Σ acts (orthogonally) on $\mathbf{R}^{2n}$, but without any further assumptions. Then

Lemma I.1. *$\mathbf{T}$ is Σ-equivariant and $\widetilde{\mathbf{J}}$-symplectic iff $\sigma^T\widetilde{\mathbf{J}}\sigma\widetilde{\mathbf{J}}$ is a commutator of $\mathbf{DT}(z)$, $\forall\, z$ near 0.*

Proof. $\mathbf{T}$ $\widetilde{\mathbf{J}}$-symplectic means that $\left(\mathbf{DT}(z)\right)^{-1} = -\widetilde{\mathbf{J}}\mathbf{DT}(z)\widetilde{\mathbf{J}}$, as $\mathbf{T}$ Σ-equivariant means that $\mathbf{DT}(\sigma z) = \sigma\mathbf{DT}(z)\sigma^T$ and so putting the two conditions together $\sigma^T\widetilde{\mathbf{J}}\sigma\widetilde{\mathbf{J}}$ is a commutator of $\mathbf{DT}(z)$, $\forall\, z$ near 0. ■

In particular $\pm\mathbf{I}_n$ are always commutators and so, in principle, one cannot rule out that elements of Σ can act symplectically (the usual situation in practice) *or* anti-symplectically or even both ways on different subspaces!

In this work the type of symmetry we consider is restricted further. We assume that the Σ-action lifts from configuration-space to phase-space, that is, we want h to satisfy the following Σ-equivariance:

$$h(\sigma^1 x, \sigma^2 x') = \sigma^2 h(x, x'), \quad \forall\sigma \in \Sigma,\ \forall\, x, x' \text{ near } 0 \tag{I.1}$$

where σ^1, resp. σ^2, denote the image of $\sigma \in \Sigma$ by an orthogonal representation of Σ on $\mathbf{R}^n$, resp. $\mathbf{R}$. Clearly the second action on $\mathbf{R}$ can only be the identity (the case we consider in this work) or an $\mathbf{Z}_2$-action. And so we get the following relation between the symmetry (I.1) of h and the symmetry (I.2) of $\mathbf{T}$.

Proposition I.2. *The generating function h satisfies (I.1) iff the symplectic map $\mathbf{T}$ it defines satisfies*

$$\mathbf{T}(\sigma^1 x, \sigma^2\sigma^1 y) = \left(\sigma^1\mathbf{T}_1(x,y), \sigma^2\sigma^1\mathbf{T}_2(x,y)\right). \tag{I.2}$$

Proof. If h satisfies (I.1) then

$$h_x(\sigma^1 x, \sigma^1 x') = \sigma^2\sigma^1 h_x(x, x') \quad \text{and} \quad h_{x'}(\sigma^1 x, \sigma^1 x) = \sigma^2\sigma^1 h_{x'}(x, x').$$

Moreover, $\mathbf{T}$ is given by the use of the Implicit Function Theorem on (2.3). It is now a typical exercise using the uniqueness of the solution to show that $\mathbf{T}$ is Σ-equivariant with the action as defined in (I.2).

Conversely, if $\mathbf{T}$ is Σ-equivariant we get that

$$h_x(x, \mathbf{T}_1(x,y)) = \sigma^2\sigma^{1T} h_x(\sigma^1 x, \sigma^1\mathbf{T}_1(x,y)),$$
$$h_{x'}(x, \mathbf{T}_1(x,y)) = \sigma^2\sigma^{1T} h_{x'}(\sigma^1 x, \sigma^1\mathbf{T}_1(x,y)),$$

and so, as $\phi : (x,y) \mapsto (x,x')$ is a diffeomorphism

$$h_x(x,y) = \sigma^2\sigma^{1T} h_x(\sigma^1 x, \sigma^1 y),$$
$$h_{x'}(x,y) = \sigma^2\sigma^{1T} h_{x'}(\sigma^1 x, \sigma^1 y)$$

By integration from the fixed-point (the origin, say),

$$h(x,y) = \int_0^1 < h_x(tx,y)\,,\, x > + h(0,y)$$

and so

$$h(\sigma^1 x, \sigma^1 y) = \sigma^2(h(x,y) - h(0,y)) + h(0,\sigma^1 y)\,. \tag{I.3}$$

As h is defined modulo constants we can take $h^o = 0$, hence $h(\sigma^1 x, 0) = \sigma^2 h(x,0)$ and similarly $h(0,\sigma^1 y) = \sigma^2 h(0,y)$. Going back to (I.3) we see that this implies that $h(\sigma^1 x, \sigma^1 y) = \sigma^2 h(x,y)$. ■

Condition (I.1) imposes that we always get an even number of copies of irreducible representations of Σ. It is not a restriction for real blocks but certainly one for complex and quaternionic blocks. It is not exactly clear what it means in practical terms to study an equivariant map whose symmetry does not show-up explicitly in configuration space. The choice of that space has some importance.

Because there is a 1-1 correspondence between maps and generating functions the symmetry of the map is buried somewhere in the Taylor expansion of h. The question is how to read it from h.

The simplest examples we can consider are $\mathbf{S}^1$-equivariant twist maps on $\mathbb{R}^2 \approx \mathbb{C}$. The equivariant maps are of the form $\mathbf{T}(z) = e^{i\theta(z\bar{z})}\, z$, for any function $\theta : \mathbb{R}_+ \to \mathbb{R}$. The quadratic terms of the generating function are

$$-\tfrac{1}{2}\cot\theta^o x^2 + \csc\theta^o x x' - \tfrac{1}{2}\cot\theta^o x'^2\,. \tag{I.4}$$

They have no obvious linear symmetry (note that the dynamics of $\mathbf{T}$ in configuration space is the projection of the dynamics of $\mathbf{T}$ onto the x-axis). Obviously this example becomes very simple when dealt with using polar coordinates. Let

$$x = \sqrt{r}\,\cos\phi\,,$$
$$y = \sqrt{r}\,\sin\phi\,,$$

then $\mathbf{T}$ becomes a classical twist map

$$r' = r$$
$$\phi' = \phi + \theta(r)\,.$$

We can use ϕ as configuration coordinate and get the following generating function

$$g(\phi,\phi') = \Theta(\phi - \phi')$$

with $(\Theta')^{-1}(r) = -\theta(-r)$. Clearly we recover the $\mathbf{S}^1$-symmetry with a diagonal action by phase shift. However, the implication of the phase shift symmetry on the generating function *in cartesian coordinates* is unclear (cf. (I.4)).

The question, of determining the restricted action of a phase-space symmetry on configuration space, becomes rapidly more obscure as can be seen already with $\mathbb{Z}_n$-equivariant area-preserving maps.

Appendix J. (p,q)-resonances for symplectic maps

The bifurcation of periodic points for symplectic maps when the linear part has multiple resonances can be quite complicated. In this appendix we describe the group-theoretic framework for the bifurcation of periodic points when the linear map $\mathbf{DT}^o$ has two resonant multipliers, referring to this as the case of simple (p,q)-resonance. We assume throughout that $3 \leq p \leq q$. The cases $p, q = 1$ or 2 could also be included, but the group-theoretic aspect is less rich and the analysis must be separated into many special cases, in particular the different Jordan block structures must also be taken into account.

Consider a Λ-parametrized family $\mathbf{T}$ of symplectic maps such that $\mathbf{DT}^o$ has two pairs of resonant multipliers on the unit circle: $\exp(\frac{2r}{p}\pi i)$ and $\exp(\frac{2s}{q}\pi i)$ with $\{r,p\} = \{s,q\} = 1$ with $q \geq p \geq 3$) ($\{\cdot,\cdot\}$ denotes the g.c.d.). We assume that no other resonances are present and so look for periodic solutions in the space of period-$[p,q]$ points (where $[\cdot,\cdot]$ is the l.c.m.). Using the framework introduced in Section 2, locally there exists a generating function and action which generate the symplectic map $\mathbf{T}$. Restriction of the action to the space of period-$[p,q]$ points results in the functional

$$\mathrm{W}_{[p,q]}(\mathbf{x},\Lambda) = \sum_{j=1}^{[p,q]} h(x^j, x^{j+1}, \Lambda)\,.$$

Let $\mathbf{L}^o = \mathrm{Hess}_{\mathbf{x}}\mathrm{W}^o_{[p,q]}$, then the kernel of $\mathbf{L}^o$ is isomorphic to $\mathbb{C}^2 = \{\,(z_1,z_2)\,\}$ where z_1 (respectively z_2) is the coordinate for the eigenspace corresponding to the p-multipliers (respectively q-multipliers). The linear part of the map is controlled by the (multi)parameter $\hat{\lambda} = (\lambda, \alpha)$ where α is the *detuning* parameter and λ controls the distance from the rational values.

Application of the Splitting Lemma to $\mathrm{W}_{[p,q]}(\mathbf{x},\Lambda)$ reduces the problem to a gradient bifurcation problem on $\mathbb{C}^2$. The reduced potential will be $\mathbb{Z}_{[p,q]}$-equivariant with action:

$$\mathbb{Z}_{[p,q]} = \langle\omega\rangle \quad \text{for} \quad \omega = \exp(2\pi i/[p,q]) \quad \text{and} \quad \omega(z_1,z_2) = (\omega^{rq'}z_1, \omega^{sp'}z_2)$$

where $p' \stackrel{\text{def}}{=} p/\{p,q\}$ and $q' \stackrel{\text{def}}{=} q/\{p,q\}$.

Without loss of generality, we suppose also that the map is reversible. This leads to more general results in the sense that the isotropy subgroup lattices for $\mathbb{Z}_{[p,q]}$ can be simply read off the ones for $\mathbb{D}_{[p,q]}$.

The method we are going to describe, to find the $\mathbb{D}_{[p,q]}$-invariants and construct the lattice of isotropy subgroups, applies equally well to the $\mathbb{Z}_{[p,q]}$-problem. The coordinates are chosen such that the reversor acts by conjugation σ on $\mathbb{C}^2$: $\sigma \cdot (z_1, z_2) = (\overline{z}_1, \overline{z}_2)$.

J.1. $\mathsf{D}_{[p,q]}$-invariant functions on $\mathbb{C}^2$

Since $\mathsf{D}_{[p,q]}$ is compact, there exists a finite basis for the $\mathsf{D}_{[p,q]}$-invariant functions. The invariants are combinations of the monomials $z_1^{k_1}\overline{z}_1^{k_2} z_2^{l_1}\overline{z}_2^{l_2}$. It is immediate that

$$u_1 = z_1\overline{z}_1 \qquad \text{and} \qquad u_2 = z_2\overline{z}_2$$

are invariants. The above construction implies that we end up with monomials of the type $z_1^k z_2^l$ with $k, l \in \mathbf{Z}$ where $k, l < 0$ represent $\overline{z}_1^{|k|}$ or $\overline{z}_2^{|l|}$. Such a monomial is $\mathbf{Z}_{[p,q]}$-invariant if

$$\frac{r}{p}k + \frac{s}{q}l \in \mathbf{Z}, \tag{J.1}$$

that is, if

$$r\,q'\,k + s\,p'\,l \in [p,q]\mathbf{Z}\,.$$

Therefore we should study the set

$$\mathcal{L} = \{(k,l) \in \mathbf{Z}^2 \,:\, rq'k + sp'l \in [p,q]\mathbf{Z}\,\}.$$

Note that $\mathcal{L}$ is a lattice. Define the map μ by

$$\mu(k,l) = rq'k + sp'l \in \{r,s\}\mathbf{Z} \quad (\text{because } \{rq', sp'\} = \{r,s\}).$$

Then

$$\mathcal{L} = \mu^{-1}([p,q]\mathbf{Z}) = \mu^{-1}(\{r,s\}\,[p,q]\mathbf{Z}),$$

since $\{\,\{r,s\},[p,q]\,\} = 1$. Define $r' = r/\{r,s\}$ and $s' = s/\{r,s\}$.

Lemma J.1. *$\mathcal{L}$ is generated by*

$$(-s'p', r'q') \qquad (\textit{line of } 0)$$

and the solutions (k_o, l_o) of

$$(r'q')\,k_o + (s'p')\,l_o = [p,q]\,. \tag{J.2}$$

Proof. From Bezout Theorem (J.2) has solutions. Let $(k,l) \in \mathcal{L}$ and use (J.2) to get down to the 0-line. Then solve on the 0-line. ■

Continuing our construction of the invariants, we note that σ-invariance means that the sum of the monomials corresponding to (k,l) and $(-k,-l)$ gives an invariant for $\mathsf{D}_{[p,q]}$. Therefore we need to find the generators of $\mathcal{L}_+ \stackrel{\text{def}}{=} \mathcal{L} \cap \{\ell \geq 0\}$ under positive addition (multiplication of invariants), that is $\mathbf{Z}_+$. (For $\mathbf{Z}_{[p,q]}$ we need only to find additive generators for $\mathcal{L}_- \stackrel{\text{def}}{=} \mathcal{L} \cap \{\ell \leq 0\}$ also.)

Lemma J.2. *$\mathcal{L}$ is generated by $(p,0)$ and $(-\tilde{s}p', q')$ for some $\tilde{s}$ (given below in (J.3)).*

Proof. $\mathcal{L}$ is generated by $(-s'p', r'q')$ and (k_o, l_o) such that $(r'q')k_o+(s'p')l_o = [p,q]$. Let $r'\overline{k} + s'\overline{l} = 1$, $\nu_o = r'$ and $\mu_o = -\{p,q\}\,\overline{l}$, then we can take $k_o = p\overline{k}$ and $l_o = q\overline{l}$ and so

$$\left.\begin{aligned}\mu_o(-s'p') + \nu_o k_o &= s'p\overline{l} + r'p\overline{k} = p\\ \mu_o(r'q') + \nu_o l_o &= -r'q\overline{l} + r'q\overline{l} = 0\end{aligned}\right\} \quad \Rightarrow \quad (p,0) \in \mathcal{L}.$$

Because $\{r', \{p,q\}\overline{l}\} = \{r', \overline{l}\} = 1$ we can find μ_1 and ν_1 such that $r'\mu_1 + \nu_1\{p,q\}\overline{l} = 1$. Then $\mu_1(r'q') + \nu_1 l_o = q'$,

$$\mu_1(-s'q') + \nu_1 k_o = -p'(s'\mu_1 - \nu_1\{p,q\}\overline{k}) = -\tilde{s}p'$$

and

$$\tilde{s} = s'\mu_1 - \nu_1\{p,q\}\overline{k}\,. \tag{J.3}$$

Note that the matrix of change of coordinates is

$$T = \begin{pmatrix} \mu_o & \nu_o \\ \mu_1 & \nu_1 \end{pmatrix}$$

and that

$$|T| = \mu_0\nu_1 - \mu_1\nu_o = -\{p,q\}\,\overline{l}\nu_1 - \mu_1 r' = -1\,,$$

consequently T is invertible. ■

Corollary J.3. *If $(p,q) = 1$ then the generators of $\mathcal{L}_+$ are $(p,0)$ and $(0,q)$, with $(-p,0)$ and $(0,-q)$ for $\mathcal{L}_-$.*

Proof. From the previous lemma, the generators of $\mathcal{L}$ are $(p,0)$ and $(-\tilde{s}p, q)$ and so we can conclude for $\mathcal{L}_+$ and $\mathcal{L}_-$. ■

J.2. Isotropy subgroups

We finish by giving the isotropy subgroup lattices for some typical cases. First, when p and q are odd with q not a multiple of p we find

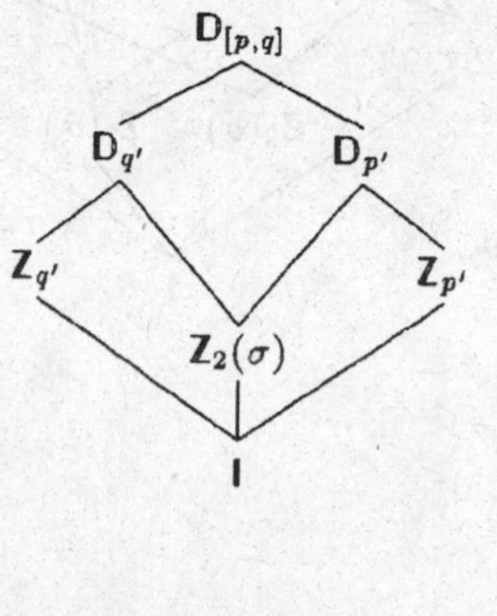

The fixed-point subspaces of $\mathbf{D}_{q'}$ ($\mathbf{D}_{p'}$) correspond to reversible orbits of period-p (q) points, $\operatorname{Fix}\mathbf{Z}_{q'}$ ($\mathbf{Z}_{p'}$) correspond to general period-p (q) orbits and $\operatorname{Fix}\mathbf{Z}_2(\sigma)$ contain the reversible period-$[p,q]$ orbits.

When p is even and q odd we get

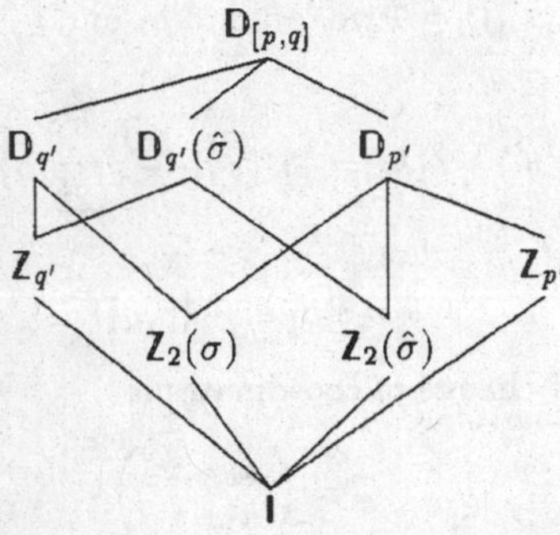

where the new fixed point subspaces correspond to period-p antireversible orbits (that is, if $\{z_i\}_{i=1}^{p}$ is such an orbit, $z_{i+\frac{1}{2}p} = \mathcal{R}z_i$) and period-$[p,q]$ antireversible orbits, respectively.

When p and q are even but $[p,q]/\{p,q\}$ is odd, we have

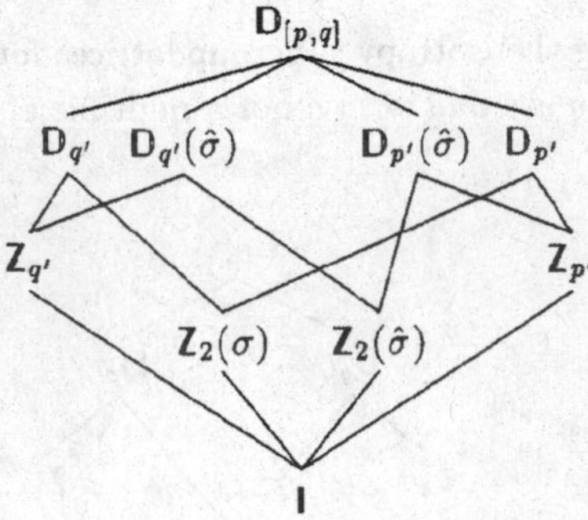

and when q is a multiple of p, say $q = np$ for some n we find

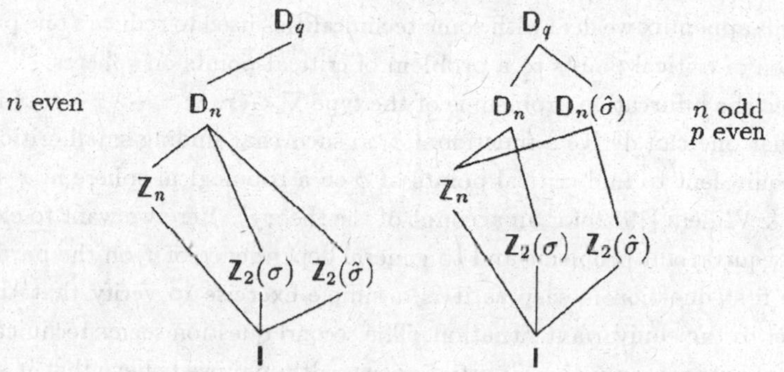

Appendix K. About reversible equivariant symplectic maps

In this appendix we are going to limit our analysis to reversible symplectic maps $\widehat{\mathbf{T}}$ where the reversor $\widehat{\mathcal{R}}$ acts *anti-symplectically* and the group Σ acts *symplectically* with $\widehat{\mathbf{J}}$ being the symplectic operator. It is of course possible to consider the same degree of generality as in Appendices F and I in which case we would end up with similar relations for $D\widehat{\mathcal{R}}^o$, $\widehat{\mathbf{J}}$ and $\mathbf{D}\widehat{\mathbf{T}}^o$. But, as we cannot yet exploit this generality very efficiently, we consider only the simpler situation.

From the Equivariant Darboux Theorem there exists a Σ-equivariant change of coordinates bringing the symplectic operator into canonical form $\widetilde{\mathbf{J}}$ (cf. Appendix I) and $\widehat{\mathcal{R}}$ is transformed into another $\widetilde{\mathbf{J}}$-antisymplectic involution, say $\mathcal{R}$, reversing $\mathbf{T}$, the transformation of $\widehat{\mathbf{T}}$.

An immediate question that arises is whether or not $\mathcal{R}$ can be linearized? We do not know the answer in all cases, but certainly yes if $\mathcal{R}$ and Σ commute. This is obtained by averaging in the proof of Meyer [1981] where such linearization was performed in the non-equivariant situation.

Now suppose that such a linearization has been done, that is, $\mathcal{R}$ is now a matrix. There are examples where $\mathcal{R}$ and Σ do not commute but one can make them skew-commute by enlarging Σ by including $\mathcal{R}\Sigma\mathcal{R}$. We suppose that this has been done ($\widehat{\mathbf{J}}$ and $\mathbf{T}$ are still Σ-equivariant even for the larger group) and so $\forall\,\sigma$, $\exists\,\sigma'$ with $\mathcal{R}\sigma = \sigma'\mathcal{R}$. We have thus constructed a $\Sigma \ltimes \mathbf{D}_q$-action on $\mathbf{X}_q^n$. When $\mathcal{R}$ and Σ commute then we have an action of the direct product $\Sigma \times \mathbf{D}_q$ with $\mathcal{R}$ a nice antisymplectic involution.

Appendix L. Bifurcations and critical points of equivariant functionals

In this appendix we deal with some technicalities used to reduce a one parameter bifurcation of critical points to a problem of critical points on spheres.

When the bifurcation problem is of the type $\nabla_x G(x,\lambda) = -\lambda\, x + f(x)$ it is well-known that one can derive a functional $\phi(x)$ such that finding small critical points of G is equivalent to find critical points of ϕ on a topological sphere in x-space (cf. Mawhin & Willem [1989] for an account of the theory). Here we want to extend the result to equivariant problems and to general dependence of f on the parameter λ.

The first question is easy as it is a simple exercise to verify that the proofs carry over to the equivariant situation. The second question seems technically more delicate and we can only give a partial answer although we believe that it should be true in all the cases we consider in this work.

To attack the problem we are going to show that subject to reasonable assumptions it is possible to cast the problem into its traditional shape where the usual results will apply.

We consider an orthogonal action of a compact group Γ on $\mathbf{R}^n$. We say that a Γ-equivariant bifurcation functional $G : \mathbf{R}^{n+1} \to \mathbf{R}$ is *finite* if one of the following holds true:

(a) $G(x,\lambda) = -\frac{1}{2}\lambda\,||x||^2 + F(x)$ for some $F : \mathbf{R}^n \to \mathbf{R}$, or

(b) there exists a number $l \in \mathbf{N}$ such that for all Γ-equivariant functionals H such that $j^{l+1}G = j^{l+1}H$ we have $(\nabla_x G \sim \nabla_x H\, ;\, \mathcal{K}^{\Gamma}_{\lambda})$ where $j^l G$ denotes the jet of order l of G.

Note that (b) corresponds to finite gradient determinacy. If (b) is satisfied then $\nabla_x G$ is $\mathcal{K}^{\Gamma}_{\lambda}$-equivalent to its k-jet. We have to use $\mathcal{K}^{\Gamma}_{\lambda}$-equivalence for the gradients as we have seen in the introduction of Chapter 3 that $\mathcal{R}^{\Gamma}_{\lambda}$-equivalence for functionals would in general not give finite determined (hence finite) problems.

We need the following preparatory lemma.

Lemma L.1. *Let $G : \mathbf{R}^{n+1} \to \mathbf{R}$ be a finite Σ-equivariant smooth functional such that $j^3 G(x,\lambda) = G_3(x) - \frac{1}{2}\lambda\,||x||^2$ for some G_3 then $(\nabla_x G \sim \nabla_x \overline{G}\, ;\, \mathcal{K}^{\Gamma}_{\lambda})$ where $\overline{G}$ is of the form $-\frac{1}{2}\lambda\,||x||^2 + \overline{F}(x)$ for some smooth $\overline{F}$.*

Proof. If G satisfies (a) then the result is immediate. Otherwise, let l be the order of finiteness of G. We proceed by induction to show that we can remove all terms with a λ-dependence up to order $l+1$, then use (b) to remove the higher order terms.

For any integer $k \geq 1$, consider the following sequence of functionals $\{G^k_i\}_{i=0}^{k-1}$ such that

$$j^{k+3} G^k_i(x,\lambda) = G_o(x) + \sum_{j=1}^{k-i} \lambda^j G_j(x) - \tfrac{1}{2}\lambda\,||x||^2\,, \tag{L.1}$$

where the degrees of $G_o \geq 3$ and of $G_j \geq k+3$. With a diagonal trick, we are going to construct Γ-equivariant changes of coordinates ψ_i^k such that

$$j^{k+3}G_{i+1}^k = j^{k+3}(G_i^k \circ \psi_i^k), \quad 0 \leq i \leq k-1,$$

and $j^{k+3}G_0^{k+1} = j^{k+3}(G_{k-1}^k \circ \psi_{k-1}^k)$. Then taking $G_0^1 = G$ and $\psi = \psi_{l-1}^l \circ \cdots \circ \psi_0^1$ we have cast $G_{l+1} = G \circ \psi$ into the right shape to apply (b) and then to conclude.

To show the induction step, write $\psi_i^k(x,\lambda) = x + \lambda^{k-i-1}u_i(x)$ where u_i is to be determined.

First we consider $0 \leq i \leq k-2$ for $k \geq 2$. With u_i of order at least $i+2$ in x, then

$$\begin{aligned} j^{k+3}G_i^k(\psi_i^k(x,\lambda),\lambda) =& G_o(x) + \lambda^{k-i-1}D(j^3G_o)(x,x,u_i) \\ &+ \sum_{j=1}^{k-i} \lambda^j G_j(x) - \tfrac{1}{2}\lambda\|x\|^2 - \lambda^{k-i}(x,u_i). \end{aligned} \tag{L.2}$$

Clearly we are done if

$$G_{k-i}(x) = (x, u_i). \tag{L.3}$$

As G_{k-i} is of order at least $i+3$ we can use Hadamard's Lemma (Gibson [1979]) to get a smooth $\hat{u}_i$ of order at least $i+2$ in x satisfying (L.3).

The last thing to do is to take care of the equivariance. Note that G_{k-i} is Γ-equivariant and so we can average $\hat{u}_i$ via

$$u_i(x) = \int_\Gamma \gamma^{-1}\hat{u}_i(\gamma x)\, d\gamma\,,$$

where $d\gamma$ is the Haar measure of Γ. An easy calculation shows that u_i is equivariant and moreover satisfies (L.3) (recall that Γ is acting orthogonally):

$$(x, u_i(x)) = (x, \int_\Gamma \gamma^T \hat{u}_i(\gamma x)\, d\gamma) = \int_\Gamma (\gamma x, \hat{u}_i(\gamma x))\, d\gamma = \int_\Gamma G_{k-i}(\gamma x)\, d\gamma = G_{k-i}(x)\,.$$

For $i = k-1$, the computations are simpler as ψ_{k-1}^k can be now taken as independent of λ. For u_{k+1} of order at least $k+1$,

$$j^{k+3}G_{k-1}^k(\psi_{k-1}^k(x),\lambda) = G_o(x+u_{k-1}) + \lambda G_1(x) - \tfrac{1}{2}\lambda\|x\|^2 - \lambda(x, u_{k-1})\,.$$

Again it is enough to solve $G_i(x) = (x, u_{k-1})$. This shows that we can push the additional λ-dependence to any order we want and then apply (b). ■

And so we get the main result.

Proposition L.2. *In the context of the Equivariant Splitting Lemma with a one dimensional bifurcation parameter* λ, *let us denote the two functionals* W_q, *resp.* $\widehat{W}_q$, *by* G , *resp.* $\hat{G}$. *Recall that* $L^o = \mathrm{Hess}_x G$.

Then if G is finite there is $\mathrm{cat}_\Gamma \mathrm{Ker}\, \mathbf{L}^o$ *Γ-orbits of solutions* $(\chi_i, \lambda_i(\chi_i))$, *for* $1 \leq i \leq \mathrm{cat}_\Gamma \mathrm{Ker}\, \mathbf{L}^o$, *on any small enough topological sphere around the origin.*

The integer $\mathrm{cat}_\Gamma \mathrm{Ker}\, \mathbf{L}^o$ *is the equivariant Lusternik-Schnirelman category of the unit sphere in* $\ker \mathbf{L}^o$.

Proof. From Lemma L.1 we can cast G into a form $\overline{G}$ to which one can apply the standard techniques (Mawhin & Willem [1989], Section 6.5) to get the equivalence between $\nabla_x G = 0$ and a critical point problem for a functional ϕ defined on any small topological sphere around the origin.

Hence the conclusion as we can average each step of the construction. ■

Remarks, corollaries. (a) One can prove that condition (b) of finiteness is satisfied for G iff it is satisfied for $\widehat{G}$.

(b) Finite $\mathcal{K}^\Gamma_\lambda$-codimension (or finite gradient $\mathcal{K}^\Gamma_\lambda$-codimension) of $\nabla_x G$ (or $\nabla_x \widehat{G}$) implies finiteness of G.

(c) As in Stuart [1979] we can extend the result to Hilbert spaces using the approach of Damon [1986] to deal with singularities of Fredholm maps.

Appendix M. Instability Lemma

In this appendix we show that one can determine loss of stability on the bifurcating branches from the reduced bifurcation equations in a general context.

Instability Lemma M.1. *Let* $\mathbf{T}$ *be a family of symplectic twist maps on* $\mathbb{R}^{2n}$ *parametrized by* $\lambda \in \mathbb{R}$. *For some* $\epsilon > 0$, *consider the following intervals* $\mathcal{U}_\epsilon$ *for* $\lambda = 0$: $(-\epsilon, 0)$ *or* $(0, \epsilon)$. *Suppose that* $\sigma(\mathbf{DT}(0,\lambda)^q) \subset \mathbf{S}^1 \setminus \{1\}$, $\forall \lambda \in \mathcal{U}_\epsilon$, *and that* $1 \in \sigma(\mathbf{DT}^{oq})$ *(it means that the origin is linearly stable for* $\lambda \in \mathcal{U}_\epsilon$*). Let* $\widehat{\mathrm{W}}_q$ *be the reduced functional for the bifurcating period-q points.*

Then, for $(\|\chi\|, \lambda)$ *sufficiently small, the period-q point bifurcating from the origin corresponding to* (χ, λ) *is unstable if, for any* $\bar{\lambda} \in \mathcal{U}_\epsilon$,

$$|\mathrm{Hess}_\chi \widehat{\mathrm{W}}_q(0, \bar{\lambda})| \cdot |\mathrm{Hess}_\chi \widehat{\mathrm{W}}_q(\chi, \lambda)| < 0 .$$

Note that $\mathbf{T}$ *can be equivariant.*

Proof. The fundamental result we need is formula (2.36) linking the residues to the Hessian of the action. With our hypotheses we know that we can use our framework to reduce, via the Splitting Lemma, the bifurcation problem for the action W_q into a bifurcation problem on Ker $\mathbf{L}^o$ where $\mathbf{L}^o = \mathrm{Hess}_{\mathbf{x}} \mathrm{W}^o_q$. Note that its structure might be much more complicated that what we have studied so far. Then we can use the splitting (2.38) and formula (2.36) to get the value of the product of the residues for the map $\mathbf{T}$ on some branch (χ, λ) of solutions of $\nabla_\chi \widehat{\mathrm{W}}_q(\chi, \lambda) = 0$.

Now we use the hypothesis that all the multipliers for $(0,\bar{\lambda})$, with $\bar{\lambda} \in \mathcal{U}_\epsilon$, are on the unit circle but not at 1, this means that all the residues are strictly positive and so

$$1 = \operatorname{sign} \prod_{k=1}^{n} R_k = (-1)^n \operatorname{sign}(|\mathbf{B}|^q |\mathbf{QL}^o\mathbf{Q}|) \cdot \operatorname{sign}|\operatorname{Hess}_\chi \widehat{\mathrm{W}}_q(0,\lambda)|.$$

as the sign of the determinants of $\mathbf{B}$ and $\mathbf{QL}^o\mathbf{Q}$ are constant in a small neighborhood of the origin.

Now we simply want to know when a multiplier crosses to the negative axis. Clearly this is achieved if the product becomes negative. And so on a branch of solution corresponding to (χ,λ)

$$\begin{aligned}\operatorname{sign} \prod_{k=1}^{n} R_k &= (-1)^n \operatorname{sign}(|\mathbf{B}|^q |\mathbf{QL}^o\mathbf{Q}|) \cdot \operatorname{sign}|\operatorname{Hess}_\chi \widehat{\mathrm{W}}_q(\chi,\lambda)| \\ &= |\operatorname{Hess}_\chi \widehat{\mathrm{W}}_q(0,\bar{\lambda})| \cdot |\operatorname{Hess}_\chi \widehat{\mathrm{W}}_q(\chi,\lambda)|.\end{aligned}$$

And so if the product is negative one of the R_k's is negative hence the solution is unstable. ■

Appendix N. Isotropy and twisted subgroups

As $\mathbf{Z}_q$ is abelian and $\Sigma\times\mathbf{Z}_q$ is a direct product we can transfer to our context the discussions in GSS II [1988, Section XVI.7/8, p.300-307] and the subsequent papers mentioned in this appendix about the action of $\Gamma\times\mathbf{S}^1$ in the context of Equivariant Hopf bifurcation. The key concepts transfer very well, but we obtain sometimes only a weaker version of some results due to the discrete symmetry brought by $\mathbf{Z}_q$. Let us elaborate on the basic results, the references providing more information.

Let $\Pi \subset \Sigma\times\mathbf{Z}_q$ be an isotropy subgroup. For any r, we denote by $\mathbf{Z}_r$ the subgroup $I\times\mathbf{Z}_r \subset \Sigma\times\mathbf{Z}_q$. In general $\mathbf{Z}_q$ might not act freely, but we can factor out the *temporal subgroup* of Π, $\mathbf{Z}_q^r \overset{\text{def}}{=} \Pi\cap\mathbf{Z}_q$ where r is given by $\mathbf{Z}_q^r = \mathbf{Z}_{q/r}$. As every element of $\mathbf{Z}_q$ commutes with Π we can define $\mathbf{Z}_r(\mathrm{W}\Pi) = \mathbf{Z}_q/\mathbf{Z}_q^r$ (cf. Section 8.1). Now we can show the equivalent to Proposition XVI.7.3, that is, all isotropy subgroups of $\Sigma\times\mathbf{Z}_q$ are twisted subgroups.

Proposition N.1. *Let $\Pi \subset \Sigma\times\mathbf{Z}_q$ be an isotropy subgroup, then there exits a subgroup $\Xi\subset\Sigma$ and a group homomorphism $\theta:\Xi\to\mathbf{Z}_r(\mathrm{W}\Pi)$ such that*

$$\Pi = \Xi^\theta\times\mathbf{Z}_q^r = \{\,(\sigma,\theta(\sigma)) \mid \sigma\in\Xi\,\}\times\mathbf{Z}_q^r.$$

Proof. Consider $\Pi' = \Pi/\mathbb{Z}_q^r \subset \Sigma\times\mathbb{Z}_q$ and the natural projection $\pi' : \Sigma\times\mathbb{Z}_r(\mathrm{W}\Pi) \to \Sigma$. As $\Pi' \cap \mathbb{Z}_r(\mathrm{W}\Pi) = \{1\}$, π' is an isomorphism from Π' onto $\pi'(\Pi')$. Define $\Xi = \pi'(\Pi')$ and we can conclude like in GSS II. ■

Note that the subgroup of *spatial* symmetries $\mathcal{K} \stackrel{\text{def}}{=} \mathrm{Ker}\,\theta$ is a normal subgroup of Σ. Following Golubitsky & Stewart [1993] we can also characterize the normalizer $\mathrm{N}\Pi$ and the Weyl subgroup $\mathrm{W}\Pi = \mathrm{N}\Pi/\Pi$ of Π in $\Sigma\times\mathbb{Z}_q$ by following their line of proof.

Proposition N.2.

$$\mathrm{N}\Pi = C(\Xi,\mathcal{K})\times\mathbb{Z}_q$$

where $C(\Xi,\mathcal{K}) = \{\ \sigma \mid \sigma\pi\sigma^{-1}\pi^{-1} \in \mathcal{K},\ \forall\pi\in\Xi\ \}$, *and*

$$\mathrm{W}\Pi = (C(\Xi,\mathcal{K})/\Xi)\times\mathbb{Z}_r(\mathrm{W}\Pi)\,.$$

Note that the example of the $\mathbf{O}(2)$-equivariant maps on $\mathbf{R}^4$ show that the main conclusion of the Golubitsky & Schaeffer paper do not hold in the discrete map setting.

Finally, in our context the dimension of the fixed-point subspaces is of prime interest for *all* the isotropy subgroups. Without resonances $\mathbb{Z}_q^r = \mathbf{I}$ and we can readily adapt Proposition XVI.8.3 in GSS II to get the following.

Given a twisted subgroup $\Xi^\theta \subset \Sigma\times\mathbb{Z}_q$ with spatial subgroup $\mathcal{K}$, then, one can show that

(a) If $\theta(\Xi) = \mathbf{I}$ then dim (Fix Ξ^θ)=2 dim (Fix Ξ),
(b) If $\theta(\Xi) = \mathbb{Z}_2$ then dim (Fix Ξ^θ)=2 [dim (Fix $\mathcal{K}$)-dim (Fix Ξ)],
(c) If $\theta(\Xi) = \mathbb{Z}_3$ then dim (Fix Ξ^θ)=dim (Fix $\mathcal{K}$)-dim (Fix Ξ),
(d) If $\theta(\Xi) = \mathbb{Z}_4$ then dim (Fix Ξ^θ)=dim (Fix $\mathcal{K}$)-dim (Fix $\mathcal{L}$),
(e) If $\theta(\Xi) = \mathbb{Z}_6$ then

$$\dim(\mathrm{Fix}\,\Xi^\theta) = \dim(\mathrm{Fix}\,\Xi) + \dim(\mathrm{Fix}\,\mathcal{K}) - \dim(\mathrm{Fix}\,\mathcal{L}) - \dim(\mathrm{Fix}\,\mathcal{M})\,,$$

where $\mathcal{L}$ and $\mathcal{M}$ are the unique subgroups satisfying $\mathcal{K}\subset\mathcal{L}, \mathcal{M}\subset\Xi$ with $|\Xi/\mathcal{L}| = 3$ and $|\Xi/\mathcal{M}| = 2$.

List of References

[1] V. I. ARNOLD [1972] *Normal forms for functions near degenerate critical points, the Weyl groups A_k, D_k, E_k and Lagrangian singularities*, Func. Anal. Appl. **6**, pp. 254-72

[2] V. I. ARNOLD [1988] *Geometrical Methods in the Theory of Ordinary Differential Equations*, Second Edition, Springer-Verlag: New York

[3] V. I. ARNOLD, V. V. KOZLOV & A. I. NEISHTADT [1988] *Mathematical aspects of classical and celestial mechanics*, in Enc. of Math. Sci., Dynamical Systems III (Ed. V. Arnold), Springer-Verlag

[4] S. AUBRY [1983] *The twist map, the extended Frenkel-Kontorova model and the devil's staircase*, Physica **7D**, pp. 240-58

[5] T. BARTSCH [1989] *Critical orbits of symmetric functionals*, Manuscr. Math. **66**, pp. 129-152

[6] T. BARTSCH [1992a] *The Conley index over a space*, Math. Zeit **209**, pp. 167-77

[7] T. BARTSCH [1992b] *Topological methods for Variational Problems with Symmetries*, Lecture Notes in Mathematics **1560** (1993), Springer Verlag

[8] T. BARTSCH & M. CLAPP [1990] *Bifurcation theory for symmetric potential operators and the equivariant cup-length*, Math. Z. **204**, pp. 341-356

[9] D. BERNSTEIN & A. KATOK [1987] *Birkhoff periodic orbits for small perturbations of completely integrable Hamiltonian systems with convex Hamiltonian*, Inv. Math. **88**, pp. 225-41

[10] A. BHOWAL, T. K. ROY & A. LAHIRI [1993] *Small-angle Krein collisions in a family of four-dimensional reversible maps*, Physics Rev. A, to appear

[11] T. J. BRIDGES [1990] *Bifurcation of periodic solutions near a collision of eigenvalues of opposite signature*, Math. Proc. Camb. Phil. Soc. **108**, pp. 575-601

[12] T. J. BRIDGES [1991] *Stability of periodic solutions near a collision of eigenvalues of opposite signature*, Math. Proc. Camb. Phil. Soc. **109**, pp. 375-403

[13] T. J. BRIDGES & R. H. CUSHMAN [1993] *Unipotent normal forms for symplectic maps*, PhysicaD **65**, pp. 211-41

[14] T. J. BRIDGES, R. H. CUSHMAN & R. S. MACKAY [1993] *Dynamics near an irrational collision of eigenvalues for symplectic maps*, Proceedings of the Field's Institute Workshop on Normal Forms and Homoclinic Chaos (edited by W. F. LANGFORD)

[15] T. J. BRIDGES, J. E. FURTER & A. LAHIRI [1991] *Collision of multipliers at $\pm i$ in reversible-symplectic maps*, Preprint, University of Warwick

[16] J. BRUCE, A.A. DU-PLESSIS & C.T.C. WALL [1987] *Determinacy and unipotency*, Invent. Math. **88**, pp. 521-534

[17] R. CUSHMAN [1983] *Geometry of the energy-momentum mapping of the spherical pendulum*, C. W. I. Newsletter, **1**, pp. 4-18

[18] J. DAMON [1984] *The unfolding and determinacy theorems for subgroups $\mathcal{A}$ and $\mathcal{K}$*, Memoirs AMS, **306**, Providence

[19] J. DAMON [1986] *A theorem of Mather and the local structure of nonlinear Fredholm maps*, Proceedings of Symposium on Pure Mathematics **45**, Part I, pp. 339-52

[20] J. DAMON [1987] *Deformations of sections of singularities and Gorenstein surface singularities*, Am. J. Math. **109**, pp. 695-722

[21] J. DAMON [1989] *Topological triviality and versality for subgroups of $\mathcal{A}$ and $\mathcal{K}$*, Memoirs AMS, **389**, Providence

[22] J. DAMON & D. MOND [1991] *$\mathcal{A}$-codimension and the vanishing topology of the discriminants*, Inv. Math. **106**, pp. 217-42

[23] P.J. DAVIS [1979] *Circulant Matrices*, John Wiley & Sons: New York

[24] M. DELLNITZ & I. MELBOURNE [1992] *The equivariant Darboux Theorem*, preprint 59, Hamburger Beiträge zur Angewandten Mathematik

[25] M. DELLNITZ, I. MELBOURNE & J. E. MARSDEN [1992] *Generic bifurcation of Hamiltonian vectorfields with symmetry*, Nonlinearity **5**, pp. 979-96

[26] R. L. DEWAR & J. D. MEISS [1993] *Flux-minimizing curves for reversible area-preserving maps*, PhysicaD **57**, pp. 476-506

[27] M. FIELD [1976] *Transversality in G-Manifolds*, Trans. AMS. **231**, pp. 429-450

[28] J. FURTER [1990] *Bifurcation of subharmonics in reversible systems and the classification of bifurcation diagrams equivariant under the dihedral groups I. Period tripling*, preprint, University of Warwick

[29] J. FURTER [1991] *Bifurcation of subharmonics in reversible systems and the classification of bifurcation diagrams equivariant under the dihedral groups II. High resonances*, preprint, University of Warwick

[30] I. M. GEL'FAND & V. B. LIDSKII [1955] *On the structure of the regions of stability of linear canonical systems of differential equations with periodic coefficients*, AMS Trans. (2)**8**, pp. 143-81

[31] C. G. GIBSON [1979] *Singular Points of Smooth Mappings*, Research Notes in Math. **25**, Pitman: London

[32] M. GOLUBITSKY & M. ROBERTS [1987] *A classification of degenerate Hopf bifurcations with* **O***(2) symmetry*, J. Diff. Eqns. **69**, pp. 216-64

[33] M. GOLUBITSKY & D. SCHAEFFER [1979] *A theory for imperfect bifurcation theory via singularity theory*, Comm. Pure & Applied Math. **32**, pp. 21-98

[34] M. GOLUBITSKY & I. STEWART [1993] *Algebraic criterion for symmetric Hopf bifurcation*, Proc. Royal Soc. Lond. **A440**, pp. 727-32

[35] M. GOLUBITSKY, I. STEWART & D. SCHAEFFER [1988] *Singularities and groups in bifurcation theory, Vol. II, Appl. Math. Sci. no. 69*, Springer-Verlag

[36] J. E. HOWARD & R. S. MACKAY [1987] *Linear stability of symplectic maps*, J. Math. Phys. **25**(5), pp. 1036-1051

[37] A. HUMMEL [1979] *Bifurcation of Periodic Points*, Thesis, University of Groningen

[38] H. KIELHÖFER [1988] *A bifurcation theorem for potential operators*, J. Func. Analysis **77**, pp. 1-8

[39] H. KOOK & J. D. MEISS [1989] *Periodic orbits for reversible-symplectic mappings*, Physica **35D**, pp. 65-86

[40] W. KRAWCEWICZ & W. MARZANTOWICZ [1990] *Ljusternik-Schnirelman methods for functionals invariant with respect to a finite group action*, J. Diff. Eqns. **85**, pp. 105-24

[41] A. LAHIRI, A. BHOWAL, T. K. ROY & M. B. SEVRYUK [1993] *Stability of invariant curves in four-dimensional reversible mappings near a 1:1 resonance*, PhysicaD **63**, pp. 99-116

[42] E. J. N. LOOIJENGA [1984] *Isolated singular points on complete intersections*, Lecture notes in Maths. **77**, London Math. Soc.

[43] R. S. MACKAY [1986] *Introduction to the dynamics of area-preserving maps*, report, University of Warwick

[44] R. S. MACKAY & J. D. MEISS [1983] *Linear stability of periodic orbits in Lagrangian systems*, Phys. Lett. **98A**, pp. 92-4

[45] R. S. MACKAY, J. D. MEISS & J. STARK [1989] *Converse KAM theory for symplectic twist maps*, Nonlinearity **2**, pp. 555-70

[46] R. J. MAGNUS [1977] *On universal unfoldings of certain real functions on a Banach space*, Math. Proc. Camb. Phil. Soc. **81**, pp. 91-5

[47] J. MATHER [1968] *Stability of C^∞ mappings III, finitely determined map-germs*, Pub. Math. IHES. **36**, pp. 127-156

[48] J. MATHER [1982] *Existence of quasiperiodic orbits for twist homeomorphisms of the annulus*, Topology **21**, pp. 457-67

[49] J. MAWHIN & M. WILLEM [1989] *Critical Point Theory and Hamiltonian Systems*, Applied Math. Sci. **74**, Springer-Verlag: New York

[50] J. C. VAN DER MEER [1985] *The Hamiltonian Hopf Bifurcation*, LNM **#1160**, Springer-Verlag: New York

[51] J. C. VAN DER MEER [1990] *Hamiltonian Hopf bifurcation with symmetry*, Nonlinearity, **3**, pp. 1041-56

[52] J. D. MEISS [1991] *Symplectic maps, variational principles and transport*, Report No. 101, Applied Math., University of Colorado-Boulder

[53] K. R. MEYER [1970] *Generic bifurcation of periodic points*, Trans. AMS **149**, pp. 95-107

[54] K. R. MEYER [1971] *Generic stability properties of periodic points*, Trans. AMS **154**, pp. 273-77

[55] K. R. MEYER [1981] *Hamiltonian systems with discrete symmetry*, J. Diff. Eqns. **41**, pp. 228-38

[56] K. R. MEYER & D. S. SCHMIDT [1971] *Periodic orbits near L_4 for mass ratios near the critical mass ratio of Routh*, Cel. Mech. **4**, pp. 99-109

[57] D. MOND & J. MONTALDI [1991] *Deformations of maps on complete intersections, Damon's K_V-equivalence and bifurcations* , Preprint 21/1991, University of Warwick

[58] J. MONTALDI, M. ROBERTS & I. STEWART [1988] *Periodic solutions near equilibria of symmetric Hamiltonian systems*, Phil. Trans. **A325**, pp. 237-93

[59] J. MONTALDI, M. ROBERTS & I. STEWART [1990] *Existence of nonlinear normal modes of symmetric Hamiltonian systems*, Nonlinearity, **3**, pp. 695-730

[60] R. PALAIS [1979] *The principle of symmetric criticality*, Comm. Math. Physics **69**, pp. 19-39

[61] I. C. PERCIVAL [1980] *Variational principles for invariant tori and cantori*, in Nonl. Dyn. and the Beam-Beam Interaction, Edited by M. Month & J. Herrera, AIP CP #**57**, pp. 302-10

[62] D. PFENNIGER [1985] *Numerical study of complex instability 1. Mappings*, Astron. & Astrophys. **150**, pp.97-111

[63] D. PFENNIGER [1987] *Complex instability around the rotation axis of stellar systems II. Rotating oscillators* **180**, pp. 79-93

[64] F. RANNOU [1974] *Numerical study of discrete area-preserving mappings*, Astron. and Astrophy. **31**, pp. 289-301

[65] R. J. RIMMER [1978] *Symmetry and bifurcation of fixed points of area-preserving maps*, J. Diff. Eqns. **29**, pp. 329-344

[66] R. J. RIMMER [1983] *Generic bifurcations for involutory area-preserving maps*, Memoirs AMS, **272**, Providence

[67] R. M. ROBERTS [1986] *Characterisations of finitely determined equivariant map-germs*, Math. Ann. **275**, pp. 583-597

[68] I. SCHOENBERG [1950] *The finite Fourier series and elementary geometry*, Amer. Math. Monthly **57**, pp. 390-404

[69] M. B. SEVRYUK & A. LAHIRI [1991] *Bifurcation of families of invariant curves in four-dimensional reversible mappings*, Physics Letters A **154**, pp. 104-110

[70] P. SLODOWY [1978] *Einige Bemerkungen zur Entfaltung symmetrischer Funktionnen*, Math. Z. **158**, pp. 157-170

[71] A. G. SOKOL'SKIJ [1974] *On the stability of an autonomous Hamiltonian system with 2-DOF in the case of equal frequencies*, J. Appl. Math. Mech. **38**, pp. 791-9

[72] C. A. STUART [1979] *An introduction to bifurcation theory based on differential calculus*, in *Nonlinear Analysis and Mechanics*, Heriot-Watt Symposium **IV**, R. J. KNOPS ed., Pitman: London, pp. 76-137

[73] H. TERAO [1983] *The bifurcation set and logarithmic vector fields*, Math. Ann. **263**, pp. 313-321

[74] A. VANDERBAUWHEDE [1982] *Local Bifurcation and Symmetry*, Res. Notes in Math. **75**, Pitman, Boston

[75] C. T. C. WALL [1981] *Finite determinacy of smooth mappings*, Bull. Lond. Math. Soc. **13**, pp. 481-539

[76] J. WILLIAMSON [1937] *On the normal forms of linear canonical transformations in dynamics*, Amer. J. Math. **59**, pp. 599-617

[77] V. A. YAKUBOVICH & V. M. STARZHINSKII [1975] *Linear differential equations with periodic coefficients, Vol 1*, Keter Publ. House Jerusalem Ltd

[78] S. S. T. YAU [1983] *Milnor algebras and equivalence relations among holomorphic functions*, Bulletin of AMS **9**, pp. 235-239

[79] C. ZUPPA [1984] *Bifurcations imparfaites de type potentiel et D-equivalence de deformations des fonctions*, Duke. Math. J. **51**, pp. 729-763

Printing: Weihert-Druck GmbH, Darmstadt
Binding: Buchbinderei Schäffer, Grünstadt

Author Index

Subject Index